Advances in Independent Component Analysis and Learning Machines

Advances in Independent Component Analysis and Learning Machines

Edited by

Ella Bingham

Samuel Kaski

Jorma Laaksonen

Jouko Lampinen

ELSEVIER

AMSTERDAM • BOSTON • HEIDELBERG • LONDON
NEW YORK • OXFORD • PARIS • SAN DIEGO
SAN FRANCISCO • SINGAPORE • SYDNEY • TOKYO
Academic Press is an imprint of Elsevier

AP

Academic Press is an imprint of Elsevier
125 London Wall, London, EC2Y 5AS, UK
525 B Street, Suite 1800, San Diego, CA 92101-4495, USA
225 Wyman Street, Waltham, MA 02451, USA
The Boulevard, Langford Lane, Kidlington, Oxford OX5 1GB, UK

Library of Congress Cataloging-in-Publication Data
A catalog record for this book is available from the Library of Congress

British Library Cataloguing in Publication Data
A catalogue record for this book is available from the British Library

ISBN: 978-0-12-802806-3

For information on all Academic Press publications
visit our website at http://store.elsevier.com/

Publisher: Matthew Deans
Acquisition Editor: Tim Pitts
Editorial Project Manager: Charlie Kent
Production Project Manager: Melissa Read
Designer: Greg Harris

Contents

Preface

This book is dedicated to Prof. Erkki Oja on the occasion of his retirement in January 2015. It contains contributions by his former students and collaborators.

Erkki Oja is Professor of Computer Science and Engineering at Aalto University School of Science. Until the end of 2014, he was also Director of the Finnish Centre of Excellence in Computational Inference Research (COIN) and its predecessors at Aalto University and Helsinki University of Technology: Adaptive Informatics Research Centre and Neural Networks Research Centre. Since 2000, he has been an IEEE Fellow, and is also an IAPR Founding Fellow, past President of ENNS, founding member of the Pattern Recognition Society of Finland (Hatutus), and a member of several other prominent scientific societies. He is past chairman of the Finnish Research Council for Natural Sciences and Engineering. Erkki is the author or coauthor of more than 330 papers and he has written three textbooks. He has about 35,000 citations, his h index is 61, and he has supervised over 50 doctoral theses. Erkki's research interests include principal and independent component analysis, self-organization, statistical pattern recognition, and applying machine learning to computer vision and signal processing. The contributions in this book are inspired by and build upon the seminal results that Erkki has achieved.

The book begins with an Introduction where several distinguished scholars describe the collaboration they have had with Erkki. The first part of the book is a collection of methodological advancements and surveys on independent component analysis and machine learning algorithms and applications. The latter half of the book contains case examples of applying pattern recognition and latent variable methods to several domains.

We would like to thank all authors for their interesting chapters.

Ella Bingham, Samuel Kaski, Jorma Laaksonen, and Jouko Lampinen
Espoo, November 2014

About the Editors

Ella Bingham received her Doctor of Science (Ph.D.) degree in Computer Science in 2003 and M.Sc. degree in Systems and Operations Research in 1998, both at Helsinki University of Technology. Her main research field has been statistical data analysis. She works at the Helsinki Institute for Information Technology (HIIT) at Aalto University and University of Helsinki. In addition, she is the Executive Director of the Foundation for Aalto University Science and Technology. Her professional interests include science policy, research administration, research assessments, and research funding.

Samuel Kaski received the Doctor of Science (Ph.D.) degree in Computer Science from Helsinki University of Technology, Finland, in 1997. He is currently a Professor at Aalto University, the Director of the Helsinki Institute for Information Technology (HIIT), Aalto University and University of Helsinki, Finland, and the Director of the Finnish Centre of Excellence in Computational Inference Research (COIN). He is an action editor of the *Journal of Machine Learning Research* and has chaired several conferences including AISTATS 2014. He has published over 200 peer-reviewed papers and supervised 18 Ph.D. theses. His current research interests include statistical machine learning, computational biology and medicine, information visualization, and exploratory information retrieval.

Jorma Laaksonen has worked with Prof. Erkki Oja since 1994 and received his Doctor of Science in Technology degree in 1997 from Helsinki University of Technology, Finland. Presently, he is a permanent teaching research scientist at the Department of Computer Science, Aalto School of Science, where he has instructed eight doctoral theses in the supervision of Prof. Oja. He is an author of 200 scientific journal, conference and edited book papers on pattern recognition, statistical classification, machine learning, and neural networks, and has a Google Scholar *h*-index of 27. His research interests are in content-based multimodal information retrieval and computer vision. Dr. Laaksonen is an associate editor of Pattern Recognition Letters, IEEE senior member, and a founding member of the SOM and LVQ Programming Teams and the PicSOM Development Group.

Jouko Lampinen obtained his Doctor of Science (Ph.D.) degree in Information Technology from Lappeenranta University of Technology in 1993. He is currently a Professor and the Head of Department at the Department of Computer Science, Aalto University School of Science. He is the Director of the Aalto M.Sc. programme in Life Science Technologies. He has published over 100 peer-reviewed papers and supervised or co-supervised over 20 Ph.D. theses. His current research interests include probabilistic modeling, and data analysis in systemic neuroscience.

List of Contributors

Traian Abrudan
Department of Computer Science, University of Oxford, Oxford, UK

Ella Bingham
Helsinki Institute for Information Technology, Aalto University and University of Helsinki, Helsinki, Finland

Guangyong Chen
Department of Computer Science and Engineering, The Chinese University of Hong Kong, Shatin, N.T., Hong Kong, China

KyungHyun Cho
University of Montreal, Montreal, Canada

Scott C. Douglas
Department of Electrical Engineering, Southern Methodist University, Dallas, Texas, USA

Markku Hauta-Kasari
School of Computing, and Institute of Photonics, University of Eastern Finland, Joensuu, Finland

Pheng Ann Heng
Department of Computer Science and Engineering, The Chinese University of Hong Kong, Shatin, N.T., Hong Kong, China

Aapo Hyvärinen
Department of Computer Science, Helsinki Institute for Information Technology, University of Helsinki, Helsinki, Finland

Satoru Ishikawa
Department of Computer Science, Aalto University, Espoo, Finland

Heikki Kälviäinen
Machine Vision and Pattern Recognition Laboratory, Department of Mathematics and Physics, Lappeenranta University of Technology, Lappeenranta, Finland

Juha Karhunen
Department of Computer Science, Aalto University, Espoo, Finland

Irwin King
The Chinese University of Hong Kong, Shatin, Hong Kong, China

Visa Koivunen
Department of Signal Processing and Acoustics, School of Electrical Engineering, Aalto University, Espoo, Finland

Zbyněk Koldovský
The Institute of Information Theory and Automation of the Czech Academy of Sciences, Czech Republic

Markus Koskela
Department of Computer Science, University of Helsinki, Helsinki, Finland

Jorma Laaksonen
Department of Computer Science, Aalto University, Espoo, Finland

Hannu Laamanen
Institute of Photonics, University of Eastern Finland, Joensuu, Finland

Heikki Mannila
Department of Computer Science, Aalto University, Espoo, Finland

Jussi Parkkinen
School of Computing, University of Eastern Finland, Joensuu, Finland, and School of Engineering, Monash University Malaysia, Jalan Lagoon Selatan, Selangor, Malaysia

Matti Pietikäinen
Center for Machine Vision Research, Department of Computer Science and Engineering, University of Oulu, Finland

Tapani Raiko
Department of Computer Science, Aalto University, Espoo, Finland

Mats Sjöberg
Department of Computer Science, University of Helsinki, Helsinki, Finland

Petr Tichavský
The Institute of Information Theory and Automation of the Czech Academy of Sciences, Czech Republic

Harri Valpola
ZenRobotics Ltd., Helsinki, Finland

Ricardo Vigário
Department of Computer Science, Aalto University, Espoo, Finland

Ville Viitaniemi
Department of Computer Science, Aalto University, Espoo, Finland

Lei Xu
Department of Computer Science and Engineering, The Chinese University of Hong Kong, Shatin, N.T., Hong Kong, China

Zhirong Yang
Department of Computer Science, Aalto University, Espoo, Finland

Guoying Zhao
Center for Machine Vision Research, Department of Computer Science and Engineering, University of Oulu, Finland

Fengyuan Zhu
Department of Computer Science and Engineering, The Chinese University of Hong Kong, Shatin, N.T., Hong Kong, China

Introduction

Ricardo Vigário

As this book celebrates some great scientific avenues inspired by visionary researchers of the likes of Erkki Oja, I believe it is of particular relevance to remember the person, his career, and the shoulders from where several such amazing prospects sprung forth.

Many are the topics of research where Erkki has brought a marked contribution. Many more are the researchers, young and older, who shared and shaped paths of their working careers with him. For the sake of conciseness and a certain focus, I took the liberty to invite only a few of Erkki's distinguished friends and colleagues to contribute with some words describing their relation to him. We will see in all of them the same admiration and recognition for a true scholar, but also some echoes of the person and the communicator, who often transcends the limits of the "work rules," and ventures into the friendship realm.

This chapter is admittedly "the" nonscientific contribution in the book. Therefore, instead of thoroughly editing its text, in search for a rather cohesive structure, I will opt for a more platonic approach: reality will always exceed all finite projections we may find for it. In the cave wall that is this book, I will therefore let each of the participants voice their own views about Erkki. Those projections will certainly not be uncorrelated, let alone independent, as they address and characterize a common multi-dimensional source. Yet, the subspace spanned by the following words will hopefully draw an impressionist sketch of the person, the scholar, and the friend.

Using the liberty that assists the convener of such a delightful forum, I will take the opportunity to open with my own, short contribution.

Erkki Oja and his learning rule. Original photo by Anni Hanén.

A STUDENT AND A CO-WORKER

In the early 1990s, I contacted a professor in Portugal on the subject of finding a suitable supervisor for a postgraduate research work. I thought I knew exactly what I wanted to do, but was rather uncertain where to go, and whom to work with. That professor, also in this chapter (LBA), suggested two names he knew who were leading researchers in the area I wanted to study – unsupervised learning, with biologically plausible computer vision goals. One worked in a remote city in Finland, far away from its capital. The other, Ralph Linsker, had a research group at IBM Thomas J. Watson Res. Center, in New York, USA. I thought: "maybe I can spend this year in Finland, and then move to IBM for the doctoral degree." This decision was made even easier because Erkki Oja had just moved to Otaniemi, a campus in the thriving capital region.

Over 20 years later, after experiencing research at several other leading machine learning groups, I find myself still working in the same department, idealized by academician Teuvo Kohonen, and developed by Erkki Oja to very high international standards. One could wrongfully mistake this persistence for inertia. None could be farther from the truth. It was very clear, from my very first encounter with Erkki and his research group, that his department was a perfect place to germinate new ideas, and even propose new research directions. At the time, it had a recognized group of experts in Neural Networks, carrying out leading research in unsupervised learning. Equally important in my decision was the fact that the university campus comprised, among many other excellent research areas, an internationally recognized brain research unit, led by the expert hands of researchers, such as the academicians Olli V. Lounasmaa and Riitta Hari.

One cannot say that neuroinformatics was then at the core of Erkki's research interests. Yet, he had always a finger on the pulse of science, and saw that a bridge between the aforementioned areas of research excellence was a valuable asset in the development of his own research endeavors. As one example, under his mentoring, pioneering research on biomedical applications of independent component analysis (ICA) was proposed. This is still, to date, one of the most accomplished areas of applied research for ICA.

The two anecdotal stories above reflect well Erkki's nature: an acute scholar, with an amiable nature; a visionary in science, as well as a true mentor, empowering his junior colleagues, leading and supporting them to take independent responsibility in science; and with a door always open, and a word of guidance at all times.

With permission, I will end these lines with the words of academician Riitta Hari: "over the years, I have had the privilege of interacting with Erkki at a scientific, academic, and personal level, and the interaction has always been smooth and effective. I highly appreciate Erkki as a scientist and a colleague. I am sure that his outstanding research and pedagogical mentoring will continue to promote science in multiple disciplines."

Prof. SIMON HAYKIN

A novel property of Neural Networks and Learning Machines is their inherent ability to learn from the environment; and through learning, improve their performance in some statistical sense. Work in such research fields, right from their inception, has been motivated by the recognition that the human brain is a powerful information processing machine, which distinguishes itself, in a remarkable manner, from the digital computer.

Professor Erkki Oja's influential scientific work spans over four highly prolific decades, from early research on associative memories in the late 1970s and early 1980s. To elaborate, Hebbian learning and principles of subspace analysis are basic to pattern recognition and machine vision, as well as blind source separation (BSS) and ICA, fields in which Prof. Oja researched throughout the 1990s and early 2000s. More recently, nonnegative matrix factorization and computational inference came into prominence. Throughout all that time, the points made herein apply to an insatiable thirst for knowledge, and an exceptional ability to detect and discover new trends in Neural Computation. Simply put, all of the above are the hallmark of a distinguished academic, namely, Prof. Erkki Oja.

A highly remarkable learning rule, known as Oja's rule, so-called in recognition of the work done by Prof. Oja, was published in 1982. The rule was motivated by Hebb's postulate of learning, which was first described in a book written by the neuropsychologist Donald Hebb in 1949. For the record, Oja's rule may be described as follows:

> A single linear neuron with a Hebbian-type adaptation rule for its synaptic weights can evolve into a filter for the first principal component of the input distribution.

The rule is simple to state, yet it is very rich in its mathematical exposé.

Furthermore, Prof. Oja and his research teams, over the years, went on to expand his learning rule for the identification of eigensubspaces, nonlinear principal component decompositions, and ICA algorithms. In addition to very elegant and efficient theoretical advances in Neural Computation and related topics, Prof. Oja always sought their use in practice, supporting pioneering research in many ambitious application areas: computer vision and pattern recognition; neuroinformatics and biomedical engineering applications; as well as proactive information retrieval and inference.

To conclude, Prof. Oja is an innovator par excellence. Knowing him as I do, he will continue to impact the world of Neural Computation through his pioneering contributions in years to come.

Prof. JOSÉ PRÍNCIPE

In the 1970s, Erkki Oja opened up roads to many discoveries in the late portion of the twentieth century and beyond. His first works on PCA provided the tone which blended a very solid grounding in mathematics with the amazing power of online adaptation that we still are grasping to fully understand. They require imagination, intuition, and transcend the aseptic world of mathematics. Erkki predicted and solidified many important applications of his simple adaptation rule (now with his name), and the applications of FastICA to imaging will always be connoted with him and his group in Finland.

His leadership in adaptive informatics, and more recently in computational inference, has propelled many young scientists in Finland, and all around the world to this exciting domain so important for big data. But on top of it all, Erkki's legacy transcends science and engineering. He is a true scholar, a gentleman, a wonderful person, and I am very fortunate to call him my friend.

Prof. TÜLAY ADALI

Erkki Oja has been a true leader in the field, and paved the way to much of the exciting work going on today in the machine learning field, particularly in nonlinear adaptive processing. He has provided the essential bridge between neural computation and adaptive optimization theory, and has not only provided the tools to address many of today's challenging problems but also offered different and fruitful ways to visualize them, making his impact particularly long lasting. The continuing strong rate in citations to his work from all periods, including those from the early 1980s, is a simple testament to the influence of his work and its continuing importance.

Beyond all this, he does inspire those around him, not only to achieve their best technically but also to enjoy life. It has been always a pleasure to be in his company, which might include sharing a fine meal with a nice glass of wine, or following the rhythm of the music, as he demonstrates to those around him how to truly be in the moment and enjoy life.

Prof. LUÍS BORGES DE ALMEIDA

I first met Erkki Oja in 1990 at a workshop on Neural Networks that I helped to organize in Sesimbra, Portugal. Over the years, I have had the opportunity to collaborate with him in a number of circumstances, the major one having been participation in the BLISS research project, which extended from 2000 to 2003.

I have come to deeply appreciate his scientific competence and good judgment, as well as his good disposition and sense of humor. He has made important scientific contributions to several fields, especially Neural Networks, ICA, and BSS. He has attracted many people of great quality to work with him, having given a great contribution to their scientific training, and having formed a renowned scientific lab.

Among all these facts that I truly value, there is one that I value the most: Erkki Oja has become a very good friend, beyond all the scientific and professional cooperation. I wish to express here my best wishes for his future.

Prof. CHRISTIAN JUTTEN

My Dear Erkki,

In a few lines, I would like to explain how you have been and you are an important person for me, at the scientific as well as human levels.

After my PhD in 1981, devoted to investigating how information is transmitted and modified through actual and formal neural networks, I considered studying learning and I have been strongly attracted by unsupervised learning, perhaps as a reaction against the trend of supervised learning.

Especially, I was very interested by Hebb's rule and its extension, Oja's rule. I was actually impassioned by the ideas, principles, and algorithms of Self-Organizing Maps (SOM) pioneered by T. Kohonen, and for which Erkki Oja did many important contributions.

At that time (and still now), I was wondering how living systems could be able to do so powerful processings, so difficult for a computer, and my researches were inspired by studying SOM, PCA, and other of your contributions in pattern recognition. I believe that probably the first algorithm of BSS, designed in 1983 with B. Ans and J. Hérault, for modeling how vertebrate can control joint motion, has been the fruit of the core question of unsupervised learning: how and what is it possible to learn without any supervisor?

I believe that I met you for the first time in 1986 in Snowbird.... Probably, you do not remember since I was a very young researcher. Later, I had a stronger contact with you in the 1990s, when our two groups submitted a proposal for a European project: unfortunately, it was not accepted. However, during the meetings for preparing the project, I have had the chance to meet you and discuss with you longer. Certainly, we became closer with the development of BSS, and the similarity between your nonlinear PCA and our first blind separation algorithms. Our two groups in Helsinki and in Grenoble contributed to design some building blocks of source separation methods, with various applications. Great moments were the first and the second ICA conference, in Aussois and then in Helsinski, and the European project BLISS in which we were partners.

In addition to the scientific respect I have for you, Erkki, I believe it was the beginning of a strong friendship. Finally, in 2008 and later in 2013, I applied for a very selective position in Institut Universitaire de France, and I asked you if you would like to act as one of my reference scientists: I am really honored that you accepted with pleasure, and each time you do the job with a perfect efficacy! Thank you a lot.

Erkki, you are now a distinguished Professor of Aalto University and you deserve it. Since 1981, you and your work have guided me and I am sure that you have been such a beacon for many other scientists, worldwide. You are my friend, and you are welcome in my lab and in my home. I wish you the best.

Prof. MARK PLUMBLEY

My own research has been influenced by the pioneering work of Erkki Oja almost ever since I began my research career. Like many others who began their PhD research in artificial neural networks, or connectionist models, in the late 1980s, I had to decide between supervised and unsupervised learning as a research topic. Having worked a little on information theory and coding in video, I wondered if this could be used as a "driver" for unsupervised learning, for a suitable algorithm.

Like many others, I had come across the famous unsupervised "Kohonen network," or SOM, in lectures in a Masters course, and I read up more about this in a text book by Kohonen. The SOM was perhaps too complex for what I was trying to do, but buried in the middle of Kohonen's text book was an apparently simple Hebb-like learning algorithm for a linear neuron, which learned to find the normalized principal component of its input data set.

This was the "Oja rule," of Erkki Oja's classic 1982 paper. That rule was simple, beautiful, and intriguing. I had found the starting point for my PhD.

My PhD work investigated the generalization to multiple outputs, building on the work by Oja and Karhunen in 1985, which found either several ordered principal components, or the space spanned by the principal components (the "principal subspace"). I found ways to use Shannon information as a motivation for the algorithm (related to Ralph Linsker's "Infomax" principle), and also to use information as an "energy" measure (Lyapunov function) to prove convergence of these Oja-like algorithms. Convergence analysis by others also demonstrated that the deceptively simple algorithm performed two complementary operations. For the original single neuron version, the weight vector finds the direction of the principal component, while simultaneously the length of the weight vector converges to unity. For the multiple-neuron version, the space spanned by the weight vectors converged to the principal subspace, while simultaneously the weight vectors themselves converged to an orthonormal set.

This "magic" self-regulation took a while to be understood properly by researchers. It meant, for example, that you cannot simply change the signs to obtain a minor-component analysis algorithm. While you would find the direction of the minor component, the self-regulation part would now be unstable and the length of the weight vector would diverge. So this apparently simple algorithm has given many researchers a lot of material to work on! In this world of Big Data, Oja's algorithm is still finding interesting applications. In the recent Machine Learning for Signal Processing conference (MLSP 2014), it was used to explore PCA estimation in distributed networks of processors.

After my PhD, as a new lecturer at King's College London, I helped establish an EU-funded "Network of Excellence" in neural networks, "NEuroNet," where Erkki Oja was one of the key partners. As part of this I participated in a small specialist workshop on Independent and Principal Component Analysis Methods in Thessaloniki in 1996, hosted by Kostas Diamantaras. At the time I was still concentrated on linear PCA-related methods, but Erkki was by then already a pioneer in the emerging field of ICA, using nonlinear neural networks.

By 2000 I was reviewing my research direction, and finally realized that ICA was going to be an interesting area to work in. My first proper ICA conference was ICA 2001, hosted by Erkki on a small island near to Helsinki University of Technology.

I knew I had a lot of catching up to do in the field of ICA, and Erkki agreed to host me for a 3-month visit in early 2002, funded by Leverhulme Trust. This visit gave me the chance to work closely with Erkki and his lab, and was instrumental in refocusing my research into ICA and source separation. It gave me the chance to work with Erkki on nonnegative versions of his nonlinear PCA model: this became the nonnegative PCA algorithm for nonnegative ICA (NN-ICA). It also gave me the chance to see how well Erkki is regarded as a research leader by colleagues, collaborators, and other staff in his lab, as well as his support and kindness to me. I still have the pair of Moomin mugs that he gave me as a souvenir of my visit!

Erkki has continued to be an inspiration to my research career. With the support of Erkki and others I hosted the International Conference on Independent Component Analysis and Signal Separation (ICA 2007) in London a few years later, and chaired the ICA Steering Committee. The work with Erkki on nonnegative ICA also led to an interest in geometrical methods of source separations, sparse representations, and compressed sensing, which still drives my current research. I wish Erkki all the best for his forthcoming retirement, and I am sure that he will continue to inspire me and many other researchers for many years to come.

Prof. KLAUS-ROBERT MÜLLER AND Dr. ANDREAS ZIEHE

Erkki, whom one of us (KRM) already met in 1991, at the occasion of the first ICANN Conference, in Helsinki, was already then an idol and my hero in neural networks. Definitely one of the great pioneers of our field. I respectfully called him Prof. Oja then, as a PhD student with still more than 1 year to go, and experienced his wonderful kindness, open mind, and his friendly advice. ICANN'91 – my first international conference where I gave a talk – certainly was pivotal in my career, since I met many other researchers there for the first time, for example, Geoff Hinton, Shun-Ichi Amari, Teuvo Kohonen, John Hertz, Werner von Seelen, and also the then "younger folks" Klaus Obermayer, Helge Ritter, Thomas Martinetz, Klaus Pawelzik, among others.

After 1991, and over the years, Erkki and I met many inspiring times, collaborated and became friends, and often so I profited from his experience and helpful advice.

In one period, our collaboration (now KRM and AZ[a]) became very intense, namely, during the EU's BLISS project, where our consortium (Erkki Oja, HUT, Helsinki; Christian Jutten, INPG, Grenoble; Luis Almeida, INESC ID, Lisbon; Simon Haykin, Mc Master, Hamilton; KRM, Fraunhofer FIRST, Berlin) furthered theoretical and practical research in BSS and ICA.

In particular, at that time, we jointly laid the very foundations to a number of novel algorithms for nonlinear ICA methods.

[a]AZ gratefully acknowledges that Erkki was a reviewer of his doctoral thesis, submitted to the University of Potsdam in 2005.

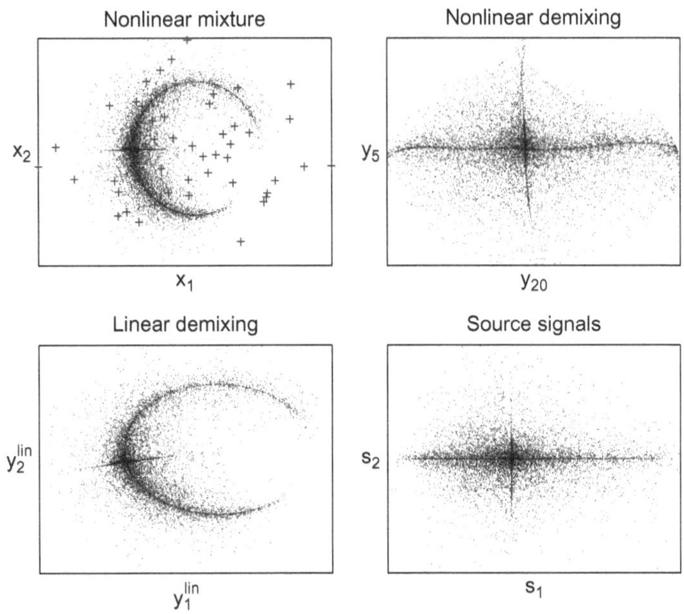

FIGURE I.1

Nonlinear blind source separation, and the "value" of some mixed (-up) speech.

While our research focused on possibilities to unmix highly nonlinear mixtures of complex sources (see Figure I.1, with an example using speech signals), this happened at the same time as the monetary union of the EU (in January 2002). In fact, to our project meeting in February 2002 in Vietri sul Mare,[b] all of us brought bags of coins to exchange and learn about the other EU nation's Euros. Erkki brought the very rare and popular Finnish Euro coins.

Recently we learned that Erkki is not only a world-class researcher, but that he is a gourmet and also – not uncommon for Finns – a passionate friend of Sauna. He shared with us that, in winter, his hobby is to enter an ice clad lake, through a hole in the ice, after Sauna. One of us (KRM) definitely looks forward to a winter visit to Finland, to also join this extraordinary experience; he is currently preparing his circulation for this event....

[b]See http://wayback.archive.org/web/20050306062356/http://ica.sa.infn.it/photos.htm.

Chapter Abstracts

CHAPTER 1

THE INITIAL CONVERGENCE RATE OF THE FastICA ALGORITHM: THE "ONE-THIRD RULE"

Scott C. Douglas

When estimating parameters adaptively using an iterative algorithm, the rate of convergence is an important gauge of the algorithm's overall performance. This chapter provides numerous results on the initial convergence rate of the well-known FastICA algorithm by Hyvärinen and Oja for independent component analysis. Particular attention is paid to the kurtosis-based form of the algorithm due to its analytical tractability. Through various analyses, it is shown that the convergence rate of the inter-channel interference (ICI) for the single-unit FastICA algorithm in an ideal linear mixing scenario is linear with a rate value of $1/3$. Thus, the algorithm reduces the average ICI by 4.77 dB at each step. Results for two-source, three-source, four-source, and general m-source linear mixtures are considered. Simulations verify the analytical results.

CHAPTER 2

IMPROVED VARIANTS OF THE FastICA ALGORITHM

Zbynvek Koldovský and Petr Tichavský

The article presents a survey of improved variants of the famous FastICA algorithm for Independent Component Analysis. Variants of the algorithm tailored to separate mixtures of stationary non-Gaussian signals and mixtures of nonstationary (block-wise stationary) non-Gaussian signals are described. Performance analyses of the algorithms are given and compared to the respective Cramér-Rao lower bounds. The behavior of FastICA variants when additive noise is present in the signal mixture is studied through a bias analysis.

CHAPTER 3

A UNIFIED PROBABILISTIC MODEL FOR INDEPENDENT AND PRINCIPAL COMPONENT ANALYSIS

Aapo Hyvärinen

Principal component analysis (PCA) and independent component analysis (ICA) are both based on a linear model of multivariate data. They are often seen as complementary tools, PCA providing dimension reduction and ICA separating underlying components or sources. In practice, a two-stage approach is often followed, where first PCA and then ICA are applied. Here, we show how PCA and ICA can be seen as special cases of the same probabilistic generative model. In contrast to conventional ICA theory, we model the variances of the components as further parameters. Such variance parameters can be integrated out in a Bayesian framework, or estimated in a more classic framework. In both cases, we find a simple objective function whose maximization enables estimation of PCA and ICA. Specifically, maximization of the objective under Gaussian assumption performs PCA, while its maximization for whitened data, under assumption of non-Gaussianity, performs ICA.

CHAPTER 4

RIEMANNIAN OPTIMIZATION IN COMPLEX-VALUED ICA

Visa Koivunen and Traian Abrudan

In many engineering applications such as beamforming, signal separation, and multiantenna communications, we are facing a constrained optimization problem w.r.t. complex-valued matrices. A prime example is the case of complex-valued independent component analysis (ICA) algorithms that require prewhitening of the data. Complex-valued observation vectors are encountered in fMRI data, radar, wireless communication, and remote-sensing applications, for example. In this chapter, we present a Riemannian geometry approach for optimization of a real-valued ICA cost function $\mathcal{J}$ of complex-valued matrix argument $\mathbf{W}$, under the constraint that $\mathbf{W}$ is an $n \times n$ unitary matrix. We present a steepest descent algorithm on the Lie group of unitary matrices $U(n)$ for finding a solution to the complex ICA problem. This algorithm moves toward the optimum along the geodesics; that is, the locally shortest paths. The developed algorithm is applied to blind source separation in multiantenna (MIMO) wireless systems where multiple datastreams are transmitted simultaneously using the same frequency resources. A well-known joint diagonalization method (Joint Approximate Diagonalization of Eigenmatrices) is employed in source separation with the developed optimization algorithm.

CHAPTER 5

NONADDITIVE OPTIMIZATION

Zhirong Yang and Irwin King

In optimization of a learning objective, additive updates that use line search along a given gradient-based direction seem to be a paradigm in the existing algorithms. However, additive updates are not necessarily the most convenient and efficient choice, especially for constrained problems. Here, we review several such problems and their nonadditive optimization algorithms, which have shown significant convenience and efficiency in practice.

CHAPTER 6

IMAGE DENOISING, LOCAL FACTOR ANALYSIS, BAYESIAN YING-YANG HARMONY LEARNING

Guangyong Chen, Fengyuan Zhu, Pheng Ann Heng and Lei Xu

A new nonlocal-filtering method LFA-BYY is proposed for image demonizing via learning a local factor analysis (LFA) model from a polluted image under processing and then denoising the image by the learned LFA model. With the help of the Bayesian Ying-Yang (BYY) harmony learning, LFA-BYY can appropriately control the dictionary complexity and learn the noise intensity from the present image under processing, while the existing state-of-the-art methods either use a pretrained dictionary or a general basis, and require an accurate noise intensity estimation provided in advance. In comparison with BM3D, K-SVD, EPLL, and Msi on the benchmark Kodak dataset and additional medical data, experiments have shown that LFA-BYY has not only obtained competitive results on images polluted by a small noise but also outperformed these competing methods when the noise intensity increases beyond a point, especially with significant improvements as the noise intensity becomes large.

CHAPTER 7

UNSUPERVISED DEEP LEARNING: A SHORT REVIEW

Juha Karhunen, Tapani Raiko and KyungHyun Cho

Deep neural networks with several layers have recently become a highly successful and popular research topic in machine learning due to their excellent performance in many benchmark problems and applications. A key idea in deep learning is to learn not only the nonlinear mapping between the inputs and outputs but also the underlying structure of the data (input) vectors. In this chapter, we first consider problems with training deep networks using backpropagation-type algorithms. After this, we consider various structures used in deep learning, including restricted Boltzmann machines, deep belief networks, deep Boltzmann machines, and nonlinear autoencoders. In the latter part of this chapter, we discuss in more detail the recently developed neural autoregressive distribution estimator and its variants.

CHAPTER 8

FROM NEURAL PCA TO DEEP UNSUPERVISED LEARNING

Harri Valpola

A network supporting deep unsupervised learning is presented. The network is an autoencoder with lateral shortcut connections from the encoder to the decoder at each level of the hierarchy. The lateral shortcut connections allow the higher levels of the hierarchy to focus on abstract invariant features. Whereas autoencoders are analogous to latent variable models with a single layer of stochastic variables, the proposed network is analogous to hierarchical latent variable models.

Learning combines denoising autoencoder and denoising sources separation frameworks. Each layer of the network contributes to the cost function a term which measures the distance of the representations produced by the encoder and the decoder. Since training signals originate from all levels of the network, all layers can learn efficiently even in deep networks.

The speedup offered by cost terms from higher levels of the hierarchy and the ability to learn invariant features are demonstrated in experiments.

CHAPTER 9

TWO DECADES OF LOCAL BINARY PATTERNS: A SURVEY

Matti Pietikäinen and Guoying Zhao

Texture is an important characteristic for many types of images. In recent years, very discriminative and computationally efficient local texture descriptors based on local binary patterns (LBPs) have been developed, which has led to significant progress in applying texture methods to different problems and applications. Due to this progress, the division between texture descriptors and more generic image or video descriptors has been disappearing. A large number of different variants of LBP have been developed to improve its robustness and to increase its discriminative power and applicability to different types of problems. In this chapter, the most recent and important variants of LBP in 2-D, spatiotemporal, 3-D, and 4-D domains are surveyed. Interesting new developments of LBP in 1-D signal analysis are also considered. Finally, some future challenges for research are presented.

CHAPTER 10

SUBSPACE APPROACH IN SPECTRAL COLOR SCIENCE

Jussi Parkkinen, Hannu Laamanen and Markku Hauta-Kasari

The use of wavelength spectrum to represent color in color processing has recently seen an increase in popularity in color and color image analysis. This is partly due to the need for more accurate color information, the development of spectral imaging technologies, and the availability of efficient spectral processing techniques. The need for accurate color information processing arises in a variety of industries. Using color spectrum also gives new possibilities, for example in medical diagnostics. The principal component analysis (PCA) has become a standard method in spectral color compression and spectrum reconstruction. Varieties of PCA and other similar expansions like independent component analysis (ICA) have also been studied in spectral color science. In this chapter, we give a short overview to the development of the PCA in spectral color science, introduce the use of PCA and ICA, and list some of the applications, where PCA has been utilized in the field of spectral color science.

CHAPTER 11

FROM PATTERN RECOGNITION METHODS TO MACHINE VISION APPLICATIONS

Heikki Kälviäinen

This chapter considers scientific challenges in developing machine vision applications based on pattern recognition methods. The goal of such solutions is to create useful and significant applications, especially using digital image processing and analysis. The focus is on visual inspection in the industry, computational vision, medical imaging and processing, and biosensors. It is important to recognize relevant phenomena, to measure them, and to understand how humans define their experience based on the measured data. This knowledge enables us to develop intelligent and robust methods for practical applications, with smart feature selection and parameter sensitivity analysis. Real-world applications for industrial machine vision, psychometric quality assessment, medical image processing, and traffic sign condition monitoring are shown. Annotation tools and expert databases are also discussed.

CHAPTER 12

ADVANCES IN VISUAL CONCEPT DETECTION: TEN YEARS OF TRECVID

Ville Viitaniemi, Mats Sjöberg, Markus Koskela, Satoru Ishikawa and Jorma Laaksonen

In this chapter, we describe the structure and operation of the visual concept-detection subsystem of the PicSOM multimedia retrieval system. We evaluate several alternative techniques used for implementing this component and show the essential results of a series of experiments in the large-scale setups of the TRECVID video retrieval evaluation campaigns in 2005, 2009, and 2014. During these years, the PicSOM system has gone through substantial evolution in both the statistical features and the detection algorithms employed. Transition from global image features to the bag-of-visual-words features and recently further to convolutional deep neural network-based features is also justified in the light of our results. Overall, during the 10 years of participation in TRECVID, the PicSOM system has shown close to the state-of-the-art performance in this very rapidly developing field of research.

CHAPTER 13

ON THE APPLICABILITY OF LATENT VARIABLE MODELING TO RESEARCH SYSTEM DATA

Ella Bingham and Heikki Mannila

In this note, we discuss the applicability of latent variable models as a tool in analyzing the structure of a research system. We consider whether tensor methods, especially Parallel Factor Analysis, are appropriate for the description of the personnel structure and publication results of different scientific disciplines in different universities. As the measured variables (personnel structure and publications) interact with both the universities and the disciplines, it is useful to view the data as a tensor. Our preliminary results suggest that tensor methods are indeed able to find meaningful structure in such data.

PART

I

Methods

CHAPTER

The initial convergence rate of the FastICA algorithm: The "One-Third Rule"

1

Scott C. Douglas

Department of Electrical Engineering, Southern Methodist University, Dallas, Texas, USA

1.1 INTRODUCTION

Research in the fields of blind source separation (BSS) and independent component analysis (ICA) has uncovered numerous algorithms and procedures for performing linear decompositions of spatial and spatio-temporal information based on their underlying statistical structures. One of the most-used procedures for both BSS and ICA is the FastICA algorithm of Hyvärinen and Oja [1,2]. This technique was derived in [1] as an approximate Newton method applied to a measure of non-Gaussianity as a separation criterion. The most-common implementation of this block-based procedure is a two-stage approach. Given a sequence of m-dimensional measurements $\mathbf{x}(k) = [x_1(k) \ \cdots \ x_m(k)]^T$, $1 \leq k \leq N$ for which a linear transformation $\mathbf{y}(k) = \mathbf{Bx}(k)$ is desired, the following steps are taken:

1. These data are first prewhitened by an $(m \times m)$ prewhitening matrix $\mathbf{P}$ such that the prewhitened measurements given by

$$\mathbf{v}(k) = \mathbf{Px}(k) \tag{1.1}$$

have uncorrelated and unit variance elements; that is

$$\frac{1}{N}\sum_{k=1}^{N} \mathbf{v}(k)\mathbf{v}^T(k) = \mathbf{I}, \tag{1.2}$$

where $\mathbf{I}$ is the identity matrix.

2. An iterative procedure is employed to adjust an $(m \times m)$ orthonormal matrix $\mathbf{W}_t = [\mathbf{w}_{1k} \ \cdots \ \mathbf{w}_{mk}]^T$ such that

$$\mathbf{y}_t(k) = \mathbf{W}_t\mathbf{v}(k) \tag{1.3}$$

contains the estimates of the independent sources. Each $\mathbf{w}_{it} = [w_{i1k} \cdots w_{imk}]^T$ in $\mathbf{W}_t$ is adjusted using

$$y_{it}(k) = \mathbf{w}_{it}^T(k)\mathbf{v}(k) \tag{1.4}$$

E. Bingham, S. Kaski, J. Laaksonen and J. Lampinen (Eds) Advances in Independent Component Analysis and Learning Machines.
http://dx.doi.org/10.1016/B978-0-12-802806-3.00001-4

$$\tilde{\mathbf{w}}_{it} = \frac{1}{N}\sum_{k=1}^{N} f(y_{it}(k))\mathbf{v}(k) - f'(y_{it}(k))\mathbf{w}_{it} \tag{1.5}$$

$$\hat{\mathbf{w}}_{i(t+1)} = \frac{\tilde{\mathbf{w}}_{it}}{\|\tilde{\mathbf{w}}_{it}\|}, \tag{1.6}$$

where $f(y)$ is a nonlinearity, $f'(y)$ is its derivative, and $\|\mathbf{w}_{it}\| = (\sum_{j=1}^{m} w_{ijt}^2)^{1/2}$ is the Euclidean norm of the vector $\mathbf{w}_{it}$. The vectors $\hat{\mathbf{w}}_{i(t+1)}$ are then orthogonalized with respect to each other to obtain the updated matrix $\mathbf{W}_{t+1}$. Although many choices for $f(y)$ are possible, two popular choices are $f(y) = y^3$ and $f(y) = \tanh(\beta y)$, where $\beta > 0$.

The second stage of this two-stage procedure is iterated until some convergence condition is met, at which point $\mathbf{B} = \mathbf{W}_{\text{final}}\mathbf{P}$ is the resulting transformation.

Assume that the source mixtures satisfy the linear model

$$\mathbf{x}(k) = \mathbf{A}\mathbf{s}(k), \tag{1.7}$$

where $\mathbf{A}$ is an unknown $(m \times m)$ mixing matrix, $\mathbf{s}(k) = [s_1(k) \ \cdots \ s_m(k)]^T$ is the source signal vector, the elements of $\mathbf{s}(k)$ are statistically independent with zero means and unit variances, and the amplitude statistics of the elements of $\mathbf{s}(k)$ satisfy

$$E\{s_i(k)f(s_i(k))\} - E\{s_i^2(k)\}E\{f'(s_i(k))\} \neq 0 \tag{1.8}$$

for $(m-1)$ of the sources, where $E\{\cdot\}$ denotes statistical expectation. Then, it can be shown in [2] that this procedure is locally stable about a separating solution yielding $y_{it}(k) = c_{ijt}s_j(k)$ as $N \to \infty$ and $t \to \infty$, where c_{ijt} is the (i,j)th element of the combined system coefficient matrix

$$\mathbf{C}_t = \mathbf{W}_t\mathbf{P}\mathbf{A}, \tag{1.9}$$

and the limiting value of $\mathbf{C}_t$ forms a permutation matrix

$$\lim_{t\to\infty} \mathbf{C}_t = \mathbf{\Phi}, \tag{1.10}$$

where $\mathbf{\Phi}$ has only one unity element along any row or column.

The convergence behavior of the FastICA algorithm given by Eqs. (1.4)–(1.6) for the cubic nonlinearity $f(y) = y^3$ was further analyzed in [1]. This analysis developed averaged evolution equations for the coefficient ratios c_{ijt}/c_{ilk} given their initial values c_{ij0}/c_{il0} and the kurtoses κ_i of the sources defined for $1 \leq i \leq m$ as

$$\kappa_i = E\{s_i^4(k)\} - 3(E\{s_i^2(k)\})^2. \tag{1.11}$$

This analysis cleverly avoids the use of the normalization condition by employing coefficient ratios.

A useful measure of separation quality is the total inter-channel interference (ICI) given by

$$\text{ICI}_{\text{tot},t} = \sum_{i=1}^{m} \text{ICI}_{i,t} \tag{1.12}$$

$$\mathrm{ICI}_{i,t} = \frac{\sum_{j=1}^{m} c_{ijt}^2}{\max_{1\le j\le m} c_{ijt}^2} - 1, \tag{1.13}$$

where $\mathrm{ICI}_{i,t}$ is the ICI of the ith output channel. The output channel ICI directly measures the degree of isolation of a single independent source at the output of a single-unit FastICA procedure. Similarly, assuming that unique sources are extracted at each single-unit output,[a] the total ICI directly measures the degree of separation of the independent sources by assessing how much the combined system matrix deviates from a scaled permutation matrix. In the sequel, we shall focus on the average value of the output channel ICI without regard to the source being extracted, given by

$$E\{\mathrm{ICI}_t\} = E\{\mathrm{ICI}_{i,t}\}, \tag{1.14}$$

where the expectation operator $E\{\cdot\}$ is taken with respect to both the input data vectors $\{\mathbf{x}(k)\}$ and the statistical distribution of a single initial coefficient vector $\mathbf{c}_0 = \mathbf{c}_{i0}$.

Superb algorithms are deserving of further study and explanation of why they work. The behavior of the FastICA algorithm has been studied in several ways. Hyvärinen has given a justification for the algorithm's fast convergence behavior when a kurtosis-based independence measure is used [1,2]. Regalia et al. [3] have related the FastICA procedure to a gradient method and have considered the convergence of the algorithm in the undermodeled case. Oja and Yuan have studied the global fixed-point behavior of the two-source FastICA algorithm with kurtosis contrast and symmetric orthogonalization of the separation matrix, with additional extensions to the arbitrary contrast case [4,5]. Cramér-Rao lower bounds for ICA and BSS are given in [6,7], and the statistical efficiency of the FastICA algorithm under finite sample sizes is explored and compared to other approaches. The local convergence analysis of FastICA taking into account the sign-flipping nature of the coefficient updates is presented in [8]. Ollila has examined the performance of the deflation-based FastICA algorithm using influence functions, and identifies situations where the performance of the algorithm can vary due to the ordering of the extracted sources [9].

While the above studies all give reasonable explanations as to when and why the FastICA algorithm separates signal mixtures, the question still remains: *What makes the FastICA algorithm converge quickly?* Hyvärinen and Oja show through the structure of the single-unit update with a kurtosis-based entropy measure that the ICI converges cubically locally at a separating solution. This analysis does not consider the global behavior of the update, nor does it define the region of locality for the cubic convergence behavior. An interesting observation about the FastICA algorithm for noise-free mixtures and a kurtosis-based separation criterion was made [10]: the average convergence of the total ICI is well-described by the following empirical rule at iteration t:

$$E\{\mathrm{ICI}_{\mathrm{tot},t+1}\} \approx \left(\frac{1}{3}\right) E\{\mathrm{ICI}_{\mathrm{tot},t}\}. \tag{1.15}$$

In this context, averaging is done over ensembles of source mixtures with fixed statistical distributions in each run, but with random initial conditions. This result can also be expressed in terms of a single-unit average ICI as

$$E\{\text{ICI}_t\} \approx \left(\frac{1}{3}\right) E\{\text{ICI}_{t-1}\} \tag{1.16}$$

$$\approx \left(\frac{1}{3}\right)^t E\{\text{ICI}_0\}. \tag{1.17}$$

That is, the FastICA algorithm with a kurtosis-based separation criterion causes the ICI to decrease by $1/3$ or 4.77 dB at each iteration, independent of the source distributions and the number of sources. In subsequent discussions, we shall refer to the relationship in Eq. (1.17) as the "(1/3)rd Rule."

In this chapter, we provide a review of existing [16,17] and additional new results on the convergence of the FastICA algorithm's behavior for the case $f(y) = y^3$ under the linear mixing model in Eq. (1.7) with statistically independent sources. We first review the proof in [10] that all locally stable stationary points of an analysis equation for the FastICA algorithm under unit-norm constraints correspond to desirable separating solutions. We then consider the initial convergence rate of various measures of the ICI of the single-unit FastICA algorithm in various scenarios, including two-source, three-source, four-source, and general m-source mixtures. In *every* case, we find that the "(1/3)rd Rule" emerges as the dominating term in the initial convergence of the algorithm, independent of the algorithm initialization and the values of the source kurtoses in the mixture. This body of work provides conclusive evidence that this particular form of the FastICA algorithm's convergence is typically observed to be exponential, constant speed, and independent of input signal statistics. Simulations verify the accuracy of the analytical results.

The organization of the chapter is as follows. In Section 1.2, the statistical analysis of the single-unit FastICA algorithm with $f(y) = y^3$ is performed, and the evolutionary equation defining the average ICI at each iteration is given. In Section 1.3, a stationary point analysis of the single-unit FastICA algorithm is provided, showing that the only stable stationary points extract a unique independent source. In Section 1.4, initial convergence rate results for the single-unit FastICA algorithm for two-source mixtures are provided for both equal-kurtosis sources and arbitrary kurtosis sources, verifying the "(1/3)rd Rule" in different ways. Section 1.5 provides analyses of the three-source, four-source, and arbitrary m-source cases for arbitrary kurtosis source mixtures under a uniform prior distribution for the combined system vector, as well as a unique analysis employing order statistics in the equal-kurtosis source case. In all of these analyses, the "(1/3)rd Rule" emerges as the main descriptor for the convergence of the algorithm. Simulations in Section 1.6 show the viabilities of the various analyses in describing the average behavior of the FastICA algorithm in various scenarios. Conclusions are drawn in Section 1.7.

1.2 STATISTICAL ANALYSIS OF THE FastICA ALGORITHM

The technique used to analyze the behavior of the FastICA algorithm is well known in the adaptive signal processing field [11]. The evolutionary behaviors of the parameters of any adaptive algorithm can be determined by developing an equivalent set of coefficient updates whose forms depend on the statistics of the signals being processed. Because the FastICA algorithm is a block-based procedure, the corresponding coefficient updates derived from the analysis are equivalent to those of the FastICA algorithm as the block size N tends to infinity.

This analysis of the FastICA algorithm uses the expectation operator defined for $\mathbf{M}(k)$ dependent on $\mathbf{s}(k)$ as

$$E\{\mathbf{M}(k)\} = \lim_{N\to\infty} \frac{1}{N}\sum_{k=1}^{N} \mathbf{M}(k). \tag{1.18}$$

Let $\mathbf{s}(k)$ contain sources that are statistically independent with zero means and unit variances, such that for any four nonequal integers $\{i,j,l,p\} \in [1,m]$,

$$E\{s_i(k)\} = 0, \quad E\{s_i(k)s_j(k)\} = 0, \tag{1.19}$$

$$E\{s_i^2(k)\} = 1, \quad E\{s_i(k)s_j(k)s_l^2(k)\} = 0, \tag{1.20}$$

$$E\{s_i(k)s_j^3(k)\} = 0, \quad E\{s_i(k)s_j(k)s_l(k)s_p(k)\} = 0. \tag{1.21}$$

Define the kurtosis of each $s_i(k)$ as in Eq. (1.11).

The first stage of the FastICA algorithm is a prewhitening step. As $N \to \infty$, we have that

$$\mathbf{P}E\{\mathbf{x}(k)\mathbf{x}^T(k)\}\mathbf{P}^T = \mathbf{PA}E\{\mathbf{s}(k)\mathbf{s}^T(k)\}\mathbf{A}^T\mathbf{P}^T \tag{1.22}$$

$$= \mathbf{PAA}^T\mathbf{P}^T = \mathbf{I}, \tag{1.23}$$

such that $\mathbf{PA} = \mathbf{\Gamma}$, where $\mathbf{\Gamma}$ is orthonormal; that is, $\mathbf{\Gamma}\mathbf{\Gamma}^T = \mathbf{\Gamma}^T\mathbf{\Gamma} = \mathbf{I}$. Thus, we have

$$\mathbf{v}(k) = \mathbf{\Gamma}\mathbf{s}(k), \tag{1.24}$$

where $E\{\mathbf{v}(k)\mathbf{v}^T(k)\} = \mathbf{I}$.

Assume a single-unit system in which $\mathbf{w}_t = \mathbf{w}_{it}$, where we have suppressed the ith index for notational simplicity. Define the transformed coefficient vector

$$\mathbf{c}_t = \mathbf{\Gamma}^T\mathbf{w}_t, \tag{1.25}$$

such that

$$y_t(k) = \mathbf{w}_t^T\mathbf{v}(k) = \mathbf{c}_t^T\mathbf{s}(k). \tag{1.26}$$

Then, by premultiplying both sides of Eq. (1.5) by $\mathbf{\Gamma}^T$, we can write the FastICA algorithm update as

$$\tilde{\mathbf{c}}_{t+1} = E\{y_t^3(k)\mathbf{s}(k)\} - 3\mathbf{c}_t \tag{1.27}$$

$$\mathbf{c}_{t+1} = \frac{\tilde{\mathbf{c}}_{t+1}}{\|\tilde{\mathbf{c}}_{t+1}\|}, \tag{1.28}$$

where $E\{y_t^2(k)\} = \mathbf{c}_t^T E\{\mathbf{s}(k)\mathbf{s}^T(k)\}\mathbf{c}_t = 1$ due to the unit-norm constraint on $\mathbf{c}_t$.

To proceed further, we require an expression for $E\{y_t^3(k)\mathbf{s}(k)\}$. The following theorem yields the necessary result and the resulting expression for the update of $\mathbf{c}_t$.

Theorem 1. *Let $\mathbf{c}_t$ be an m-dimensional vector, and suppose the elements of $\mathbf{s}(k)$ satisfy Eqs.* (1.19) *and* (1.20). *Define $y_t(k) = \mathbf{c}_t^T\mathbf{s}(k)$ for $1 \leq n \leq N$. Then,*

$$E\{y_t^3(k)\mathbf{s}(k)\} = \mathbf{Kf}(\mathbf{c}_t) + 3\|\mathbf{c}_t\|^2\mathbf{c}_t, \tag{1.29}$$

where $\mathbf{K}$ is a diagonal matrix whose (i,i)th element is κ_i and the ith element of $\mathbf{f}(\mathbf{c}_t)$ is c_{it}^3.

Corollary 1 (First shown in [1]). *As $N \to \infty$, the single-unit FastICA algorithm update in Eqs.* (1.4)–(1.6) *is equivalent to the vector update*

$$\mathbf{c}_{t+1} = \frac{\mathbf{Kf}(\mathbf{c}_t)}{\|\mathbf{Kf}(\mathbf{c}_t)\|}, \tag{1.30}$$

where $\mathbf{w}_t = \mathbf{PAc}_t$, or the scalar update for $1 \leq i \leq m$:

$$c_{i(t+1)} = \frac{\kappa_i c_{it}^3}{\sqrt{\sum_{j=1}^m \kappa_j^2 c_{jt}^6}}. \tag{1.31}$$

Proofs of these results are in the Appendix.

Discussion. The result in Eq. (1.30) yields several insights into the structure and convergence behavior of the FastICA algorithm. Specifically,

1. The convergence of each element of the combined system vector $\mathbf{c}_t = \mathbf{A}^T\mathbf{P}^T\mathbf{w}_t$ is largely uncoupled with respect to every other element of $\mathbf{c}_t$. The only coupling is due to the normalization constraint $\|\mathbf{c}_t\| = 1$.
2. The average behavior of the FastICA algorithm can be easily simulated for any initial combined system vector $\mathbf{c}_0$ if the kurtoses $\{\kappa_i\}$ of the sources are known and the sources satisfy the moment conditions in Eqs. (1.19)–(1.21). Thus, complicated Monte Carlo simulations involving source signal mixtures $\mathbf{x}(k)$ are not required.
3. The update in Eq. (1.30) is identical to the power method for finding the principal eigenvector [12] of the *time-varying* diagonal matrix $\mathbf{R}_t$ whose (i,i)th element is $\kappa_i c_{it}^2$.

This last result can be used to develop a simple implementation of the FastICA algorithm that is similar in structure to the method of orthogonal iterations for principal component analysis [12]. Table 1.1 shows a MATLAB-based implementation of the FastICA algorithm that employs the built-in functions `chol` and `qr` to implement the prewhitening and orthogonalization stages of the algorithm, respectively. In this program, `iter` determines the total number of iterations of the algorithm, and the

Table 1.1 The FastICA Algorithm in MATLAB

```
[N,m] = size(x);
W = eye(m);
v = x/chol((x'*x)/N);
for t=1:iter
        y = v*W;
        [W,R] = qr(v'*y.^3 - 3*W*N);
end
```

`for` loop can be replaced by a `while` loop with an appropriately chosen stopping criterion. At convergence, the *i* diagonal entry of `R` is equal to the kurtosis of the *i*th extracted source.

1.3 STATIONARY POINT ANALYSIS OF THE FastICA ALGORITHM

The results of the last section can be used to determine the stationary points of the FastICA iteration. Our analysis extends the work in [1,2] by finding *all* of the possible stationary points of the algorithm under the normalization condition and not just the desirable separating ones. Similar analytical results for a related prewhitened algorithm are described in [13]. For our analysis, we shall analyze the situation where the signs of each c_{it} are ignored, such that Eq. (1.31) is

$$|c_{i(t+1)}| = \frac{|\kappa_i c_{it}^3|}{\sqrt{\sum_{j=1}^{m} \kappa_j^2 c_{jt}^6}}. \tag{1.32}$$

Clearly, the cubic nature of the update in Eq. (1.31) causes the sign of each c_{it} to alternate back and forth as k is incremented if $\kappa_i < 0$.

Assume without loss of generality that the $|\kappa_i|$ values are ordered, such that $|\kappa_1| \geq |\kappa_2| \geq \cdots \geq |\kappa_m|$. Moreover, let m_p denote the smallest integer for which $|\kappa_j| = 0$ for $m_p + 1 \leq j \leq m$, and define $\mathcal{J}$ as any subset of the elements of $\mathcal{I} = \{1, 2, \ldots, m_p\}$. Then, we have the following.

Theorem 2. *The set of potential stationary points of Eq.* (1.32) *for* $\mathbf{c}_t = \mathbf{c}_s = [c_{1,s} \; \cdots \; c_{m,s}]^T$ *are given by all possible subsets of* $\mathcal{J}$ *with*

$$|c_{i,s}| = \sqrt{\frac{|\kappa_i|^{-1}}{\sum_{j\in\mathcal{J}} |\kappa_j|^{-1}}}, \quad i \in \mathcal{J} c_{j,s} = 0, j \notin \mathcal{J}. \tag{1.33}$$

Theorem 3. *The set of stable stationary points of Eq.* (1.32) *are the separating solutions defined for each* $i \in \mathcal{I}$ *as*

$$|c_{i,s}| = 1, \quad c_{j,s} = 0, \quad 1 \leq j \leq m, j \neq i. \tag{1.34}$$

Proofs of these results are in the Appendix.

Discussion. The above two theorems indicate that, if the FastICA algorithm converges, the only possible stationary points are those corresponding to separating solutions, such that the system output $y(k)$ contains one of the independent sources. Moreover, due to the unit variance constraint, the scaling of the source at the system output is unity. All of these results assume infinite data, such that $N \to \infty$, and thus numerical issues due to estimation quality are not considered here.

1.4 INITIAL CONVERGENCE OF THE FastICA ALGORITHM FOR TWO-SOURCE MIXTURES

1.4.1 OVERVIEW OF RESULTS

In this section, we provide our first analytical justifications of the convergence rate of the FastICA algorithm as predicted by Eq. (1.17) for $f(y) = y^3$. Our initial study considers the simplest mixing situation possible – a single-unit FastICA procedure applied to a noiseless two-source mixture of non-Gaussian-distributed sources. Although one may think that such a situation is overly simple and therefore not useful for practical performance predictions, we contend that (a) the simplest scenarios often allow one to exactly describe particular algorithmic characteristics that provide a qualitative description of the algorithm's behavior in more complex and more realistic scenarios, and (b) the two-source separation case is exactly encountered at the $(m-1)$th stage of source extraction in the m-dimensional FastICA algorithm.

For a two-source mixture, we show the following technical results:

- For equal-(magnitude)-kurtosis source mixtures, we determine
  - an exact expression for the probability density function (p.d.f.) of the ICI given an arbitrary initial distribution of the ICI;
  - a bound on the average ICI at iteration t for an arbitrary smooth initial distribution of the ICI;
  - an exact expression for the average ICI assuming a uniformly distributed look direction for the initial separating weight vector;
  - an approximate but exponentially tight expression for the average ICI for an arbitrary smooth initial distribution of the ICI over the interval [0, 1].

  In addition, an exploration of the limiting p.d.f. $p_t(u)$ of the ICI as a function of iteration number shows that its form is largely insensitive to the initial distribution of the ICI.

- For mixtures of two sources having *arbitrary* kurtoses κ_1 and κ_2, we derive exact and limiting expressions for the average ICI at iteration t assuming a uniformly distributed look direction for the initial separating weight vector.

In all of the cases above, we show that the "(1/3)rd Rule" describes the evolutionary behavior of the algorithm.

1.4.2 PRELIMINARIES

Our analysis of the average value of the ICI for the FastICA algorithm is performed in a coefficient-stochastic setting, in which

1. the number of measurements used to compute the averages in Eq. (1.18) is infinite, so that the evolutionary behavior described by the update in Eq. (1.30) is an accurate description of the FastICA algorithm given an initial combined system coefficient vector $\mathbf{c}_0$; and
2. the initial vector $\mathbf{c}_0$ possesses some distribution on the unit hypersphere, and thus the average performance of the algorithm depends on how this initial condition affects the evolution of the ICI performance measure in Eq. (1.14).

This type of analysis technique is well known in the adaptive filtering community and has led to a number of important results concerning the convergence behaviors of various algorithms in that field [11].

Because we are concerned with two source mixtures in this section, it will be additionally useful to consider an *intrinsic parametrization* of $\mathbf{c}_t$ that guarantees a minimal representation of the coefficient vector within its constraint space. Such a minimal representation is one-dimensional and clearly polar in form, which we choose to be

$$\mathbf{c}_t = [\cos(\theta_t) \quad \sin(\theta_t)]^T , \tag{1.35}$$

where θ_t is effectively "adjusted" by the FastICA procedure. The goal of the source separation task in terms of this procedure is to adjust θ_t such that it converges to one of the four values $\theta_{\text{opt}} \in \{0, \pi/2, \pi, 3\pi/2\}$. For purposes of descriptive simplicity, we shall only consider the convergent points $\theta_{\text{opt}} \in \{0, \pi/2\}$, as the behaviors in the other three quadrants in $\mathbf{c}_t$ are identical due to symmetry. We shall find the representation in Eq. (1.35) useful for understanding the behavior of FastICA in the two-source case.

In the two-coefficient case, the FastICA algorithm could be described in terms of a single-variable update on θ_t, an issue we take up in Section 1.7. In this section, we instead consider the value of ICI_t directly in terms of the probability density of the initial angle θ_0 defining the vector $\mathbf{c}(0)$. Without loss of generality, our discussion will be restricted to the angular range $0 \leq \theta_t \leq \pi/2$ due to four-quadrant symmetry, and the kurtoses κ_1 and κ_2 are assumed positive. Considering for the moment the arbitrary source kurtosis case, suppose $\mathbf{c}_0$ lies on the unit circle between $\theta = 0$ and $\theta = \pi/2$. Then, using the relationship in Eq. (1.30), it can be shown that

$$\frac{c_{2,t}^2}{c_{1,t}^2} = \left(\frac{\kappa_2}{\kappa_1}\right)^{3^t-1}\left(\frac{c_{2,0}^2}{c_{1,0}^2}\right)^{3^t} \tag{1.36}$$

$$= \frac{\kappa_1}{\kappa_2}\left(\sqrt{\frac{\kappa_2}{\kappa_1}}\tan(\theta_0)\right)^{2(3^t)}, \tag{1.37}$$

where $\theta_0 = \arctan(c_{1,0}/c_{2,0})$. Define the constant

$$a = \sqrt{\frac{\kappa_1}{\kappa_2}}. \tag{1.38}$$

Then, the ICI can be written as

$$\mathrm{ICI}_t = \max\left[\frac{c_{2,t}^2}{c_{1,t}^2}, \frac{c_{1,t}^2}{c_{2,t}^2}\right], \tag{1.39}$$

which translates into the relationship

$$\mathrm{ICI}_t = \begin{cases} \frac{\kappa_1}{\kappa_2}\left(\sqrt{\frac{\kappa_2}{\kappa_1}}\tan(\theta_0)\right)^{2(3^t)}, & 0 \le \theta \le \alpha_t \\ \frac{\kappa_2}{\kappa_1}\left(\sqrt{\frac{\kappa_1}{\kappa_2}}\cot(\theta_0)\right)^{2(3^t)}, & \alpha_t < \theta \le \pi/2 \end{cases}, \tag{1.40}$$

where

$$\alpha_t = \arctan\left(a \cdot (a^{-1})^{\frac{1}{3^t}}\right). \tag{1.41}$$

From Eq. (1.40), we see that ICI_t depends on θ_0, κ_2/κ_1, and t. For a given distribution of $\mathbf{c}_0$, denoted in the angle variable θ_0 as $\overline{p}_0(\theta)$, the relationship in Eq. (1.40) induces a distribution $p_t(u)$ of the ICI at iteration t. This distribution is best expressed in terms of the distribution of θ_0, as the ICI cost combines the distribution of θ_0 over the disjoint regions $[0, \alpha_t)$ and $[\alpha_t, \pi/2]$ in a nonlinear way. For this reason, we consider the following two cases separately:

1. *Equal-(Magnitude)-Kurtosis Sources.* For $\kappa_1 = \kappa_2$, we have $\alpha_t = \pi/4$ for all t. In such cases, it is reasonable to assume that θ_0 is distributed symmetrically about $\theta_0 = \pi/4$, such that ICI_t obeys the *scalar evolutionary equations* given by

$$\mathrm{ICI}_{t+1} = (\mathrm{ICI}_t)^3 \tag{1.42}$$

$$\mathrm{ICI}_t = (\mathrm{ICI}_0)^{3^t}. \tag{1.43}$$

   In this situation, the behavior of ICI_t is completely determined by the distribution $p_0(u)$ of ICI_0.
2. *Arbitrary-(Magnitude)-Kurtosis Sources.* In this case, it is easiest to explore the distribution of ICI_t and its moments, such as $E\{\mathrm{ICI}_t\}$ through the distribution of θ_0, denoted as $\overline{p}_0(\theta)$.

In either case, the distributions $p_0(u)$ or $\overline{p}_0(\theta)$ represent our uncertainty about the rows of the mixing matrix, and setting $p_0(u) = \delta(u)$, $\overline{p}_0(\theta) = \delta(\theta)$ or $\overline{p}_0(\theta - \pi/2) = \delta(\theta)$ would make $E\{\mathrm{ICI}_t\} = 0$ for all t.

1.4.3 EQUAL-KURTOSIS SOURCES CASE

For two-source mixtures with equal-kurtosis sources, the scalar evolutionary equation in Eq. (1.43) for the ICI is cubically convergent *globally* as long as the saddle point $\text{ICI}_0 = 1$ occurs with zero finite probability. This situation is perhaps the best one could hope for from a Newton-based procedure. In what follows, we determine several characteristics concerning the *average* convergence performance of the algorithm in the situation where the p.d.f. of ICI_0, denoted as $p_0(u)$, is restricted in some way.

1.4.3.1 A bound on the average ICI

Our first result is a bound on the average ICI at iteration t for weak assumptions on the p.d.f. of the initial ICI, as stated in the following theorem.

Theorem 4. *Let ICI_0 be arbitrarily distributed on $[0, ICI_{\max}]$ with distribution $p_0(u)$, where $0 < ICI_{\max} \leq 1$, subject to the additional condition that the probability density of ICI_0 has no point masses, or equivalently, the cumulative distribution function of ICI_0 is continuous over the interval $[0, 1]$. Then, an upper bound on the average ICI of the FastICA algorithm at iteration t in the two-source case is*

$$E\{ICI_t\} \leq \frac{K(ICI_{\max})^{3^t}}{3^t + 1}, \tag{1.44}$$

where

$$K = p_{\max} ICI_{\max}, \tag{1.45}$$

and $p_{\max} = \max_{0 \leq u \leq ICI_{\max}} p_0(u)$. Furthermore, this bound is tight in the case where ICI_0 is uniformly distributed, in which case $K = 1$.

The proof of this theorem is shown in the Appendix.

Discussion. This theorem takes that for reasonable distribution assumptions on the initial ICI, the ICI at time t is bounded by a function consisting of the product of a linear-converging term and a cubically converging term. Cubic convergence is the described behavior of two-source FastICA in a deterministic setting, and it is ultimately attained under stochastic initial conditions of the separation system vector if the initial distribution of the ICI is bounded away from unity, although it may take a number of iterations before this cubically converging term dominates the expression. During the initial convergence period, the bound is *linear* with rate $(1/3)$, in an expression that resembles Eq. (1.17). Moreover, if the uncertainty about the mixing system prevents one from bounding the ICI away from unity – a likely scenario in practice – then the bound predicts only linear convergence.

1.4.3.2 The probability density function of the ICI

Since ICI_t is related to ICI_0 through a monotonically increasing function in the equal-kurtosis source case, it is straightforward to determine the p.d.f. of ICI_t in terms of the p.d.f. of ICI_0. The following theorem and associated two corollaries relate to this p.d.f.

Theorem 5. *Let the initial inter-channel interference ICI_0 have p.d.f. $p_0(u)$. Then, the p.d.f. of ICI_t is*

$$p_t(u) = \left(\frac{1}{3}\right)^t p_0\left(u^{\left(\frac{1}{3^t}\right)}\right) \frac{1}{u^{1-\left(\frac{1}{3^t}\right)}}. \tag{1.46}$$

Corollary 2. *Suppose the initial separating weight vector has an angular distribution $\overline{p}_0(\theta)$ over the interval $0 \leq \theta \leq \pi/4$. Then, the p.d.f. of ICI_t is*

$$p_t(u) = \left(\frac{1}{3}\right)^t \overline{p}_0(\arctan(u^{\frac{1}{2}\frac{1}{3^t}}))\frac{1}{2}\frac{1}{1+u^{\frac{1}{3^t}}}\frac{1}{u^{1-\frac{1}{2}\frac{1}{3^t}}}. \tag{1.47}$$

Corollary 3. *Suppose the initial separating weight vector is uniformly distributed on the unit circle. Then, the p.d.f. of ICI_t is*

$$p_t(u) = \left(\frac{1}{3}\right)^t \frac{2}{\pi}\frac{1}{1+u^{\frac{1}{3^t}}}\frac{1}{u^{1-\frac{1}{2}\frac{1}{3^t}}}. \tag{1.48}$$

The proofs of these results are shown in the Appendix.

Discussion. The above theorem and corollaries indicate that, in the equal-kurtosis case, the distribution of the ICI quickly approaches a function that is approximately given by

$$p_t(u) \sim \left(\frac{1}{3}\right)^t \frac{K_t}{u}, \tag{1.49}$$

where

$$\lim_{t\to\infty} K_t = \lim_{u\to 1^-} p_0(u). \tag{1.50}$$

Thus, the distribution of ICI_0 only appears to affect the distribution of ICI_t through its limiting value near $\text{ICI}_0 = 1$, as the function $p_0(u^{(1/3^t)})$ quickly approaches $\lim_{u\to 1^-} p_0(u)$ for all $0 < u \leq 1$ and values of t greater than 4. Thus, for even small values of t, the p.d.f. of ICI_t is highly skewed toward zero, indicating fast convergence of the ICI. Moreover, if $\lim_{u\to 1^-} p_0(u) = 0$, then convergence is clearly faster than linear.

1.4.3.3 The average value of the ICI

Our attention now turns to the average ICI of the FastICA algorithm, denoted as $E\{\text{ICI}_t\}$, for a two-source mixture with equal-kurtosis sources. From our previous derivations, it is clear that we require a distribution for the initial value of θ_t, denoted as θ_0, to obtain an exact result. The behavior of the p.d.f. of ICI_t observed in the previous subsection, however, suggests that we may be able to obtain a relationship between $p_0(u)$ and $E\{\text{ICI}_t\}$ in more general circumstances. The following theorem shows that, in fact, this is exactly the case.

Theorem 6. *Suppose the p.d.f. $p_0(u)$ of ICI_0 is nonzero and well-behaved at $ICI_0 = 1$, such that*

$$\lim_{u\to 1^-} p_0(u) = K, \tag{1.51}$$

where $K > 0$ *and the left-sided derivatives* $p_0^{(i)}(u) = \lim_{\Delta \to 0}[p_0^{(i-1)}(u - \Delta) - p_0^{(i-1)}(u)]/\Delta$ *with* $p_0^{(0)}(u) = p_0(u)$ *exist and are finite at* $u = 1$ *for* $i \in \{1, 2\}$. *Then, the average ICI at iteration t obeys the relation*

$$E\{ICI_t\} = \frac{K}{3^t + 1} + \mathcal{O}\left(\left[\frac{1}{9}\right]^t\right), \tag{1.52}$$

where the second term on the right-hand side of the above relation contains terms that are converging exponentially with at least a $(1/9)$*th rate.*

Corollary 4. *An approximate expression for* $E\{ICI_t\}$ *for moderate values of t in such situations is*

$$\widehat{E\{ICI_t\}} = \left(\frac{1}{3}\right)^t K. \tag{1.53}$$

The proof of this theorem and corollary is shown in the Appendix.

Discussion. The above theorem and corollary indicate that it is the distribution of the initial ICI near unity which determines the absolute ICI level at iteration t. Convergence again is linear with a rate of $(1/3)$. In practice, this result means that one is basically guaranteed no better and no worse average performance from the FastICA algorithm than the "(1/3)rd Rule" predicts, because one cannot guarantee that the chosen look direction is bounded away from a saddle point by a finite interval.

The above result suggests that the "(1/3)rd Rule" applies for a general initial condition for the equal-kurtosis two-source case, but it is useful as a confirmation to verify that this behavior is obtained under a reasonable distribution choice for $\mathbf{w}_0$ or $\mathbf{c}_0$. In this regard, it seems natural to assume that θ_0 is uniformly distributed in the range $[0, \pi/4]$, as this would indicate no preferred direction for the initial weight vector. The following theorem relates to this assumed prior distribution on the separation system vector.

Theorem 7. *Assume that the sources in a two-source mixture have equal-magnitude kurtoses, and let the direction of the separation system weight vector be uniformly distributed in angular space. Then, the average value of the ICI at iteration t is exactly*

$$E\{ICI_t\} = -1 + \frac{4}{\pi}\sum_{j=0}^{3^t-1}\frac{1}{2j+1}(-1)^j. \tag{1.54}$$

The proof of this theorem is shown in the Appendix.

Discussion. The series expansion in Eq. (1.54) can be evaluated for any given k, and since we are most interested in small values of k (e.g., $0 < k \leq 15$), this is perhaps the most straightforward way to numerically evaluate $E\{\text{ICI}_t\}$. Another method is to use the beta function as described in [14]. Numerical evaluation shows that

$$E\{\text{ICI}_1\} = 0.10347427\ldots \approx 0.3787\, E\{\text{ICI}_0\} \tag{1.55}$$

$$E\{\text{ICI}_2\} = 0.03526023\ldots \approx 0.3408\, E\{\text{ICI}_1\} \tag{1.56}$$

$$E\{\text{ICI}_3\} = 0.01178522\ldots \approx 0.3342\, E\{\text{ICI}_2\} \tag{1.57}$$

$$E\{\text{ICI}_4\} = 0.00392960\ldots \approx 0.3334\, E\{\text{ICI}_3\} \tag{1.58}$$

$$E\{\text{ICI}_5\} = 0.00130991\ldots \approx 0.3333\, E\{\text{ICI}_4\}. \tag{1.59}$$

Thus, the average ICI is converging to zero, and the rate of convergence appears to approach $(1/3)$, again giving justification to the claim in Eq. (1.17).

Given that we have an exact expression for $E\{\text{ICI}_t\}$ under a particular initial condition, it is interesting to see how accurate the approximate expression for the ICI in Eq. (1.53) is. For a uniformly distributed look direction for $\mathbf{w}_0$ or $\mathbf{c}_0$ in angular space, it is straightforward to show by setting $u = 1$ in Eq. (1.122) that $K = 1/\pi$, such that Eq. (1.53) predicts an average ICI at iteration t of approximately

$$\widehat{E\{\text{ICI}_t\}} \approx \frac{1}{\pi}\left(\frac{1}{3}\right)^t. \tag{1.60}$$

Numerical evaluation of Eq. (1.60) indicates that it is quite accurate in predicting the value of $E\{\text{ICI}_t\}$ in Eq. (1.54) for all but the smallest values of t, as shown below:

$$\widehat{E\{\text{ICI}_0\}} = 0.31830989\ldots \approx 1.16494809\, E\{\text{ICI}_0\} \tag{1.61}$$

$$\widehat{E\{\text{ICI}_1\}} = 0.10610330\ldots \approx 1.02540751\, E\{\text{ICI}_1\} \tag{1.62}$$

$$\widehat{E\{\text{ICI}_2\}} = 0.03536777\ldots \approx 1.00304974\, E\{\text{ICI}_2\} \tag{1.63}$$

$$\widehat{E\{\text{ICI}_3\}} = 0.01178926\ldots \approx 1.00034247\, E\{\text{ICI}_3\} \tag{1.64}$$

$$\widehat{E\{\text{ICI}_4\}} = 0.00392975\ldots \approx 1.00003810\, E\{\text{ICI}_4\} \tag{1.65}$$

$$\widehat{E\{\text{ICI}_5\}} = 0.00130992\ldots \approx 1.00000423\, E\{\text{ICI}_5\}. \tag{1.66}$$

1.4.4 ARBITRARY-KURTOSIS SOURCES CASE

In the previous subsection, we focused on the behavior of FastICA applied to two-source mixtures assuming that the magnitudes of the source kurtosis were equal. In practice, the distributions of the sources will be unknown and different, and thus they will likely not have identical kurtosis magnitudes. The study in this subsection focuses on this scenario, deriving exact and limiting expressions for the average ICI in such cases.

The starting point for our study is the expression for ICI_t in Eq. (1.40), for which $\overline{p}_0(\theta)$ is the angular distribution of $\mathbf{c}_0$ over the first quadrant in the two-dimensional plane. We require an assumption for the distribution $\overline{p}_0(\theta)$, and in the lack of any prior knowledge of the mixing system, we choose a uniform distribution given by $\overline{p}_0(\theta) = 2/\pi, 0 \leq \theta \leq \pi/2$. Under this assumption, we have the following results.

Theorem 8. *Assuming a uniform distribution for the unknown separation weight vector, a limiting expression for the average ICI for a mixture of two sources with kurtoses κ_1 and κ_2 is*

$$E\{\widehat{ICI}_t\} = \frac{2}{\pi\left(\sqrt{\frac{\kappa_1}{\kappa_2}} + \sqrt{\frac{\kappa_2}{\kappa_1}}\right)} \left(\frac{1}{3}\right)^t. \tag{1.67}$$

Corollary 5. *Under the above initial separation weight vector distribution and source assumptions, an exact expression for the average ICI at iteration t is*

$$E\{ICI_t\} = a^{-2(3^t-1)}\left[-\arctan\left(a\cdot(a^{-1})^{\frac{1}{3^t}}\right) + \sum_{j=0}^{3^t-1}\frac{(-1)^j}{2j+1}\left(a\cdot(a^{-1})^{\frac{1}{3^t}}\right)^{2j+1}\right]$$

$$+\, a^{2(3^t-1)}\left[-\arctan\left(a^{-1}\left(a^{\frac{1}{3^t}}\right)\right) + \sum_{j=0}^{3^t-1}\frac{(-1)^j}{2j+1}\left(a^{-1}\left(a^{\frac{1}{3^t}}\right)\right)^{2j+1}\right], \tag{1.68}$$

where $a = \sqrt{\kappa_2/\kappa_1}$.

The proof of the theorem is shown in the Appendix.

Discussion. Several points can be made from the above results:

1. The result in Eq. (1.67) generalizes that of Eq. (1.60) to the case of arbitrary-kurtosis mixtures. It also follows the "(1/3)rd Rule," that is the asymptotic convergence rate of the FastICA algorithm is *unaffected* by the kurtosis magnitudes.
2. The expression in Eq. (1.67) clearly cannot apply at $t = 0$, so one should use $E\{\mathrm{ICI}_0\} = (4/\pi) - 1$ as derived in Eq. (1.132) for the initial value of the sequence.
3. We also conclude from this analysis that the FastICA algorithm is more likely to extract large-magnitude-kurtosis sources than smaller-magnitude-kurtosis ones. This fact is motivated by the integral in the proof in Eq. (1.135), in which $\alpha_t > \pi/4$ when $\kappa_2 > \kappa_1$. In other words, sources with larger-magnitude kurtoses have larger "capture" regions in angular space as compared to those for smaller-magnitude ones.
4. The average ICI is uniformly lower when the kurtosis magnitudes of the sources differ. For example, the average single-unit ICI for a two-source mixture where $\kappa_1/\kappa_2 = 10$ or $\kappa_1/\kappa_2 = 0.1$ will be 2.4 dB lower than that for a mixture where $\kappa_1/\kappa_2 = 1$ at every iteration $t \geq 1$.
5. If one of the sources in a two-source mixture has a zero kurtosis, the FastICA algorithm provides one-step convergence to $E\{\mathrm{ICI}_1\} = 0$ with infinite data. In practice, numerical effects limit the convergence speed in these cases.

Finally, although numerical checks have verified that $E\{\mathrm{ICI}_t\}$ in Eq. (1.68) is approximately equal to $E\{\widehat{\mathrm{ICI}}_t\}$ in Eq. (1.67) for $1 \leq t \leq 3$, the expression in Eq. (1.68) involves differences of terms that become exponentially large, so that exact numerical evaluation of the expression is difficult with typical computational tools (e.g., MATLAB) for $t \geq 4$. Thus, while it should be possible to apply the same ideas used to derive the p.d.f. of ICI_t in Eq. (1.48) to the arbitrary-kurtosis case, numerical issues would likely limit its usefulness for performance prediction.

1.5 INITIAL CONVERGENCE OF THE FastICA ALGORITHM FOR THREE OR MORE SOURCE MIXTURES

1.5.1 OVERVIEW OF RESULTS

In this section, we provide results on the convergence analysis of the single-unit FastICA algorithm operating on mixtures of three or more sources. The analytical tools involved change somewhat from the previous two-source case, as the combined system coefficient vector is inherently multi-dimensional in these cases. Even so, we find through our analyses additional confirmation for the "(1/3)rd Rule" described in Eq. (1.17).

The results presented include the following:

- A limiting expression for $E\{\mathrm{ICI}_t\}$ for an arbitrary-kurtoses three-source mixture assuming a uniformly distributed direction for the initial weight vector as k gets large.
- A limiting expression for $E\{\mathrm{ICI}_t\}$ for an arbitrary-kurtoses four-source mixture assuming a uniformly distributed direction for the initial weight vector as k gets large.
- An approximate expression for $E\{\mathrm{ICI}_t\}$ with bounding term for an arbitrary-kurtoses m-source mixture assuming a uniformly distributed direction for the initial weight vector as k gets large.
- An exact expression for $E\{\mathrm{ICI}_t\}$ for an equal-kurtosis m-source mixture for a particular assumption on the initial distribution of the combined system coefficient vector.

In each case, it is shown that the rule in Eq. (1.17) emerges as the dominant descriptor of the algorithm's convergence behavior.

1.5.2 PRELIMINARIES

The mathematical tools used to show results for greater than two-source mixtures are different from those used previously. To illustrate the approach, consider a single-unit three-source FastICA procedure with cubic nonlinearity, in which the unconstrained combined system vector at iteration k is given by

$$\mathbf{c}_t = \begin{bmatrix} \kappa_1^{\frac{q}{2}} x^{\frac{p}{2}} & \kappa_2^{\frac{q}{2}} y^{\frac{p}{2}} & \kappa_3^{\frac{q}{2}} z^{\frac{p}{2}} \end{bmatrix}^T, \tag{1.69}$$

where $p = 2(3^t)$ and $q = 3^t - 1$. Note that both p and q are functions of k. At time $k = 0$, we have

$$\mathbf{c}_0 = [x \quad y \quad z]^T, \tag{1.70}$$

where x, y, and z are random variables due to the coefficient-stochastic setting being employed.

Due to the unit-norm constraint imposed by the FastICA algorithm, one would normally constrain x, y, and z such that $x^2 + y^2 + z^2 = 1$ and attempt to generate a distribution across x, y, and z so that the constraint space is spanned with some probability, such as a uniform probability. To obtain a uniform distribution on the unit sphere, however, another concept can be used. Suppose x, y, and z are zero mean, uncorrelated, and jointly Gaussian, such that

$$p_{xyz}(x, y, z) = \frac{1}{(2\pi)^{3/2}} \exp\left(-\frac{x^2 + y^2 + z^2}{2}\right). \tag{1.71}$$

Then, the variables $x/\sqrt{x^2 + y^2 + z^2}$, $y/\sqrt{x^2 + y^2 + z^2}$, and $z/\sqrt{x^2 + y^2 + z^2}$ are uniformly distributed on the unit sphere. More importantly, we can express the ICI by considering ratios of powers of x, y, and z *without normalization*. The expectations that result from such an assumption involve Gaussian kernels, but they do not depend on trigonometric functions, angular distributions on the unit sphere, or challenging integration regions. This strategy is used to provide many of the results of this section.

1.5.3 THREE-SOURCE CASE

Consider the FastICA algorithm operating on $m = 3$ source linear mixtures with arbitrary kurtosis κ_1, κ_2, and κ_3. Moreover, assume that the prior distribution of the initial combined system coefficient vector $\mathbf{c}_0$ is uniform on the unit three-sphere. The following theorem describes the approximate evolution of the average ICI in this situation.

Theorem 9. *A limiting expression for the average ICI for a mixture of three sources with kurtoses κ_1, κ_2, and κ_3 for a uniform prior on $\mathbf{c}_0$ is*

$$\widehat{E\{ICI_t\}} \approx K_3 \left(\frac{1}{3}\right)^t, \tag{1.72}$$

where

$$K_3 = \frac{2}{\pi} \frac{1}{\sqrt{\kappa_1\kappa_2 + \kappa_1\kappa_3 + \kappa_2\kappa_3}} \left[\frac{\kappa_1\kappa_2}{\kappa_1 + \kappa_2} + \frac{\kappa_1\kappa_3}{\kappa_1 + \kappa_3} + \frac{\kappa_2\kappa_3}{\kappa_2 + \kappa_3}\right]. \tag{1.73}$$

The proof of the theorem is shown in the Appendix.

Discussion. A couple of points can be made from the result in Eq. (1.72):

1. When $\kappa_3 = 0$, the result in Eq. (1.72) reduces to the expression for the ICI for the two-source arbitrary-kurtoses case in Eq. (1.67).
2. The maximum average ICI occurs when $\kappa_1 = \kappa_2 = \kappa_3$, in which case

$$E\{\widehat{\mathrm{ICI}_t}\} = \frac{\sqrt{3}}{\pi} \left(\frac{1}{3}\right)^t, \tag{1.74}$$

which is $\sqrt{3}$ larger than the maximum average ICI in the two-source case ($\kappa_1 = \kappa_2$).

The result also clearly shows that the "(1/3)rd Rule" is reasonable for arbitrary three-source mixtures.

1.5.4 FOUR-SOURCE CASE

Consider the FastICA algorithm operating on $m = 4$ source linear mixtures with arbitrary kurtosis κ_1, κ_2, κ_3, and κ_4. Moreover, assume that the prior distribution of the initial combined system coefficient vector $\mathbf{c}_0$ is uniform on the unit four-sphere. The following theorem describes the approximate evolution of the average ICI in this situation.

Theorem 10. *A limiting expression for the average ICI for a mixture of four sources with kurtoses κ_1, κ_2, κ_3, and κ_4 for a uniform prior on $\mathbf{c}_0$ is*

$$\widehat{E\{ICI_t\}} \approx K_4 \left(\frac{1}{3}\right)^t, \tag{1.75}$$

where

$$\begin{aligned}
K_4 = \frac{4}{\pi^2} \Bigg\{ & \frac{1}{\sqrt{\kappa_1/\kappa_2} + \sqrt{\kappa_2/\kappa_1}} \left[\frac{\sqrt{\kappa_3^{-1}}}{\sqrt{\kappa_1^{-1} + \kappa_2^{-1} + \kappa_3^{-1}}} \arctan\left(\frac{\sqrt{\kappa_4^{-1}}}{\sqrt{\kappa_1^{-1} + \kappa_2^{-1} + \kappa_3^{-1}}} \right) \right. \\
& \left. + \frac{\sqrt{\kappa_4^{-1}}}{\sqrt{\kappa_1^{-1} + \kappa_2^{-1} + \kappa_4^{-1}}} \arctan\left(\frac{\sqrt{\kappa_3^{-1}}}{\sqrt{\kappa_1^{-1} + \kappa_2^{-1} + \kappa_4^{-1}}} \right) \right] \\
& + \frac{1}{\sqrt{\kappa_1/\kappa_3} + \sqrt{\kappa_3/\kappa_1}} \left[\frac{\sqrt{\kappa_2^{-1}}}{\sqrt{\kappa_1^{-1} + \kappa_2^{-1} + \kappa_3^{-1}}} \arctan\left(\frac{\sqrt{\kappa_4^{-1}}}{\sqrt{\kappa_1^{-1} + \kappa_2^{-1} + \kappa_3^{-1}}} \right) \right. \\
& \left. + \frac{\sqrt{\kappa_4^{-1}}}{\sqrt{\kappa_1^{-1} + \kappa_3^{-1} + \kappa_4^{-1}}} \arctan\left(\frac{\sqrt{\kappa_2^{-1}}}{\sqrt{\kappa_1^{-1} + \kappa_3^{-1} + \kappa_4^{-1}}} \right) \right] \\
& + \frac{1}{\sqrt{\kappa_1/\kappa_4} + \sqrt{\kappa_4/\kappa_1}} \left[\frac{\sqrt{\kappa_2^{-1}}}{\sqrt{\kappa_1^{-1} + \kappa_2^{-1} + \kappa_4^{-1}}} \arctan\left(\frac{\sqrt{\kappa_3^{-1}}}{\sqrt{\kappa_1^{-1} + \kappa_2^{-1} + \kappa_4^{-1}}} \right) \right. \\
& \left. + \frac{\sqrt{\kappa_3^{-1}}}{\sqrt{\kappa_1^{-1} + \kappa_3^{-1} + \kappa_4^{-1}}} \arctan\left(\frac{\sqrt{\kappa_2^{-1}}}{\sqrt{\kappa_1^{-1} + \kappa_3^{-1} + \kappa_4^{-1}}} \right) \right] \\
& + \frac{1}{\sqrt{\kappa_2/\kappa_3} + \sqrt{\kappa_3/\kappa_2}} \left[\frac{\sqrt{\kappa_1^{-1}}}{\sqrt{\kappa_1^{-1} + \kappa_2^{-1} + \kappa_3^{-1}}} \arctan\left(\frac{\sqrt{\kappa_4^{-1}}}{\sqrt{\kappa_1^{-1} + \kappa_2^{-1} + \kappa_3^{-1}}} \right) \right. \\
& \left. + \frac{\sqrt{\kappa_4^{-1}}}{\sqrt{\kappa_2^{-1} + \kappa_3^{-1} + \kappa_4^{-1}}} \arctan\left(\frac{\sqrt{\kappa_1^{-1}}}{\sqrt{\kappa_2^{-1} + \kappa_3^{-1} + \kappa_4^{-1}}} \right) \right]
\end{aligned}$$

$$
\begin{aligned}
&+\frac{1}{\sqrt{\kappa_2/\kappa_4}+\sqrt{\kappa_4/\kappa_2}}\left[\frac{\sqrt{\kappa_1^{-1}}}{\sqrt{\kappa_1^{-1}+\kappa_2^{-1}+\kappa_4^{-1}}}\arctan\left(\frac{\sqrt{\kappa_3^{-1}}}{\sqrt{\kappa_1^{-1}+\kappa_2^{-1}+\kappa_4^{-1}}}\right)\right.\\
&\left.+\frac{\sqrt{\kappa_3^{-1}}}{\sqrt{\kappa_2^{-1}+\kappa_3^{-1}+\kappa_4^{-1}}}\arctan\left(\frac{\sqrt{\kappa_1^{-1}}}{\sqrt{\kappa_2^{-1}+\kappa_3^{-1}+\kappa_4^{-1}}}\right)\right]\\
&+\frac{1}{\sqrt{\kappa_3/\kappa_4}+\sqrt{\kappa_4/\kappa_3}}\left[\frac{\sqrt{\kappa_1^{-1}}}{\sqrt{\kappa_1^{-1}+\kappa_3^{-1}+\kappa_4^{-1}}}\arctan\left(\frac{\sqrt{\kappa_2^{-1}}}{\sqrt{\kappa_1^{-1}+\kappa_3^{-1}+\kappa_4^{-1}}}\right)\right.\\
&\left.\left.+\frac{\sqrt{\kappa_2^{-1}}}{\sqrt{\kappa_2^{-1}+\kappa_3^{-1}+\kappa_4^{-1}}}\arctan\left(\frac{\sqrt{\kappa_1^{-1}}}{\sqrt{\kappa_2^{-1}+\kappa_3^{-1}+\kappa_4^{-1}}}\right)\right]\right\}.
\end{aligned}
\tag{1.76}
$$

The proof of the theorem is shown in the Appendix.

Discussion. A couple of points can be made from the result in Eq. (1.75):

1. When $\kappa_4 = 0$, the result in Eq. (1.75) reduced to the expression in Eq. (1.72) for the three-dimensional case.
2. When $\kappa_1 = \kappa_2 = \kappa_3 = \kappa_4$, we have

$$
E\{\widehat{\mathrm{ICI}}_t\} = \frac{4}{\pi\sqrt{3}}\left(\frac{1}{3}\right)^t, \tag{1.77}
$$

which is 4/3 times larger than the maximum ICI in the three-source case with $\kappa_1 = \kappa_2 = \kappa_3$ and $4/\sqrt{3} = 2.31$ times larger than the maximum ICI in the two-source case with $\kappa_1 = \kappa_2$.

Again, these results support the "(1/3)rd Rule" described earlier.

1.5.5 GENERAL *m*-SOURCE CASE

We now provide an analysis of the average ICI of the FastICA algorithm for general m-source mixtures assuming arbitrary kurtosis values. As in the previous three- and four-source cases, we assume a uniform distribution prior for $\mathbf{c}_0$ in the unit-m-hypersphere, in which the update vector at iteration t without normalization is given by

$$
\mathbf{c}_t = \begin{bmatrix} \kappa_1^{\frac{q}{2}} x_1^{\frac{p}{2}} & \cdots & \kappa_1^{\frac{q}{2}} x_n^{\frac{p}{2}} \end{bmatrix}^T, \tag{1.78}
$$

where $p = 2(3^t)$ and $q = 3^t - 1$. The following theorem provides a bounded expression for the average ICI in this case.

Theorem 11. *For a uniform prior on the initial combined system coefficient vector $\mathbf{c}_t$, the average ICI at iteration t for the FastICA algorithm with cubic nonlinearity satisfies*

$$
E\{ICI_t\} = g(\kappa_1, \ldots, \kappa_m)\left(\frac{1}{3}\right)^t + R(t, \kappa_1, \ldots, \kappa_m), \tag{1.79}
$$

where

$$g(\kappa_1, \ldots, \kappa_m) = \begin{cases} \frac{2^{m-1}}{(2\pi)^{m/2}} \sum_{n=1}^{m} \sum_{i=1, i\neq n}^{m} \int_0^{b_{n1}} dx_1 \cdots \int_0^{b_{nm}} dx_m \frac{(m-2)!!\sqrt{\pi}}{\sqrt{2}\left(1+(b_{ni})^2+\sum_{j=1, j\neq n, j\neq i}^{m} x_j^2\right)^{m/2}} & m \text{ is odd} \\ \frac{2^{m-1}}{(2\pi)^{m/2}} \sum_{n=1}^{m} \sum_{i=1, i\neq n}^{m} \int_0^{b_{n1}} dx_1 \cdots \int_0^{b_{nm}} dx_m \frac{\left[\frac{1}{2}(m-2)\right]! 2^{(m-2)/2}}{\left(1+(b_{ni})^2+\sum_{j=1, j\neq n, j\neq i}^{m} x_j^2\right)^{m/2}} & m \text{ is even} \end{cases} \tag{1.80}$$

$$b_{ni} = \sqrt{\frac{\kappa_n}{\kappa_i}}, \tag{1.81}$$

and the error term $R(t, \kappa_1, \ldots, \kappa_m)$ *is bounded by*

$$|R(t, \kappa_1, \ldots, \kappa_m)| \leq c(m) \left(\frac{1}{3}\right)^t \sum_{n=1}^{m} \left[(m-1) \prod_{i=2}^{m} b_{ni} \left(1 - \prod_{i=2}^{m} b_{ni}^{-\frac{2}{p}}\right) \right.$$
$$\left. + \sum_{i=2}^{m} \left(\left(1 - \frac{1}{\sqrt{p}}\right)^{p+1} b_{ni}^{1-\frac{2}{p}} + \frac{b_{ni}}{p} \left| 1 - \frac{p}{p+1} b_{ni}^{-\frac{4}{p}} \right| \right) \prod_{i=2}^{m} b_{ni}^{1-\frac{2}{p}} \right]. \tag{1.82}$$

Furthermore, we have

$$\lim_{t\to\infty} \frac{R(t, \kappa_1, \ldots, \kappa_m)}{\left(\frac{1}{3}\right)^t} = 0. \tag{1.83}$$

The proof of the theorem is shown in the Appendix.

Discussion. Two important points can be made regarding these results:

1. The conditions for this derivation encompass those of the three- and four-source cases considered previously. However, Eq. (1.80) is somewhat stronger mathematically, as the error in the approximate ICI expression is bounded by a term that decreases faster than $(1/3)^t$. These results provide justification for the approximations used earlier.
2. Again, the "(1/3)rd Rule" is supported by these results.

1.5.6 EQUAL-KURTOSIS *m*-SOURCE CASE USING ORDER STATISTICS

Our previous efforts to analyze the average single-unit ICI for the FastICA algorithm assume a uniform prior distribution on $\mathbf{c}_0$ that is not preferential to any direction. The developed results involve bounds on approximations, and we have described no simple, direct expression for the value of $E\{\mathrm{ICI}_t\}$ for *m*-dimensional mixtures in such cases. In this section, we consider another prior distribution for the combined system coefficient vector that leads to an *exact* expression for $E\{\mathrm{ICI}_t\}$ for mixtures of sources with equal kurtoses. The results in this subsection are based on the following simple but very nice observation: *In the convergence analysis of FastICA, ordering of the coefficients within the update relations does not matter.* Hence, the coefficients within $\mathbf{c}_t$ can be reordered to obtain a structured distribution of the coefficients. This methodology does create some challenges – for example, rank-ordering changes the integration regions, and it also creates a correlated set of evolving random vector elements from an identically and independently distributed (i.i.d.) set if $\mathbf{c}_t$ has i.i.d. elements and scaling is ignored. Despite these facts, the integrals become much simpler in at least one specific case.

Suppose the elements of the unnormalized vector $\mathbf{c}_0 = [\hat{c}_{10} \cdots \hat{c}_{m0}]^T$ are uniformly distributed on the interval [0, 1], such that

$$\overline{p}(\hat{c}_{10}, \hat{c}_{20}, \ldots, \hat{c}_{m0}) = \begin{cases} 1 & \text{if } 0 \leq \{\hat{c}_{i0}\} \leq 1,\ 1 \leq i \leq m \\ 0 & \text{otherwise} \end{cases}. \tag{1.84}$$

Of course, $\mathbf{c}_0$ is normally constrained to be unit length, but as scaling does not matter in the computation of the ICI, we choose an unscaled version of this vector instead. Assuming that each c_{i0} is positive as opposed to zero-mean (e.g., Unif$[-1, 1]$) also does not change the average convergence behavior of the FastICA algorithm.

What does this assumption mean from the standpoint of a practical initial condition? A uniform distribution throughout the m-dimensional unit hypercube centered at $[0 \cdots 0]$, when projected onto the m-dimensional unit hypersphere, tends to concentrate probability in the $m^{-1/2}[\pm 1\ \pm 1\ \cdots\ \pm 1]^T$ directions of m-dimensional space, an undesirable situation from the prospective of FastICA convergence. Moreover, the situation is further made more challenging if all of the sources have the same kurtosis, in which case convergence has been shown to be slower than in the unequal-kurtosis case. Thus, the analytical results of such an initial condition and source kurtosis distribution are likely to be a more conservative prediction of the average performance as compared to more realistic conditions in which the direction of $\mathbf{c}_0$ is uniformly distributed.

Under this situation, the value of $E\{\mathrm{ICI}_t\}$ is remarkably easy to compute because of the following result concerning the order statistics of i.i.d. Unif[0, 1] random variables [15]:

Fact. Let $c_1 \geq c_2 \geq \cdots \geq c_m$ be the ordered set of $\{\hat{c}_1, \hat{c}_2, \ldots, \hat{c}_m\}$ that has the joint p.d.f. in Eq. (1.84). Then, the joint p.d.f. of $\{c_1, c_2, \ldots, c_m\}$ is

$$\overline{p}(c_1, c_2, \ldots, c_m) = \begin{cases} m! & \text{if } 0 \leq c_m \leq c_{m-1} \leq \cdots \leq c_1 \leq 1 \\ 0 & \text{otherwise} \end{cases}. \tag{1.85}$$

Based on this result, we can develop an exact expression for the average ICI of the single-unit FastICA algorithm with this particular initial condition prior.

Theorem 12. *Assume that the unnormalized initial combined system coefficient vector* $\mathbf{c}_0$ *has the distribution in Eq. (1.85). Then, if the source mixture contains equal-kurtosis sources, the average value of the ICI at iteration t is exactly*

$$E\{ICI_t\} = \frac{m-1}{2(3^t)+1}. \tag{1.86}$$

The proof of this theorem is shown in the Appendix.

Discussion. Three remarks are in order here:

1. The result in Eq. (1.86) supports the "(1/3)rd Rule" described previously.
2. Unlike the previously presented results for three or more source mixtures, the expression in Eq. (1.86) is exact, although the analyses are for different initial conditions.
3. Because of the relations

$$0.5 = \frac{1}{2} \geq \frac{1}{\pi} \approx 0.3183 \tag{1.87}$$

$$1 = \frac{2}{2} \geq \frac{\sqrt{3}}{\pi} \approx 0.5513 \tag{1.88}$$

$$1.5 = \frac{3}{2} \geq \frac{4}{\pi\sqrt{3}} \approx 0.7351, \tag{1.89}$$

the value of $E\{\text{ICI}_t\}$ in Eq. (1.86) appears always to be above that of our previous expressions for $E\{\text{ICI}_t\}$ and $E\{\widehat{\text{ICI}}_t\}$ which assumed a more conservative direction prior for $\mathbf{c}_0$. This result is in agreement with our previously made comments.

1.6 NUMERICAL EVALUATIONS

We now verify several of the results and observations of the chapter via numerical evaluations. All computations have been performed within the MATLAB technical computing environment. All sources have been generated from MATLAB's `rand` or `randn` command using linear or nonlinear transformations, where appropriate. In each case, we have computed theoretical results as predicted from equations within the chapter and compared these with ensemble averages of the output of the single-unit FastICA algorithm with $f(y) = y^3$ after data prewhitening. In some cases, the results between figures are related – for example, an averaged value is computed using the exact data points shown in a previous figure – so that observations can be directly related. In every case, we have used averages of 100,000 ensembles to compute the quantities shown, and we have chosen a data block size of $N = 10{,}000$ for the FastICA algorithm.

We first examine the accuracy of the analytical model in Eq. (1.30) in predicting the qualitative behavior of the FastICA algorithm. Figure 1.1 shows eight iterations of the relation in Eq. (1.30) for an $m = 3$ source mixture with equal-kurtosis

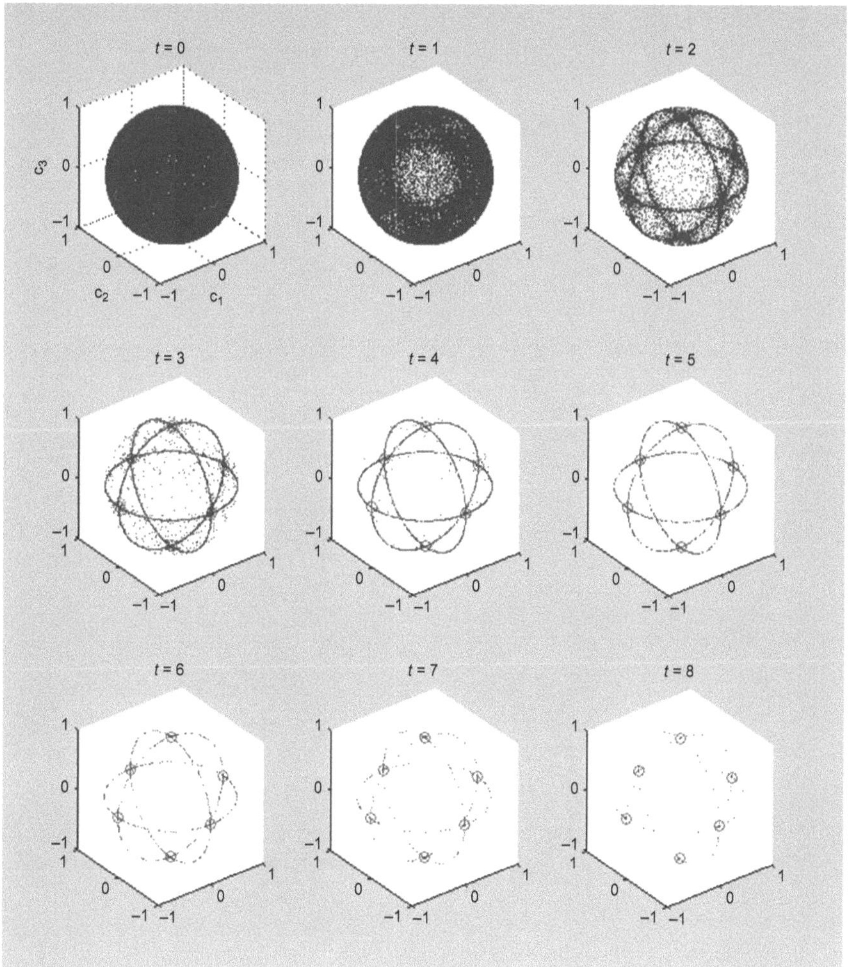

FIGURE 1.1

Convergence of the single-unit analysis equation in Eq. (1.30) for $m = 3$ and 100,000 different initial coefficient vectors spanning the unit three-sphere.

sources ($|\kappa_1| = |\kappa_2| = |\kappa_3|$) for 100,000 different initial $\mathbf{c}_0$ vectors spanning the unit three-sphere, as generated from i.i.d. Gaussian random elements for c_{10}, c_{20}, and c_{30}. After several iterations, the values of $\mathbf{c}_t$ are clustered along the circles that lie in the planes of each pair of coordinate axes. Convergence of the iteration then proceeds along these circles to one of the six stable stationary points corresponding to $[\pm 1\ 0\ 0]^T$, $[0\ \pm 1\ 0]^T$, or $[0\ 0\ \pm 1]^T$. Figure 1.2 shows eight iterations of the single-unit FastICA algorithm applied to 100,000 sets of three i.i.d. Unif$(-\sqrt{3}, \sqrt{3})$ sources with the *numerically identical* 100,000 different initial $\mathbf{c}_0$ vectors shown in the $t = 0$ graph of Figure 1.1. The similarity of the data clouds in Figures 1.1 and 1.2

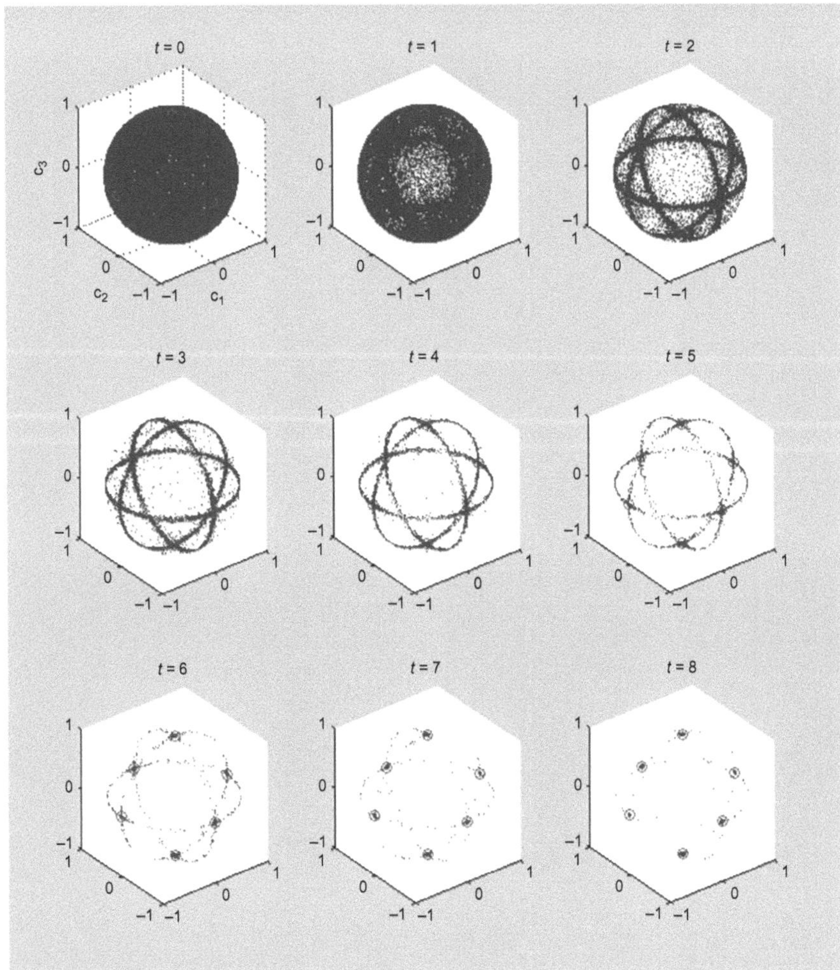

FIGURE 1.2

Convergence of the single-unit FastICA procedure for mixtures of three uniformly-distributed independent sources and 100,000 different initial coefficient vectors spanning the unit three-sphere.

indicates that the analytical model in Eq. (1.30) is accurate in predicting overall performance in practice, although the numerical behavior of the single-unit FastICA algorithm is slightly different due to the use of $N = 10{,}000$ sample data blocks to compute the numerical quantities used in the FastICA procedures.

Figure 1.3 shows the average value of the ICI, $E\{\text{ICI}_t\}$, as computed from the *numerically identical* initial conditions and simulation runs that generated the data points shown in Figure 1.2, along with the approximate predicted value $E\{\widehat{\text{ICI}}_t\}$ as

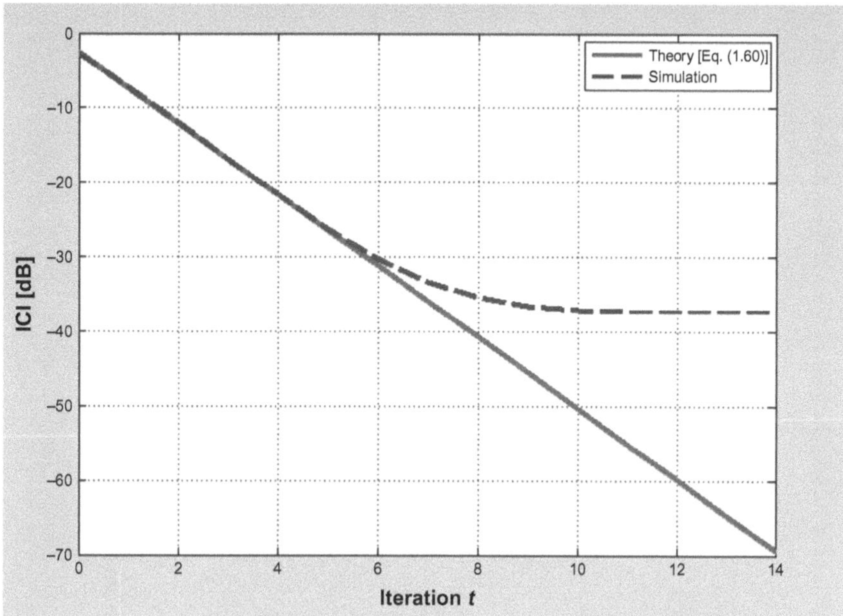

FIGURE 1.3

Evolutions of inter-channel interference from predictions and simulations for mixtures of three uniformly-distributed independent sources.

computed from Eq. (1.60). As can be seen, the "(1/3)rd Rule" is accurate in describing the initial convergence performance of the FastICA algorithm in this case. Note that the averaged value of $E\{\text{ICI}_t\}$ reaches a limiting value due to finite data block size of $N = 10{,}000$ in the algorithm, and thus there is no additional faster-than-linear convergence regime observed in practice.

We now explore the behavior of the FastICA algorithm for mixtures of different source types. Figure 1.4 shows eight iterations of the single-unit FastICA algorithm applied to 100,000 sets of three i.i.d. sources consisting of one i.i.d. Unif$(-\sqrt{3}, \sqrt{3})$ source ($|\kappa_1| = 6/5$), one unit-variance Laplacian source ($|\kappa_2| = 3$), and one binary-(± 1) source ($|\kappa_3| = 2$). In this case, we observe preferential convergence along the c_2 axis corresponding to the Laplacian source, which verifies our observation that sources with larger-magnitude kurtosis are more likely to be extracted by the FastICA algorithm. Examining the 100,000 simulation runs, we find that the Laplacian source was extracted 45.25% of the time, the binary source was extracted 33.26% of the time, and the uniform-distributed source was extracted 21.49% of the time.

Figure 1.5 shows the average value of the ICI, $E\{\text{ICI}_t\}$, as computed from the *numerically identical* initial conditions and simulation runs that generated the data points shown in Figure 1.4, along with the approximate predicted value $E\{\widehat{\text{ICI}}_t\}$ as

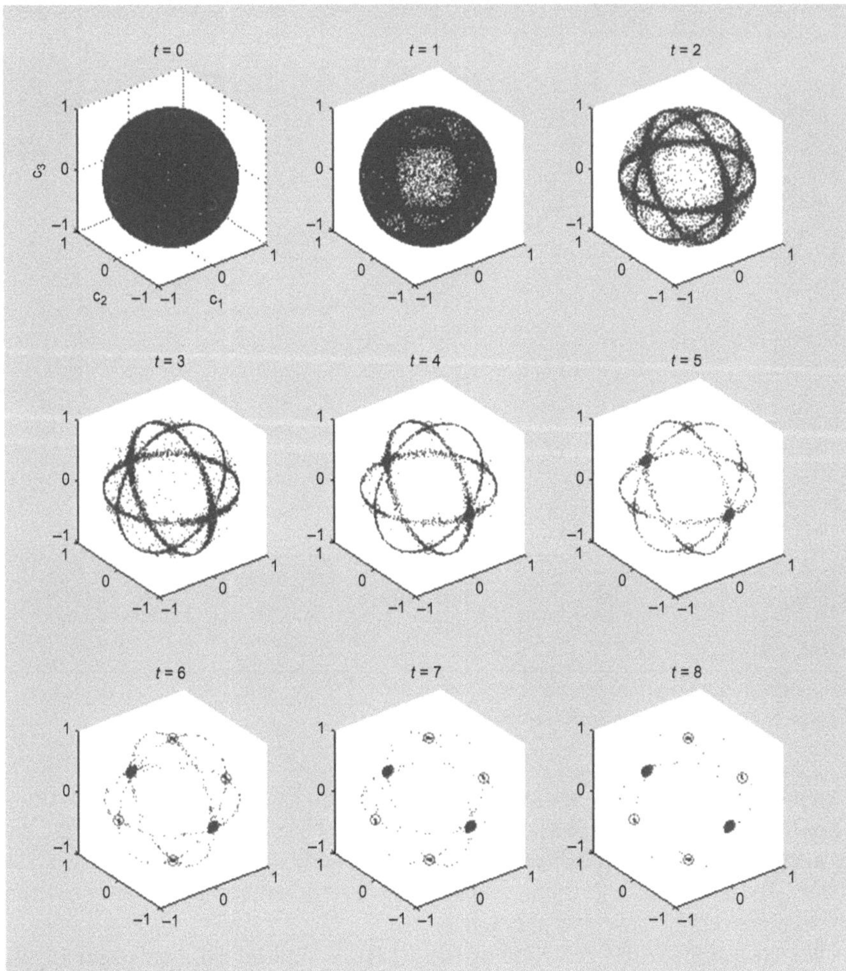

FIGURE 1.4

Convergence of the single-unit FastICA procedure for mixtures of one uniformly-distributed, one Laplacian, and one binary source and 100,000 different initial coefficient vectors spanning the unit three-sphere.

computed from Eq. (1.72). Again, the "(1/3)rd Rule" is accurate in describing the initial convergence performance of the FastICA algorithm in this case. Note that the averaged value of $E\{\mathrm{ICI}_t\}$ reaches a higher limiting value than that shown in Figure 1.3, which is likely due to the impulsive nature of the Laplacian source which has been shown to limit the statistical efficiency of the algorithm due to finite sample effects [9].

We now consider a two-source equal-kurtosis case to verify the p.d.f. results of Section 1.4. Figure 1.6 shows the values of $p_t(u)$ as computed from Eq. (1.48) for

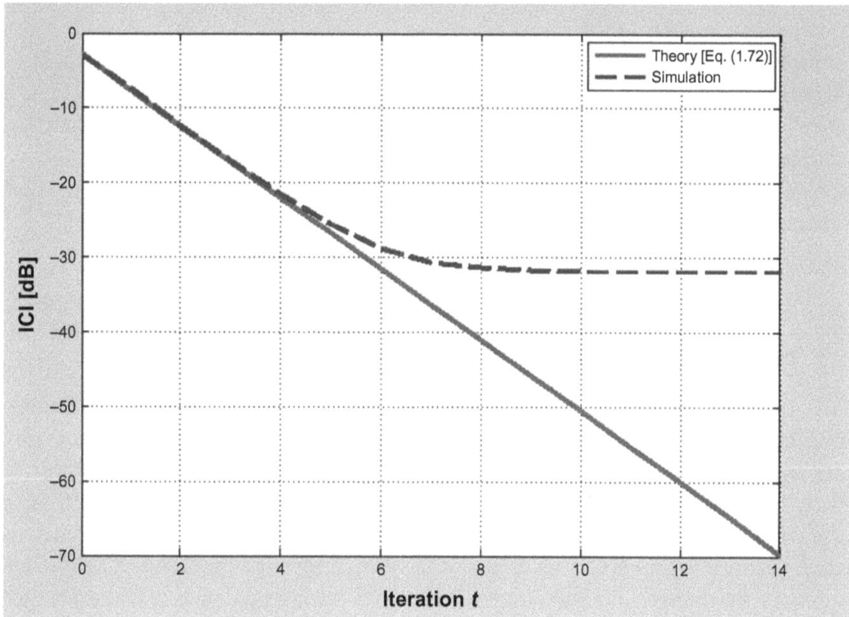

FIGURE 1.5

Evolutions of inter-channel interference from predictions and simulations for mixtures of one uniformly-distributed, one Laplacian, and one binary source.

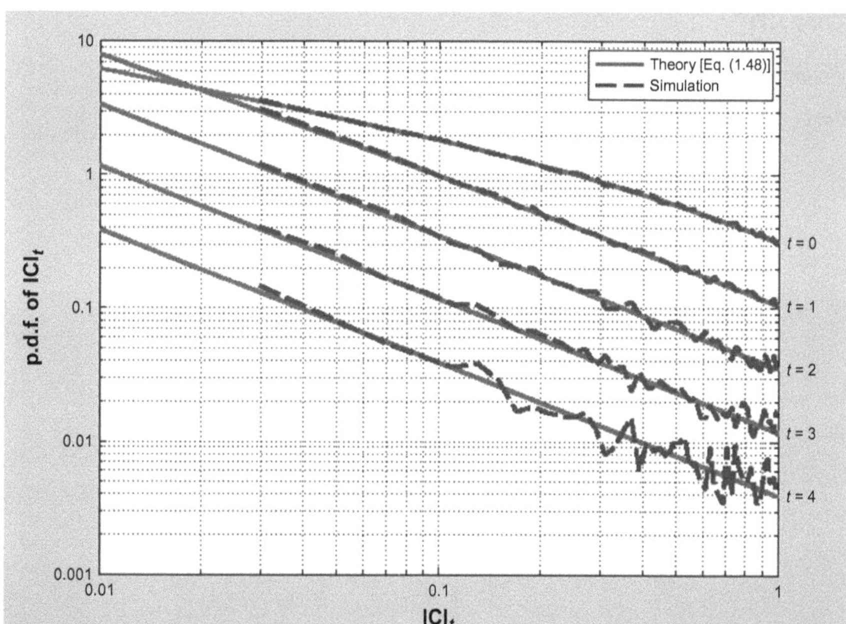

FIGURE 1.6

Evolution of the p.d.f. of the inter-channel interference for mixtures of two uniformly-distributed independent sources.

$t = \{0, 1, 2, 3, 4\}$ along with sample-based histograms of the distributions of ICI_t for 100,000 different FastICA procedures applied to sets of two i.i.d. Unif$(-\sqrt{3}, \sqrt{3})$ sources with a block size of $N = 10{,}000$. As can be seen, the theoretical expressions accurately match the histograms from the simulation results. Moreover, the functional form of the p.d.f. rapidly converges to that in Eq. (1.49), again showing a $(1/3)^t$ decrease almost everywhere except near the origin (not depicted).

1.7 CONCLUSION

In this chapter, we have analyzed the average initial convergence rate of the FastICA algorithm for the case of a cubic nonlinearity as applied to noiseless linear mixtures of independent sources. Our analysis indicates that the "(1/3)rd Rule," in which the average ICI decreases by a third or 4.77 dB at each iteration, is an accurate description of this algorithm's average performance no matter what the source kurtosis values are. This behavior is verified through numerous analyses and confirmed by numerical simulations. It provides strong evidence that the FastICA algorithm provides consistent, fast convergence for source separation and ICA tasks.

APPENDIX

PROOF OF THEOREM 1

Proof. Consider the (i,j)th element of the matrix $E\{\mathbf{s}(k)y_t^2(k)\mathbf{s}^T(k)\}$, given by

$$E\{s_i(k)s_j(k)y_t^2(k)\} = E\left\{ s_i(k)s_j(k)\sum_{l=1}^{m}\sum_{p=1}^{m} c_{lt}c_{pt}s_l(k)s_p(k)\right\} \tag{1.90}$$

$$= \sum_{l=1}^{m}\sum_{p=1}^{m} c_{lt}c_{pt}E\{s_i(k)s_j(k)s_l(k)s_p(k)\}. \tag{1.91}$$

The expectation on the right-hand side of Eq. (1.91) can be evaluated using Eqs. (1.19)–(1.21) as

$$E\{s_i(k)s_j(k)s_l(k)s_p(k)\} = \begin{cases} 1 & \text{if } i=j\neq l=p \\ & \text{or if } i=l\neq j=p \\ & \text{or if } i=p\neq j=l. \\ E\{s_i^4(k)\} & \text{if } i=j=l=p \\ 0 & \text{otherwise} \end{cases} \tag{1.92}$$

Therefore,

$$\sum_{l=1}^{m}\sum_{p=1}^{m} c_{lt}c_{pt}E\{s_i(k)s_j(k)s_l(k)s_p(k)\} = \left[\kappa_i c_{it}^2 + \|\mathbf{c}\|^2\right]\delta_{ij} + 2c_{it}c_{jt}. \tag{1.93}$$

In matrix form, we have

$$E\{\mathbf{s}(k)y_t^2(k)\mathbf{s}^T(k)\} = \mathbf{K}\mathrm{diag}\left[\mathbf{c}_t\mathbf{c}_t^T\right] + \|\mathbf{c}_t\|^2\mathbf{I} + 2\mathbf{c}_t\mathbf{c}_t^T, \tag{1.94}$$

where diag[**M**] is a diagonal matrix whose diagonal entries are the diagonal elements of **M**. Therefore,

$$E\{y_t^3(k)\mathbf{s}(k)\} = E\{\mathbf{s}(k)y_t^2(k)\mathbf{s}^T(k)\}\mathbf{c}_t \tag{1.95}$$

$$= \mathbf{K}\mathrm{diag}\left[\mathbf{c}_t\mathbf{c}_t^T\right]\mathbf{c}_t + 3\|\mathbf{c}_t\|^2\mathbf{c}_t \tag{1.96}$$

$$= \mathbf{K}\mathbf{f}(\mathbf{c}_t) + 3\|\mathbf{c}_t\|^2\mathbf{c}_t. \tag{1.97}$$

The corollary follows by substituting Eq. (1.97) into Eq. (1.27) and then recognizing that $\|\mathbf{c}_t\| = 1$ for all k due to Eq. (1.28). □

PROOF OF THEOREMS 2 AND 3

Proof. A cursory study of Eq. (1.32) shows that setting $c_{it} = 0$ results in $c_{i(t+1)} = 0$. Moreover, if $\kappa_i = 0$, then $c_{i(t+1)} = 0$ no matter what c_{it} is. Clearly, we only need to consider stationary points of Eq. (1.32) for which any subset of the $\{c_{i,s}\}$ values for $1 \le i \le m_p$ are nonzero. Call this subset of indices $\mathcal{J}$. Then, for indices $i \in \mathcal{J}$, we can simplify Eq. (1.32) to obtain

$$|\kappa_i|c_{i,s}^2 = \sqrt{\sum_{j\in\mathcal{J}} \kappa_j^2 c_{j,s}^6}. \tag{1.98}$$

By dividing both sides of Eq. (1.98) by $|\kappa_i|$ and summing across $i \in \mathcal{J}$, we have

$$\sum_{i\in\mathcal{J}} c_{i,s}^2 = \sum_{i\in\mathcal{J}} \frac{1}{|\kappa_i|}\sqrt{\sum_{j\in\mathcal{J}} \kappa_j^2 c_{j,s}^6}. \tag{1.99}$$

Since $\mathbf{c}_s$ must be of unit length, the left-hand side of Eq. (1.99) must be one, yielding the relation

$$\sqrt{\sum_{j\in\mathcal{J}} \kappa_j^2 c_{j,s}^6} = \frac{1}{\sum_{i\in\mathcal{J}} |\kappa_i|^{-1}}. \tag{1.100}$$

Substituting Eq. (1.100) into the right-hand side of Eq. (1.98) and dividing both sides of the resulting expression by $|\kappa_i|$ produces

$$c_{i,s}^2 = \frac{|\kappa_i|^{-1}}{\sum_{j\in\mathcal{J}} |\kappa_j|^{-1}}. \tag{1.101}$$

Taking square roots of both sides of Eq. (1.101) yields the condition in Eq. (1.33).

To determine the local stability of Eq. (1.32) about the solutions defined in Eq. (1.33), define the perturbed coefficient values $|c_i| = c_{i,s} + \Delta_{it}$ where $c_{i,s} = |c_{i,s}|$ in Eq. (1.33) and $\mathcal{J}$ is any valid subset of the elements of $\mathcal{I}$. Because of the unit-norm constraint in Eq. (1.28), we only need to consider perturbations $\{\Delta_i\}$ that are tangent or orthogonal to $\mathbf{c}_s$, such that

$$\sum_{j=1}^{m} c_{s,j}\Delta_j = 0. \tag{1.102}$$

Furthermore, assume that each $|\Delta_i| \ll 1$. Consider first solutions for which $c_{i,s} \neq 0$ for two or more indices $i \in \mathcal{J}$. Then, to first order in each Δ_i,

$$|c_{i(t+1)}| = \left|\frac{\kappa_i(c_{i,s}+\Delta_i)^3}{\sqrt{\sum_{j\in\mathcal{J}}\kappa_j^2(c_{j,s}+\Delta_j)^6 + \sum_{j\notin\mathcal{J}}\kappa_j^2\Delta_j^6}}\right| \tag{1.103}$$

$$\approx \left|\frac{\kappa_i\left[c_{i,s}^3 + 3\Delta_i c_{i,s}^2\right]}{\sqrt{\sum_{j\in\mathcal{J}}\kappa_j^2\left[c_{j,s}^6 + 6c_{j,s}^5\Delta_j\right]}}\right| \tag{1.104}$$

$$= \left|\frac{\kappa_i c_{i,s}^3}{\sqrt{\sum_{j\in\mathcal{J}}\kappa_j^2 c_{j,s}^6}}\right| \cdot \left|\frac{1+3\frac{\Delta_i}{c_{i,s}}}{\sqrt{1+6\frac{\sum_{l\in\mathcal{J}}\kappa_l^2 c_{l,s}^5\Delta_l}{\sum_{j\in\mathcal{J}}\kappa_j^2 c_{j,s}^6}}}\right|, \tag{1.105}$$

where we have used the fact that $|\Delta_i| \ll 1$ to simplify Eq. (1.103) to obtain Eq. (1.104). It can be shown using Eq. (1.98) that

$$\frac{\kappa_l^2 c_{l,s}^4}{\sum_{j\in\mathcal{J}}\kappa_j^2 c_{j,s}^6} = 1. \tag{1.106}$$

Substituting Eqs. (1.106) and (1.33) into the right-hand side of Eq. (1.105) and simplifying, we obtain

$$|c_{i(t+1)}| = |c_{i,s}| \cdot \left|\frac{1+3\frac{\Delta_i}{c_{i,s}}}{\sqrt{1+6\sum_{j\in J} c_{j,s}\Delta_j}}\right| \tag{1.107}$$

$$= |c_{i,s}| \cdot \left|1 + 3\frac{\Delta_i}{c_{i,s}}\right| \tag{1.108}$$

$$= |c_{i,s}| + 3\mathrm{sgn}[c_{i,s}]\Delta_i. \tag{1.109}$$

Therefore, we have

$$|c_{i(t+1)}| - |c_{i,s}| = 3\mathrm{sgn}[c_{i,s}]\Delta_i \tag{1.110}$$

$$= 3\mathrm{sgn}[c_{i,s}]\left[|c_{it}| - |c_{i,s}|\right] \tag{1.111}$$

and thus

$$\left| \frac{|c_{i(t+1)}| - |c_{i,s}|}{|c_{it}| - |c_{i,s}|} \right| = 3. \tag{1.112}$$

Equation (1.112) indicates that for indices i for which $c_{i,s} \neq 0$, any deviations of c_{it} away from $c_{i,s}$ grow at each iteration by a factor of three. This behavior is clearly unstable. Moreover, these deviations only affect the solutions for $\mathbf{c}_s$ that have two or more nonzero elements, because if $\mathbf{c}_s$ has only one nonzero element $c_{i,s} = 1$, then the corresponding perturbation in c_{it} is $\Delta_i = 0$ due to the constraint in Eq. (1.102). These results prove that all solutions of the form in Eq. (1.33) that are not of the form in Eq. (1.34) are unstable.

Now, consider the space of stationary points defined by Eq. (1.34). Due to the constraint in Eq. (1.102), we only need to consider the deviations of those coefficients $|c_{jt}| = \Delta_j$ whose stationary values are zero, as $|c_{it}| = 1$. For $j \neq i$,

$$|c_{j(t+1)}| = \left| \frac{\kappa_j \Delta_j^3}{\sqrt{\kappa_i^2 + \sum_{p \notin \mathcal{J}} \kappa_p^2 \Delta_p^6}} \right| \tag{1.113}$$

$$\approx \frac{|\kappa_j|}{|\kappa_i|} |\Delta_j|^3 \tag{1.114}$$

$$= \left(\frac{|\kappa_j|}{|\kappa_i|} |\Delta_j|^2 \right) |c_{jt}|. \tag{1.115}$$

Therefore,

$$\frac{|c_{j(t+1)}|}{|c_{jt}|} = \frac{|\kappa_j|}{|\kappa_i|} \Delta_j^2. \tag{1.116}$$

Since each $|\Delta_j| \ll 1$, the right-hand side of Eq. (1.116) is less than one. Hence, small deviations of each $|c_{jt}|$ away from zero decay to zero over time. This result proves the local stability of the update about the separating solutions in Eq. (1.34). Taken together, Eqs. (1.112) and (1.116) prove both theorems. □

PROOF OF THEOREM 4

Proof. Let $p_0(u)$ denote the p.d.f. of ICI_0. Then, we have

$$E\{\text{ICI}_t\} = \int_0^{\text{ICI}_{\max}} u^{3^t} p_0(u)\, du. \tag{1.117}$$

Using the holder inequality, we have

$$E\{\text{ICI}_t\} \leq \left(\int_0^{\text{ICI}_{\max}} u^{r3^t} \right)^{\frac{1}{r}} \left(\int_0^{\text{ICI}_{\max}} p_0^s(u)\, du \right)^{\frac{1}{s}} \tag{1.118}$$

$$\leq \left(\frac{1}{r3^t + 1} \right)^{\frac{1}{r}} (\text{ICI}_{\max})^{3^t + \frac{1}{r}} \left(\int_0^{\text{ICI}_{\max}} p_0^s(u)\, du \right)^{\frac{1}{s}}, \tag{1.119}$$

where $1/r + 1/s = 1$. Letting $s \to \infty$ and $r \to 1$, we have the inequality of the theorem. The second part of the theorem follows by direct integration of Eq. (1.117) with the p.d.f.

$$p_0(u) = \begin{cases} \dfrac{1}{\mathrm{ICI}_{\max}}, & 0 \leq u \leq \mathrm{ICI}_{\max} \\ 0, & \text{otherwise} \end{cases}. \tag{1.120}$$

□

PROOFS OF THEOREM 5 AND ASSOCIATED COROLLARIES

Proof. Let $u_t = g_t(u_0)$, where u_0 is an r.v. with distribution $p_0(u)$. Then, the p.d.f. of u_t is

$$p_t(u) = p_0\left(g_t^{-1}(u)\right) \frac{\mathrm{d}g_t^{-1}(u)}{\mathrm{d}u}, \tag{1.121}$$

where $u_0 = g_t^{-1}(u_t)$ is the inverse function of $u_t = g_t(u_0) = u_0^{3^t}$. Since $g_t^{-1}(u) = u^{\left(\frac{1}{3^t}\right)}$, we obtain the p.d.f. in Eq. (1.46). For the corollaries, consider the distribution of $p_0(u)$ that results from assuming that $\mathbf{c}_0 = [\cos(\theta_0)\ \sin(\theta_0)]^T$ is distributed on the unit arc $\theta_0 \in [0, \pi/4]$ according to the p.d.f. $\overline{p}_0(\theta)$. Using the relationship $\mathrm{ICI}_0 = \tan^2(\theta_0)$ and the p.d.f. transformation result in Eq. (1.121), it can be shown that

$$p_0(u) = \overline{p}_0\left(\arctan\left(u^{\frac{1}{2}}\right)\right) \frac{1}{2} \frac{1}{1+u} u^{-\frac{1}{2}}. \tag{1.122}$$

Substituting this expression into Eq. (1.46), we obtain the expression in Eq. (1.47). Finally, if θ_0 is uniformly distributed in $[0, \pi/4]$, we have $\overline{p}_0(\theta) = 4/\pi$, such that Eq. (1.47) simplifies to Eq. (1.48). □

PROOF OF THEOREM 6

Proof. Consider the expectation $E\{\mathrm{ICI}_t\}$ in Eq. (1.117), as given by

$$E\{\mathrm{ICI}_t\} = \int_0^1 \zeta^{3^t} p_0(\zeta)\,\mathrm{d}\zeta. \tag{1.123}$$

Define the variable transformation $\zeta = \mathrm{e}^{-x}$, such that

$$E\{\mathrm{ICI}_t\} = \int_0^\infty \mathrm{e}^{-(3^t+1)x} p_0(\mathrm{e}^{-x})\,\mathrm{d}x. \tag{1.124}$$

Integrating this integral by parts with the choices $u = p(\mathrm{e}^{-x})$ and $\mathrm{d}v = \mathrm{e}^{-(3^t+1)x}\mathrm{d}x$, we obtain

$$E\{\mathrm{ICI}_t\} = -\frac{1}{3^t+1} \mathrm{e}^{-(3^t+1)x} p_0(\mathrm{e}^{-x})\Big|_0^\infty - \frac{1}{3^t+1} \int_0^\infty \mathrm{e}^{-(3^t+2)x} p_0^{(1)}(\mathrm{e}^{-x})\mathrm{d}x \tag{1.125}$$

$$= \frac{K_1}{3^t+1} - \frac{1}{3^t+1} \int_0^\infty \mathrm{e}^{-(3^t+2)x} p_0^{(1)}(\mathrm{e}^{-x})\mathrm{d}x. \tag{1.126}$$

The integral on the right-hand side of Eq. (1.126) can be integrated by parts to obtain

$$\int_0^\infty \mathrm{e}^{-(3^t+2)x} p_0^{(1)}(\mathrm{e}^{-x}) = \frac{p_0^{(1)}(1)}{3^t+2} - \frac{1}{3^t+2}\int_0^\infty \mathrm{e}^{-3^{t+1}x} p_0^{(2)}(\mathrm{e}^{-x})\,\mathrm{d}x, \tag{1.127}$$

such that Eq. (1.126) becomes

$$E\{\mathrm{ICI}_t\} = \frac{K_1}{3^t+1} + \frac{1}{(3^t+1)(3^t+2)}\left[p_0^{(1)}(1) - \int_0^\infty \mathrm{e}^{-3^{t+1}x} p_0^{(2)}(\mathrm{e}^{-x})\,\mathrm{d}x\right]. \tag{1.128}$$

The results of the theorem and corollary follow directly from this expression. □

PROOF OF THEOREM 7

Proof. If $\mathbf{w}_0$ (or equivalently, $\mathbf{c}_0$) has a uniformly distributed direction in two-dimensional space, the p.d.f. of the angle θ_0 is uniform over the interval $[0, 2\pi)$. Due to the eight-fold symmetry of ICI_t as a function of angle θ_0 in this situation, we can restrict our study to the interval $\theta_0 \in [0, \pi/4]$. Thus, we can write ICI_0 as a random variable and evaluate its expected value as

$$E\{\mathrm{ICI}_0\} = \frac{4}{\pi}\int_0^{\pi/4} \tan^2(\theta)\,\mathrm{d}\theta. \tag{1.129}$$

Using the formula

$$\int \tan^i(\theta)\,\mathrm{d}\theta = \frac{\tan^{i-1}(\theta)}{i-1} - \int \tan^{i-2}(\theta)\,\mathrm{d}\theta, \tag{1.130}$$

we find that

$$E\{\mathrm{ICI}_0\} = \frac{4}{\pi}\left[\tan(\pi/4) - \frac{\pi}{4}\right] \tag{1.131}$$

$$= \frac{4}{\pi} - 1 \tag{1.132}$$

$$\approx 0.27323954\ldots \tag{1.133}$$

To evaluate the expected value of ICI_t, we note the relationship in Eq. (1.40), which for equal-kurtosis sources gives the integral

$$E\{\mathrm{ICI}_t\} = \frac{4}{\pi}\int_0^{\pi/4} \tan^{2(3^t)}(\theta)\,\mathrm{d}\theta. \tag{1.134}$$

By recursive application of Eq. (1.130), it is straightforward to show that $E\{\mathrm{ICI}_t\}$ has the series expansion in Eq. (1.54). □

PROOF OF THEOREM 8

Proof. The starting point for this proof is Eq. (1.40). Letting $\overline{p}_0(\theta) = 2/\pi$ for $0 \le \theta \le pi/2$, the expectation of ICI_t is

$$E\{\mathrm{ICI}_t\} = \frac{2}{\pi}\left\{\int_0^{\alpha_t} a^2\left(a^{-1}\tan\theta\right)^{2(3^t)}\,\mathrm{d}\theta + a^{-2}\int_{\alpha_t}^{\pi/2}(a\cot\theta)^{2(3^t)}\,\mathrm{d}\theta\right\}. \tag{1.135}$$

Define

$$a_t = \tan(\alpha_t) = a \cdot (a^{-1})^{\frac{1}{3^t}}. \tag{1.136}$$

Using the relationships

$$\cot\theta = \tan\left(\frac{\pi}{2} - \theta\right) \tag{1.137}$$

$$\frac{\pi}{2} - \arctan(a_t) = \arctan\left(a_t^{-1}\right), \tag{1.138}$$

we can rewrite Eq. (1.135) as

$$E\{\mathrm{ICI}_t\} = \frac{2}{\pi}\left\{\int_0^{\arctan(a_t)} a^2\left(a^{-1}\tan\theta\right)^{2(3^t)} \mathrm{d}\theta + a^{-2}\int_0^{\arctan(a_t^{-1})} (a\tan\theta)^{2(3^t)}\,\mathrm{d}\theta\right\}. \tag{1.139}$$

By the change of variables $u = b^{-1}\tan\theta$, one can show for any positive constants b and $b_t = b\cdot(b^{-1})^{1/(3^t)}$ that

$$\int_0^{\arctan(b_t)} \left(b^{-1}\tan\theta\right)^{2(3^t)} \mathrm{d}\theta = \int_0^{(b^{-1})^{1/(3^t)}} \frac{u^{2(3^t)}}{b^{-1}+bu^2}\mathrm{d}u. \tag{1.140}$$

As $k \to \infty$, the area underneath the integrand in Eq. (1.140) is concentrated at $u = (b^{-1})^{1/(3^t)}$, such that we may approximate

$$\int_0^{(b^{-1})^{1/(3^t)}} \frac{u^{2(3^t)}}{b^{-1}+bu^2}\mathrm{d}u \approx \frac{1}{b^{-1}+b\cdot(b^{-2})^{1/(3^t)}}\int_0^{(b^{-1})^{1/(3^t)}} u^{2(3^t)}\mathrm{d}u \tag{1.141}$$

$$= \left(\frac{1}{b^{-1}(b)^{\frac{1}{3^t}} + b\left(b^{-1}\right)^{\frac{1}{3^t}}}\right)\frac{b^{-2}}{2(3^t)+1}. \tag{1.142}$$

Using the result in Eq. (1.142) to approximate the integrals in Eq. (1.135) yields

$$E\{\widehat{\mathrm{ICI}}_t\} = \frac{1}{\pi}\left(\frac{2}{a^{-1}(a)^{\frac{1}{3^t}} + a(a^{-1})^{\frac{1}{3^t}}}\right)\frac{2}{2(3^t)+1}. \tag{1.143}$$

As $t \to \infty$, we find that both $a^{1/(3^t)}$ and $(a^{-1})^{1/(3^t)}$ quickly tend to unity, and $2(3^t) \gg 1$. Thus, we can simplify Eq. (1.143) to the final expression given by Eq. (1.67).

The proof of Corollary 5 is similar to that used to derive Eq. (1.54) for equal-kurtosis mixtures and is omitted. □

PROOF OF THEOREM 9

Proof. Consider the component of the ICI at iteration k in situations where the first kurtosis component is being extracted. Then, we can write this portion of the average ICI as

$$E\left\{\mathrm{ICI}_t^{(1)}\right\} = \frac{8}{(2\pi)^{3/2}} \int_0^{\infty} \mathrm{d}x \int_0^{ax} \mathrm{d}y \int_0^{bx} \mathrm{d}z \times \left[\frac{\left(\frac{y}{a}\right)^p + \left(\frac{z}{b}\right)^p}{x^p}\right] \exp\left(-\frac{x^2+y^2+z^2}{2}\right), \tag{1.144}$$

where

$$a = \left(\frac{\kappa_1}{\kappa_2}\right)^{\frac{q}{p}} \quad \text{and} \quad b = \left(\frac{\kappa_1}{\kappa_3}\right)^{\frac{q}{p}}. \tag{1.145}$$

The factor of "8" premultiplying the integral in Eq. (1.144) is due to the limits of zero for the three integrals on the right-hand side of Eq. (1.144). Rewriting this integral, we obtain

$$E\left\{\mathrm{ICI}_t^{(1)}\right\} = \frac{8}{(2\pi)^{3/2}} \int_0^{\infty} \frac{\mathrm{e}^{-\frac{x^2}{2}}}{x^p} \mathrm{d}x \left[\int_0^{ax} \left(\left(\frac{y}{a}\right)^p + \left(\frac{z}{b}\right)^p\right) \mathrm{e}^{-\frac{y^2}{2}} \mathrm{d}y\right] \int_0^{bx} \mathrm{e}^{-\frac{z^2}{2}} \mathrm{d}z. \tag{1.146}$$

The integral in brackets on the right-hand side of Eq. (1.146) can be approximated as

$$\int_0^{ax} \left(\left(\frac{y}{a}\right)^p + \left(\frac{z}{b}\right)^p\right) \mathrm{e}^{-\frac{y^2}{2}} \mathrm{d}y \approx \left[\mathrm{e}^{-\frac{a^2x^2}{2}} \int_0^{ax} \left(\frac{y}{a}\right)^p \mathrm{d}y\right] + \left(\frac{z}{b}\right)^p \int_0^{ax} \mathrm{e}^{-\frac{y^2}{2}} \mathrm{d}y \tag{1.147}$$

$$= \left[a\,\mathrm{e}^{-\frac{a^2x^2}{2}} \frac{x^{p+1}}{p+1}\right] + \left(\frac{z}{b}\right)^p \int_0^{ax} \mathrm{e}^{-\frac{y^2}{2}} \mathrm{d}y. \tag{1.148}$$

Thus, we can approximate

$$\left[\int_0^{ax} \left(\left(\frac{y}{a}\right)^p + \left(\frac{z}{b}\right)^p\right) \mathrm{e}^{-\frac{y^2}{2}} \mathrm{d}y\right] \int_0^{bx} \mathrm{e}^{-\frac{z^2}{2}} \mathrm{d}z = \frac{x^{p+1}}{p+1} \left[a\,\mathrm{e}^{-\frac{a^2x^2}{2}} \int_0^{bx} \mathrm{e}^{-\frac{z^2}{2}} \mathrm{d}z + b\,\mathrm{e}^{-\frac{b^2x^2}{2}} \int_0^{ax} \mathrm{e}^{-\frac{y^2}{2}} \mathrm{d}y\right] \tag{1.149}$$

Substituting Eq. (1.149) into Eq. (1.146), we obtain

$$E\left\{\widehat{\mathrm{ICI}}_t^{(1)}\right\} = \frac{8}{(2\pi)^{3/2}(p+1)} \int_0^{\infty} x\,\mathrm{e}^{-\frac{x^2}{2}} \times \left[a\,\mathrm{e}^{-\frac{a^2x^2}{2}} \int_0^{bx} \mathrm{e}^{-\frac{z^2}{2}} \mathrm{d}z + b\,\mathrm{e}^{-\frac{b^2x^2}{2}} \int_0^{ax} \mathrm{e}^{-\frac{y^2}{2}} \mathrm{d}y\right] \mathrm{d}x \tag{1.150}$$

$$= \frac{8}{(2\pi)^{3/2}(p+1)} \left[\int_0^{\infty} ax\,\mathrm{e}^{-\frac{x^2(1+a^2)}{2}} \int_0^{bx} \mathrm{e}^{-\frac{z^2}{2}} \mathrm{d}z\,\mathrm{d}x + \int_0^{\infty} bx\,\mathrm{e}^{-\frac{x^2(1+b^2)}{2}} \int_0^{ax} \mathrm{e}^{-\frac{y^2}{2}} \mathrm{d}y\,\mathrm{d}x\right]. \tag{1.151}$$

Now, we have

$$\int_0^{\infty} x e^{-\frac{x^2(1+a^2)}{2}}\left[\int_0^{bx} e^{-\frac{z^2}{2}}\,dz\right]dx = \frac{1}{1+a^2}\int_0^{\infty}(1+a^2)x e^{-\frac{x^2(1+a^2)}{2}}\left[\int_0^{bx} e^{-\frac{z^2}{2}}\,dz\right]dx \tag{1.152}$$

$$= \frac{1}{1+a^2}\left[-e^{-\frac{x^2(1+a^2)}{2}}\int_0^{bx} e^{-\frac{z^2}{2}}\,dz\Big|_0^{\infty} + \int_0^{\infty} b\, e^{-\frac{x^2(1+a^2+b^2)}{2}}\,dx\right] \tag{1.153}$$

$$= \frac{b}{1+a^2}\sqrt{\frac{2\pi}{1+a^2+b^2}}\int_0^{\infty}\frac{e^{-\frac{x^2}{2(1+a^2+b^2)^{-1}}}\,dx}{\sqrt{2\pi(1+a^2+b^2)^{-1}}} \tag{1.154}$$

$$= \frac{b}{1+a^2}\sqrt{\frac{\pi}{2(1+a^2+b^2)}}. \tag{1.155}$$

Therefore,

$$E\left\{\widehat{\mathrm{ICI}}_t^{(1)}\right\} = \frac{8}{(2\pi)^{3/2}(p+1)}\sqrt{\frac{\pi}{2(1+a^2+b^2)}}\left[\frac{b}{a^{-1}+a}+\frac{a}{b^{-1}+b}\right] \tag{1.156}$$

$$= \frac{2}{\pi(p+1)}\frac{1}{\sqrt{1+a^2+b^2}}\left[\frac{b}{a^{-1}+a}+\frac{a}{b^{-1}+b}\right]. \tag{1.157}$$

Now, as k increases, we have

$$\lim_{k\to\infty}\frac{q}{p} = \frac{1}{2}, \tag{1.158}$$

such that

$$\lim_{k\to\infty} a = \sqrt{\frac{\kappa_1}{\kappa_2}} \quad \lim_{k\to\infty} b = \sqrt{\frac{\kappa_1}{\kappa_3}}. \tag{1.159}$$

Substituting these results into Eq. (1.157), we obtain

$$E\left\{\widehat{\mathrm{ICI}}_t^{(1)}\right\} = \frac{2}{\pi(2(3^t)+1)}\frac{1}{\sqrt{\kappa_1^{-1}+\kappa_2^{-1}+\kappa_3^{-1}}}\left[\frac{\sqrt{\kappa_3^{-1}}}{\sqrt{\frac{\kappa_1}{\kappa_2}}+\sqrt{\frac{\kappa_2}{\kappa_1}}}+\frac{\sqrt{\kappa_2^{-1}}}{\sqrt{\frac{\kappa_1}{\kappa_3}}+\sqrt{\frac{\kappa_3}{\kappa_1}}}\right] \tag{1.160}$$

$$= \frac{2}{\pi(2(3^t)+1)}\frac{1}{\sqrt{\kappa_1\kappa_2+\kappa_1\kappa_3+\kappa_2\kappa_3}}\left[\frac{\kappa_1\kappa_2}{\kappa_1+\kappa_2}+\frac{\kappa_1\kappa_3}{\kappa_1+\kappa_3}\right]. \tag{1.161}$$

As k increases, $2(3^t) \gg 1$, such that

$$E\left\{\widehat{\mathrm{ICI}}_t^{(1)}\right\} = \frac{1}{\pi}\left(\frac{1}{3}\right)^t\frac{1}{\sqrt{\kappa_1\kappa_2+\kappa_1\kappa_3+\kappa_2\kappa_3}}\left[\frac{\kappa_1\kappa_2}{\kappa_1+\kappa_2}+\frac{\kappa_1\kappa_3}{\kappa_1+\kappa_3}\right]. \tag{1.162}$$

Invoking symmetry, we have

$$E\left\{\widehat{\mathrm{ICI}}_t^{(2)}\right\} = \frac{1}{\pi}\left(\frac{1}{3}\right)^t \frac{1}{\sqrt{\kappa_1\kappa_2+\kappa_1\kappa_3+\kappa_2\kappa_3}}\left[\frac{\kappa_2\kappa_1}{\kappa_2+\kappa_1}+\frac{\kappa_2\kappa_3}{\kappa_2+\kappa_3}\right]. \tag{1.163}$$

$$E\left\{\widehat{\mathrm{ICI}}_t^{(3)}\right\} = \frac{1}{\pi}\left(\frac{1}{3}\right)^t \frac{1}{\sqrt{\kappa_1\kappa_2+\kappa_1\kappa_3+\kappa_2\kappa_3}}\left[\frac{\kappa_3\kappa_1}{\kappa_3+\kappa_1}+\frac{\kappa_3\kappa_2}{\kappa_3+\kappa_2}\right]. \tag{1.164}$$

Finally, we have

$$E\{\widehat{\mathrm{ICI}}_t\} = E\left\{\widehat{\mathrm{ICI}}_t^{(1)}\right\} + E\left\{\widehat{\mathrm{ICI}}_t^{(2)}\right\} + E\left\{\widehat{\mathrm{ICI}}_t^{(3)}\right\}. \tag{1.165}$$

Substitution of the expressions into the above relation yields the result. □

PROOF OF THEOREM 10

Proof. The proof for the four-source case follows that for the $m = 3$ case, in which the integral of interest is

$$\begin{aligned} E\left\{\mathrm{ICI}_t^{(1)}\right\} =& \frac{16}{(2\pi)^{4/2}}\int_0^\infty \mathrm{d}x \int_0^{ax}\mathrm{d}y\int_0^{bx}\mathrm{d}z\int_0^{cx}\mathrm{d}w \\ &\times\left[\frac{\left(\frac{y}{a}\right)^p+\left(\frac{z}{b}\right)^p+\left(\frac{w}{c}\right)^p}{x^p}\right]\exp\left(-\frac{x^2+y^2+z^2+w^2}{2}\right), \end{aligned} \tag{1.166}$$

where a and b are defined as previously and

$$c = \left(\frac{\kappa_1}{\kappa_4}\right)^{\frac{q}{p}}. \tag{1.167}$$

We first focus on the ICI when the first source is dominant, in which case

$$\begin{aligned} E\left\{\mathrm{ICI}_t^{(1)}\right\} =& \frac{16}{(2\pi)^{4/2}}\int_0^\infty \frac{\mathrm{e}^{-\frac{x^2}{2}}}{x^p}\mathrm{d}x\left[\int_0^{ax}\left(\left(\frac{y}{a}\right)^p+\left(\frac{z}{b}\right)^p+\left(\frac{w}{c}\right)^p\right)\mathrm{e}^{-\frac{y^2}{2}}\mathrm{d}y\right] \\ &\times\int_0^{bx}\mathrm{e}^{-\frac{z^2}{2}}\mathrm{d}z\int_0^{cx}\mathrm{e}^{-\frac{w^2}{2}}\mathrm{d}w. \end{aligned} \tag{1.168}$$

The integrals needed to solve this problem are given on the following pages without explanation. The pattern of evaluation is very similar to the previous case.

$$\int_0^{ax}\left(\frac{y}{a}\right)^p\mathrm{e}^{-\frac{y^2}{2}}\mathrm{d}y \approx \mathrm{e}^{-\frac{a^2x^2}{2}}\int_0^{ax}\left(\frac{y}{a}\right)^p\mathrm{d}y \tag{1.169}$$

$$\begin{aligned} \int_0^{ax}\left(\left(\frac{y}{a}\right)^p+\left(\frac{z}{b}\right)^p+\left(\frac{w}{c}\right)^p\right)\mathrm{e}^{-\frac{y^2}{2}}\mathrm{d}y =& \left[a\,\mathrm{e}^{-\frac{a^2x^2}{2}}\frac{x^{p+1}}{p+1}\right] \\ &+\left[\left(\frac{z}{b}\right)^p+\left(\frac{w}{c}\right)^p\right]\int_0^{ax}\mathrm{e}^{-\frac{y^2}{2}}\mathrm{d}y. \end{aligned} \tag{1.170}$$

$$\left[\int_0^{ax}\left(\left(\frac{y}{a}\right)^p+\left(\frac{z}{b}\right)^p+\left(\frac{w}{c}\right)^p\right)\mathrm{e}^{-\frac{y^2}{2}}\mathrm{d}y\right]\int_0^{bx}\mathrm{e}^{-\frac{z^2}{2}}\mathrm{d}z\int_0^{cx}\mathrm{e}^{-\frac{w^2}{2}}\mathrm{d}w$$
$$=\frac{x^{p+1}}{p+1}\left[a\,\mathrm{e}^{-\frac{a^2x^2}{2}}\int_0^{bx}\mathrm{e}^{-\frac{z^2}{2}}\mathrm{d}z\int_0^{cx}\mathrm{e}^{-\frac{w^2}{2}}\mathrm{d}w\right.$$
$$+b\,\mathrm{e}^{-\frac{b^2x^2}{2}}\int_0^{ax}\mathrm{e}^{-\frac{y^2}{2}}\mathrm{d}y\int_0^{cx}\mathrm{e}^{-\frac{w^2}{2}}\mathrm{d}w$$
$$\left.+c\,\mathrm{e}^{-\frac{c^2x^2}{2}}\int_0^{ax}\mathrm{e}^{-\frac{y^2}{2}}\mathrm{d}y\int_0^{bx}\mathrm{e}^{-\frac{z^2}{2}}\mathrm{d}z\right]. \quad (1.171)$$

$$E\left\{\mathrm{ICI}_t^{(1)}\right\}=\frac{16}{(2\pi)^{4/2}(p+1)}\int_0^{\infty}x\,\mathrm{e}^{-\frac{x^2}{2}}\left[a\,\mathrm{e}^{-\frac{a^2x^2}{2}}\int_0^{bx}\mathrm{e}^{-\frac{z^2}{2}}\mathrm{d}z\int_0^{cx}\mathrm{e}^{-\frac{w^2}{2}}\mathrm{d}w\right.$$
$$+b\,\mathrm{e}^{-\frac{b^2x^2}{2}}\int_0^{ax}\mathrm{e}^{-\frac{y^2}{2}}\mathrm{d}y\int_0^{cx}\mathrm{e}^{-\frac{w^2}{2}}\mathrm{d}w$$
$$\left.+\,c\,\mathrm{e}^{-\frac{c^2x^2}{2}}\int_0^{ax}\mathrm{e}^{-\frac{y^2}{2}}\mathrm{d}y\int_0^{bx}\mathrm{e}^{-\frac{z^2}{2}}\mathrm{d}z\right]\mathrm{d}x \quad (1.172)$$
$$=\frac{16}{(2\pi)^{4/2}(p+1)}\left[\int_0^{\infty}ax\,\mathrm{e}^{-\frac{x^2(1+a^2)}{2}}\int_0^{bx}\mathrm{e}^{-\frac{z^2}{2}}\mathrm{d}z\int_0^{cx}\mathrm{e}^{-\frac{w^2}{2}}\mathrm{d}w\,\mathrm{d}x\right.$$
$$+\int_0^{\infty}bx\,\mathrm{e}^{-\frac{x^2(1+b^2)}{2}}\int_0^{ax}\mathrm{e}^{-\frac{y^2}{2}}\mathrm{d}y\int_0^{cx}\mathrm{e}^{-\frac{w^2}{2}}\mathrm{d}w\,\mathrm{d}x$$
$$\left.+\int_0^{\infty}cx\,\mathrm{e}^{-\frac{x^2(1+c^2)}{2}}\int_0^{ax}\mathrm{e}^{-\frac{y^2}{2}}\mathrm{d}y\int_0^{bx}\mathrm{e}^{-\frac{z^2}{2}}\mathrm{d}z\,\mathrm{d}x\right]. \quad (1.173)$$

$$\int_0^{\infty}x\,\mathrm{e}^{-\frac{x^2(1+a^2)}{2}}\left[\int_0^{bx}\mathrm{e}^{-\frac{z^2}{2}}\mathrm{d}z\int_0^{cx}\mathrm{e}^{-\frac{w^2}{2}}\mathrm{d}w\right]\mathrm{d}x$$
$$=\frac{1}{1+a^2}\int_0^{\infty}(1+a^2)x\,\mathrm{e}^{-\frac{x^2(1+a^2)}{2}}\left[\int_0^{bx}\mathrm{e}^{-\frac{z^2}{2}}\mathrm{d}z\int_0^{cx}\mathrm{e}^{-\frac{w^2}{2}}\mathrm{d}w\right]\mathrm{d}x \quad (1.174)$$
$$=\frac{1}{1+a^2}\left[-\mathrm{e}^{-\frac{x^2(1+a^2)}{2}}\int_0^{bx}\mathrm{e}^{-\frac{z^2}{2}}\mathrm{d}z\int_0^{cx}\mathrm{e}^{-\frac{w^2}{2}}\mathrm{d}w\Big|_0^{\infty}+\int_0^{\infty}b\,\mathrm{e}^{-\frac{x^2(1+a^2+b^2)}{2}}\right.$$
$$\left.\times\int_0^{cx}\mathrm{e}^{-\frac{w^2}{2}}\mathrm{d}w\,\mathrm{d}x+\int_0^{\infty}c\,\mathrm{e}^{-\frac{x^2(1+a^2+c^2)}{2}}\int_0^{bx}\mathrm{e}^{-\frac{z^2}{2}}\mathrm{d}z\,\mathrm{d}x\right] \quad (1.175)$$

$$= \frac{1}{1+a^2}\left[\int_0^\infty b\,\mathrm{e}^{-\frac{x^2(1+a^2+b^2)}{2}}\int_0^{cx}\mathrm{e}^{-\frac{w^2}{2}}\,\mathrm{d}w\,\mathrm{d}x\right.$$

$$\left. + \int_0^\infty c\,\mathrm{e}^{-\frac{x^2(1+a^2+c^2)}{2}}\int_0^{bx}\mathrm{e}^{-\frac{z^2}{2}}\,\mathrm{d}z\,\mathrm{d}x\right]. \tag{1.176}$$

$$\int_0^\infty \mathrm{e}^{-\frac{x^2}{2}}\int_0^{\alpha x}\mathrm{e}^{-\frac{w^2}{2}}\,\mathrm{d}w\,\mathrm{d}x = \arctan(a). \tag{1.177}$$

$$\int_0^\infty b\,\mathrm{e}^{-\frac{x^2(1+a^2+b^2)}{2}}\int_0^{cx}\mathrm{e}^{-\frac{w^2}{2}}\,\mathrm{d}w\,\mathrm{d}x = \frac{b}{\sqrt{1+a^2+b^2}}\arctan\left(\frac{c}{\sqrt{1+a^2+b^2}}\right). \tag{1.178}$$

$$a\int_0^\infty x\,\mathrm{e}^{-\frac{x^2(1+a^2)}{2}}\left[\int_0^{bx}\mathrm{e}^{-\frac{z^2}{2}}\,\mathrm{d}z\int_0^{cx}\mathrm{e}^{-\frac{w^2}{2}}\,\mathrm{d}w\right]\mathrm{d}x$$

$$= \frac{1}{a^{-1}+a}\left[\frac{b}{\sqrt{1+a^2+b^2}}\arctan\left(\frac{c}{\sqrt{1+a^2+b^2}}\right)\right.$$

$$\left. + \frac{c}{\sqrt{1+a^2+c^2}}\arctan\left(\frac{b}{\sqrt{1+a^2+b^2}}\right)\right] \tag{1.179}$$

$$= \frac{1}{\sqrt{\kappa_1/\kappa_2}+\sqrt{\kappa_2/\kappa_1}}\left[\frac{\sqrt{\kappa_3^{-1}}}{\sqrt{\kappa_1^{-1}+\kappa_2^{-1}+\kappa_3^{-1}}}\arctan\left(\frac{\sqrt{\kappa_4^{-1}}}{\sqrt{\kappa_1^{-1}+\kappa_2^{-1}+\kappa_3^{-1}}}\right)\right.$$

$$\left. + \frac{\sqrt{\kappa_4^{-1}}}{\sqrt{\kappa_1^{-1}+\kappa_2^{-1}+\kappa_4^{-1}}}\arctan\left(\frac{\sqrt{\kappa_3^{-1}}}{\sqrt{\kappa_1^{-1}+\kappa_2^{-1}+\kappa_4^{-1}}}\right)\right]. \tag{1.180}$$

$$E\left\{\widehat{\mathrm{ICI}}_t^{(1)}\right\} = \frac{4}{\pi^2(p+1)}\left\{\frac{1}{\sqrt{\kappa_1/\kappa_2}+\sqrt{\kappa_2/\kappa_1}}\left[\frac{\sqrt{\kappa_3^{-1}}}{\sqrt{\kappa_1^{-1}+\kappa_2^{-1}+\kappa_3^{-1}}}\right.\right.$$

$$\times \arctan\left(\frac{\sqrt{\kappa_4^{-1}}}{\sqrt{\kappa_1^{-1}+\kappa_2^{-1}+\kappa_3^{-1}}}\right)$$

$$\left. + \frac{\sqrt{\kappa_4^{-1}}}{\sqrt{\kappa_1^{-1}+\kappa_2^{-1}+\kappa_4^{-1}}}\arctan\left(\frac{\sqrt{\kappa_3^{-1}}}{\sqrt{\kappa_1^{-1}+\kappa_2^{-1}+\kappa_4^{-1}}}\right)\right]$$

$$+ \frac{1}{\sqrt{\kappa_1/\kappa_3}+\sqrt{\kappa_3/\kappa_1}}\left[\frac{\sqrt{\kappa_2^{-1}}}{\sqrt{\kappa_1^{-1}+\kappa_2^{-1}+\kappa_3^{-1}}}\arctan\left(\frac{\sqrt{\kappa_4^{-1}}}{\sqrt{\kappa_1^{-1}+\kappa_2^{-1}+\kappa_3^{-1}}}\right)\right.$$

$$
+\frac{\sqrt{\kappa_4^{-1}}}{\sqrt{\kappa_1^{-1}+\kappa_3^{-1}+\kappa_4^{-1}}}\arctan\left(\frac{\sqrt{\kappa_2^{-1}}}{\sqrt{\kappa_1^{-1}+\kappa_3^{-1}+\kappa_4^{-1}}}\right)\Bigg]
$$

$$
+\frac{1}{\sqrt{\kappa_1/\kappa_4}+\sqrt{\kappa_4/\kappa_1}}\Bigg[\frac{\sqrt{\kappa_2^{-1}}}{\sqrt{\kappa_1^{-1}+\kappa_2^{-1}+\kappa_4^{-1}}}\arctan\left(\frac{\sqrt{\kappa_3^{-1}}}{\sqrt{\kappa_1^{-1}+\kappa_2^{-1}+\kappa_4^{-1}}}\right)
$$

$$
+\frac{\sqrt{\kappa_3^{-1}}}{\sqrt{\kappa_1^{-1}+\kappa_3^{-1}+\kappa_4^{-1}}}\arctan\left(\frac{\sqrt{\kappa_2^{-1}}}{\sqrt{\kappa_1^{-1}+\kappa_3^{-1}+\kappa_4^{-1}}}\right)\Bigg]\Bigg\}. \tag{1.181}
$$

$$
E\left\{\widehat{\mathrm{ICI}}_t^{(2)}\right\}=\frac{4}{\pi^2(p+1)}\Bigg\{\frac{1}{\sqrt{\kappa_1/\kappa_2}+\sqrt{\kappa_2/\kappa_1}}\Bigg[\frac{\sqrt{\kappa_3^{-1}}}{\sqrt{\kappa_1^{-1}+\kappa_2^{-1}+\kappa_3^{-1}}}
$$

$$
\times\arctan\left(\frac{\sqrt{\kappa_4^{-1}}}{\sqrt{\kappa_1^{-1}+\kappa_2^{-1}+\kappa_3^{-1}}}\right)
$$

$$
+\frac{\sqrt{\kappa_4^{-1}}}{\sqrt{\kappa_1^{-1}+\kappa_2^{-1}+\kappa_4^{-1}}}\arctan\left(\frac{\sqrt{\kappa_3^{-1}}}{\sqrt{\kappa_1^{-1}+\kappa_2^{-1}+\kappa_4^{-1}}}\right)\Bigg]
$$

$$
+\frac{1}{\sqrt{\kappa_2/\kappa_3}+\sqrt{\kappa_3/\kappa_2}}\Bigg[\frac{\sqrt{\kappa_1^{-1}}}{\sqrt{\kappa_1^{-1}+\kappa_2^{-1}+\kappa_3^{-1}}}\arctan\left(\frac{\sqrt{\kappa_4^{-1}}}{\sqrt{\kappa_1^{-1}+\kappa_2^{-1}+\kappa_3^{-1}}}\right)
$$

$$
+\frac{\sqrt{\kappa_4^{-1}}}{\sqrt{\kappa_2^{-1}+\kappa_3^{-1}+\kappa_4^{-1}}}\arctan\left(\frac{\sqrt{\kappa_1^{-1}}}{\sqrt{\kappa_2^{-1}+\kappa_3^{-1}+\kappa_4^{-1}}}\right)\Bigg]
$$

$$
+\frac{1}{\sqrt{\kappa_2/\kappa_4}+\sqrt{\kappa_4/\kappa_2}}\Bigg[\frac{\sqrt{\kappa_1^{-1}}}{\sqrt{\kappa_1^{-1}+\kappa_2^{-1}+\kappa_4^{-1}}}\arctan\left(\frac{\sqrt{\kappa_3^{-1}}}{\sqrt{\kappa_1^{-1}+\kappa_2^{-1}+\kappa_4^{-1}}}\right)
$$

$$
+\frac{\sqrt{\kappa_3^{-1}}}{\sqrt{\kappa_2^{-1}+\kappa_3^{-1}+\kappa_4^{-1}}}\arctan\left(\frac{\sqrt{\kappa_1^{-1}}}{\sqrt{\kappa_2^{-1}+\kappa_3^{-1}+\kappa_4^{-1}}}\right)\Bigg]\Bigg\}. \tag{1.182}
$$

$$
E\left\{\widehat{\mathrm{ICI}}_t^{(3)}\right\}=\frac{4}{\pi^2(p+1)}\Bigg\{\frac{1}{\sqrt{\kappa_1/\kappa_3}+\sqrt{\kappa_3/\kappa_1}}\Bigg[\frac{\sqrt{\kappa_2^{-1}}}{\sqrt{\kappa_1^{-1}+\kappa_2^{-1}+\kappa_3^{-1}}}
$$

$$
\times\arctan\left(\frac{\sqrt{\kappa_4^{-1}}}{\sqrt{\kappa_1^{-1}+\kappa_2^{-1}+\kappa_3^{-1}}}\right)
$$

$$
\begin{aligned}
&+\frac{\sqrt{\kappa_4^{-1}}}{\sqrt{\kappa_1^{-1}+\kappa_3^{-1}+\kappa_4^{-1}}}\arctan\left(\frac{\sqrt{\kappa_2^{-1}}}{\sqrt{\kappa_1^{-1}+\kappa_3^{-1}+\kappa_4^{-1}}}\right)\Bigg]\\
&+\frac{1}{\sqrt{\kappa_2/\kappa_3}+\sqrt{\kappa_3/\kappa_2}}\left[\frac{\sqrt{\kappa_1^{-1}}}{\sqrt{\kappa_1^{-1}+\kappa_2^{-1}+\kappa_3^{-1}}}\arctan\left(\frac{\sqrt{\kappa_4^{-1}}}{\sqrt{\kappa_1^{-1}+\kappa_2^{-1}+\kappa_3^{-1}}}\right)\right.\\
&\left.+\frac{\sqrt{\kappa_4^{-1}}}{\sqrt{\kappa_2^{-1}+\kappa_3^{-1}+\kappa_4^{-1}}}\arctan\left(\frac{\sqrt{\kappa_1^{-1}}}{\sqrt{\kappa_2^{-1}+\kappa_3^{-1}+\kappa_4^{-1}}}\right)\right]\\
&+\frac{1}{\sqrt{\kappa_3/\kappa_4}+\sqrt{\kappa_4/\kappa_3}}\left[\frac{\sqrt{\kappa_1^{-1}}}{\sqrt{\kappa_1^{-1}+\kappa_3^{-1}+\kappa_4^{-1}}}\arctan\left(\frac{\sqrt{\kappa_2^{-1}}}{\sqrt{\kappa_1^{-1}+\kappa_3^{-1}+\kappa_4^{-1}}}\right)\right.\\
&\left.\left.+\frac{\sqrt{\kappa_2^{-1}}}{\sqrt{\kappa_2^{-1}+\kappa_3^{-1}+\kappa_4^{-1}}}\arctan\left(\frac{\sqrt{\kappa_1^{-1}}}{\sqrt{\kappa_2^{-1}+\kappa_3^{-1}+\kappa_4^{-1}}}\right)\right]\right\}. \qquad (1.183)
\end{aligned}
$$

$$
\begin{aligned}
E\left\{\widehat{\mathrm{ICI}}_t^{(4)}\right\} &= \frac{4}{\pi^2(p+1)}\left\{\frac{1}{\sqrt{\kappa_1/\kappa_4}+\sqrt{\kappa_4/\kappa_1}}\left[\frac{\sqrt{\kappa_2^{-1}}}{\sqrt{\kappa_1^{-1}+\kappa_2^{-1}+\kappa_4^{-1}}}\right.\right.\\
&\times\arctan\left(\frac{\sqrt{\kappa_3^{-1}}}{\sqrt{\kappa_1^{-1}+\kappa_2^{-1}+\kappa_4^{-1}}}\right)\\
&\left.+\frac{\sqrt{\kappa_3^{-1}}}{\sqrt{\kappa_1^{-1}+\kappa_3^{-1}+\kappa_4^{-1}}}\arctan\left(\frac{\sqrt{\kappa_2^{-1}}}{\sqrt{\kappa_1^{-1}+\kappa_3^{-1}+\kappa_4^{-1}}}\right)\right]\\
&+\frac{1}{\sqrt{\kappa_2/\kappa_4}+\sqrt{\kappa_4/\kappa_2}}\left[\frac{\sqrt{\kappa_1^{-1}}}{\sqrt{\kappa_1^{-1}+\kappa_2^{-1}+\kappa_4^{-1}}}\arctan\left(\frac{\sqrt{\kappa_3^{-1}}}{\sqrt{\kappa_1^{-1}+\kappa_2^{-1}+\kappa_4^{-1}}}\right)\right.\\
&\left.+\frac{\sqrt{\kappa_3^{-1}}}{\sqrt{\kappa_2^{-1}+\kappa_3^{-1}+\kappa_4^{-1}}}\arctan\left(\frac{\sqrt{\kappa_1^{-1}}}{\sqrt{\kappa_2^{-1}+\kappa_3^{-1}+\kappa_4^{-1}}}\right)\right]\\
&+\frac{1}{\sqrt{\kappa_3/\kappa_4}+\sqrt{\kappa_4/\kappa_3}}\left[\frac{\sqrt{\kappa_1^{-1}}}{\sqrt{\kappa_1^{-1}+\kappa_3^{-1}+\kappa_4^{-1}}}\arctan\left(\frac{\sqrt{\kappa_2^{-1}}}{\sqrt{\kappa_1^{-1}+\kappa_3^{-1}+\kappa_4^{-1}}}\right)\right.
\end{aligned}
$$

$$+\frac{\sqrt{\kappa_2^{-1}}}{\sqrt{\kappa_2^{-1}+\kappa_3^{-1}+\kappa_4^{-1}}}\arctan\left(\frac{\sqrt{\kappa_1^{-1}}}{\sqrt{\kappa_2^{-1}+\kappa_3^{-1}+\kappa_4^{-1}}}\right)\Bigg]\Bigg\}. \tag{1.184}$$

$$E\{\widehat{\mathrm{ICI}}_t\}=E\left\{\widehat{\mathrm{ICI}}_t^{(1)}\right\}+E\left\{\widehat{\mathrm{ICI}}_t^{(2)}\right\}+E\left\{\widehat{\mathrm{ICI}}_t^{(3)}\right\}+E\left\{\widehat{\mathrm{ICI}}_t^{(4)}\right\}. \tag{1.185}$$

The final result is found by substitution of the relations into Eq. (1.185). □

PROOF OF THEOREM 11

Proof. Consider the component of the ICI at iteration t in the situations where the first kurtosis component is being extracted. Then, we can write this portion of the average ICI as

$$E\left\{\mathrm{ICI}_t^{(1)}\right\}=\frac{2^m}{(2\pi)^{m/2}}\int_0^{\infty}\mathrm{d}x_1\int_0^{a_2x_1}\mathrm{d}x_2\cdots\int_0^{a_mx_1}\mathrm{d}x_m \times\left[\frac{\sum_{i=2}^m\left(\frac{x_i}{a_i}\right)^p}{x_1^p}\right]\exp\left(-\frac{\sum_{i=1}^m x_i^2}{2}\right) \tag{1.186}$$

$$=\frac{2^m}{(2\pi)^{m/2}}\sum_{i=2}^m\int_0^{\infty}\mathrm{d}x_1\int_0^{a_2x_1}\mathrm{d}x_2\cdots\int_0^{a_mx_1}\mathrm{d}x_m \times\left[\frac{\left(\frac{x_i}{a_i}\right)^p}{x_1^p}\right]\exp\left(-\frac{\sum_{i=1}^m x_i^2}{2}\right), \tag{1.187}$$

where

$$a_i=\left(\frac{\kappa_1}{\kappa_i}\right)^{\frac{q}{p}},\quad i=2,\ldots,m. \tag{1.188}$$

Make the transform for the variables $x_2,\ldots,x_m$ in the above integral,

$$E\left\{\mathrm{ICI}_t^{(1)}\right\}=\frac{2^m}{(2\pi)^{m/2}}\sum_{i=2}^m\int_0^{a_2}\mathrm{d}x_2\cdots\int_0^{a_m}\mathrm{d}x_m\int_0^{\infty} \times\left(\frac{x_i}{a_i}\right)^p x_1^{m-1}\exp\left(-\frac{x_1^2}{2}\left(1+\sum_{i=2}^m x_i^2\right)\right)\mathrm{d}x_1. \tag{1.189}$$

The most inside integral can be calculated as

$$\int_0^{\infty}x_1^{m-1}\exp\left(-\frac{x_1^2}{2}\left(1+\sum_{i=2}^m x_i^2\right)\right)\mathrm{d}x_1=\begin{cases}\dfrac{(m-2)!!\sqrt{\pi}}{\sqrt{2}\left(1+\sum_{i=2}^m x_i^2\right)^{m/2}} & m\text{ is odd}\\[2ex] \dfrac{\left[\frac{1}{2}(m-2)\right]!2^{(m-2)/2}}{\left(1+\sum_{i=2}^m x_i^2\right)^{m/2}} & m\text{ is even,}\end{cases}$$

where $(m-2)!! = 1 \cdot 3 \cdot 5 \cdot 7 \cdots (m-2)$. Thus,

$$E\left\{\mathrm{ICI}_t^{(1)}\right\} = \begin{cases} \frac{2^m}{(2\pi)^{m/2}} \sum_{i=2}^{m} \int_0^{a_2} \mathrm{d}x_2 \cdots \int_0^{a_m} \mathrm{d}x_m \left(\frac{x_i}{a_i}\right)^p \frac{(m-2)!!\sqrt{\pi}}{\sqrt{2}\left(1+\sum_{j=2}^m x_j^2\right)^{m/2}} & m \text{ is odd} \\ \frac{2^m}{(2\pi)^{m/2}} \sum_{i=2}^{m} \int_0^{a_2} \mathrm{d}x_2 \cdots \int_0^{a_m} \mathrm{d}x_m \left(\frac{x_i}{a_i}\right)^p \frac{\left[\frac{1}{2}(m-2)\right]!2^{(m-2)/2}}{\left(1+\sum_{j=2}^m x_j^2\right)^{m/2}} & m \text{ is even.} \end{cases}$$

When $k \to \infty$, $(\frac{x_i}{a_i})^p \to 0$ at the interval $0 \le x_i < a_i$, and $a_i \to \sqrt{\frac{\kappa_1}{\kappa_i}} =: b_{1i}$, we can approximate the integral

$$\int_0^{a_i} \left(\frac{x_i}{a_i}\right)^p \frac{1}{\left(1+\sum_{j=2}^m x_j^2\right)^{m/2}} \mathrm{d}x_i \approx \frac{1}{\left(1+a_i^2+\sum_{j=2, j\neq i}^m x_j^2\right)^{m/2}} \int_0^{a_i} \left(\frac{x_i}{a_i}\right)^p \mathrm{d}x_i \quad (1.190)$$

$$= \frac{1}{\left(1+a_i^2+\sum_{j=2, j\neq i}^m x_j^2\right)^{m/2}} \frac{a_i}{p+1} \quad (1.191)$$

$$\approx \frac{1}{\left(1+b_{1i}^2+\sum_{j=2, j\neq i}^m x_j^2\right)^{m/2}} \frac{b_{1i}}{p}. \quad (1.192)$$

Note that since $p = 2(3^t)$, we have

$$E\left\{\mathrm{ICI}_t^{(1)}\right\} \approx \begin{cases} \frac{1}{2(3)^t} \frac{2^m}{(2\pi)^{m/2}} \sum_{i=2}^{m} \int_0^{b_{12}} \mathrm{d}x_2 \cdots \int_0^{b_{1m}} \mathrm{d}x_m \frac{(m-2)!!\sqrt{\pi}}{\sqrt{2}\left(1+b_{1i}^2+\sum_{j=2, j\neq i}^m x_j^2\right)^{m/2}} & m \text{ is odd} \\ \frac{1}{2(3)^t} \frac{2^m}{(2\pi)^{m/2}} \sum_{i=2}^{m} \int_0^{b_{12}} \mathrm{d}x_2 \cdots \int_0^{b_{1m}} \mathrm{d}x_m \frac{\left[\frac{1}{2}(m-2)\right]!2^{(m-2)/2}}{\left(1+b_{1i}^2+\sum_{j=2, j\neq i}^m x_j^2\right)^{m/2}} & m \text{ is even.} \end{cases}$$

Similarly, we get

$$E\left\{\mathrm{ICI}_t^{(n)}\right\} \approx \begin{cases} \frac{1}{2(3)^t} \frac{2^m}{(2\pi)^{m/2}} \sum_{i=1, i\neq n}^{m} \int_0^{b_{n1}} \mathrm{d}x_1 \cdots \int_0^{b_{nm}} \mathrm{d}x_m \frac{(m-2)!!\sqrt{\pi}}{\sqrt{2}\left(1+(b_{ni})^2+\sum_{j=1, j\neq n, j\neq i}^m x_j^2\right)^{m/2}} & m \text{ is odd} \\ \frac{1}{2(3)^t} \frac{2^m}{(2\pi)^{m/2}} \sum_{i=1, i\neq n}^{m} \int_0^{b_{n1}} \mathrm{d}x_1 \cdots \int_0^{b_{nm}} \mathrm{d}x_m \frac{\left[\frac{1}{2}(m-2)\right]!2^{(m-2)/2}}{\left(1+(b_{ni})^2+\sum_{j=1, j\neq n, j\neq i}^m x_j^2\right)^{m/2}} & m \text{ is even} \end{cases}$$

for $n = 2, \ldots, m$ and

$$b_{ni} = \sqrt{\frac{\kappa_n}{\kappa_i}}. \tag{1.193}$$

Therefore,

$$E\{\mathrm{ICI}_t\} = \sum_{n=1}^{m} E\left\{\mathrm{ICI}_t^{(n)}\right\} \tag{1.194}$$

$$\approx g(\kappa_1, \ldots, \kappa_m)\left(\frac{1}{3}\right)^t, \tag{1.195}$$

where

$$g(\kappa_1, \ldots, \kappa_m) = \begin{cases} \frac{2^{m-1}}{(2\pi)^{m/2}} \sum_{n=1}^{m} \sum_{i=1, i\neq n}^{m} \int_0^{b_{n1}} \mathrm{d}x_1 \cdots \int_0^{b_{nm}} \mathrm{d}x_m \frac{(m-2)!!\sqrt{\pi}}{\sqrt{2}\left(1+(b_{ni})^2+\sum_{j=1, j\neq n, j\neq i}^{m} x_j^2\right)^{m/2}} & m \text{ is odd} \\ \frac{2^{m-1}}{(2\pi)^{m/2}} \sum_{n=1}^{m} \sum_{i=1, i\neq n}^{m} \int_0^{b_{n1}} \mathrm{d}x_1 \cdots \int_0^{b_{nm}} \mathrm{d}x_m \frac{\left[\frac{1}{2}(m-2)\right]! 2^{(m-2)/2}}{\left(1+(b_{ni})^2+\sum_{j=1, j\neq n, j\neq i}^{m} x_j^2\right)^{m/2}} & m \text{ is even.} \end{cases}$$

The errors have been introduced in both Eqs. (1.190) and (1.192). We first estimate the error in Eq. (1.190).

$$\int_0^{a_i} \left(\frac{x_i}{a_i}\right)^p \frac{1}{\left(1+\sum_{i=2}^{m} x_i^2\right)^{m/2}} \mathrm{d}x_i$$

$$= \left(\int_0^{\left(1-\frac{1}{\sqrt{p}}\right)a_i} + \int_{\left(1-\frac{1}{\sqrt{p}}\right)a_i}^{a_i}\right) \left(\frac{x_i}{a_i}\right)^p \frac{1}{\left(1+\sum_{i=2}^{m} x_i^2\right)^{m/2}} \mathrm{d}x_i \tag{1.196}$$

$$\leq \left(1-\frac{1}{\sqrt{p}}\right)^{p+1} a_i + \frac{1}{\left(1+\left(1-\frac{1}{\sqrt{p}}\right)^2 a_i^2 + \sum_{j=2, j\neq i}^{m} x_j^2\right)^{m/2}} \int_{\left(1-\frac{1}{\sqrt{p}}\right)a_i}^{a_i} \left(\frac{x_i}{a_i}\right)^p \mathrm{d}x_i \tag{1.197}$$

$$\leq \left(1-\frac{1}{\sqrt{p}}\right)^{p+1} a_i + \frac{1}{\left(1-\frac{1}{\sqrt{p}}\right)^m} \frac{1}{\left(1+a_i^2+\sum_{j=2, j\neq i}^{m} x_j^2\right)^{m/2}} \int_{\left(1-\frac{1}{\sqrt{p}}\right)a_i}^{a_i} \left(\frac{x_i}{a_i}\right)^p \mathrm{d}x_i \tag{1.198}$$

$$\leq \left(1-\frac{1}{\sqrt{p}}\right)^{p+1} a_i + \frac{1}{\left(1-\frac{1}{\sqrt{p}}\right)^m} \frac{1}{\left(1+a_i^2+\sum_{j=2, j\neq i}^{m} x_j^2\right)^{m/2}} \int_0^{a_i} \left(\frac{x_i}{a_i}\right)^p \mathrm{d}x_i. \tag{1.199}$$

Thus,

$$0 \le \int_0^{a_i} \left(\frac{x_i}{a_i}\right)^p \frac{1}{\left(1+\sum_{i=2}^m x_i^2\right)^{m/2}} \mathrm{d}x_i - \frac{1}{\left(1+a_i^2+\sum_{j=2,j\neq i}^m x_i^2\right)^{m/2}} \int_0^{a_i} \left(\frac{x_i}{a_i}\right)^p \mathrm{d}x_i \tag{1.200}$$

$$\le \left(1-\frac{1}{\sqrt{p}}\right)^{p+1} a_i. \tag{1.201}$$

To estimate the error in Eq. (1.192), recall that $a_i = (\frac{\kappa_1}{\kappa_i})^{\frac{q}{p}} = b_{1i}^{1-\frac{2}{p}}$, we have

$$\left| \frac{1}{\left(1+a_i^2+\sum_{j=2,j\neq i}^m x_j^2\right)^{m/2}} \frac{a_i}{p+1} - \frac{1}{\left(1+b_{1i}^2+\sum_{j=2,j\neq i}^m x_j^2\right)^{m/2}} \frac{b_{1i}}{p} \right| \tag{1.202}$$

$$\le \left| \frac{1}{\left(1+b_{1i}^2+\sum_{j=2,j\neq i}^m x_j^2\right)^{m/2}} \frac{b_{1i}^{1-\frac{4}{p}}}{p+1} - \frac{1}{\left(1+b_{1i}^2+\sum_{j=2,j\neq i}^m x_j^2\right)^{m/2}} \frac{b_{1i}}{p} \right| \tag{1.203}$$

$$\le \frac{1}{\left(1+b_{1i}^2+\sum_{j=2,j\neq i}^m x_j^2\right)^{m/2}} \frac{b_{1i}}{p} \left| 1 - \frac{p}{p+1} b_{1i}^{-\frac{4}{p}} \right| \tag{1.204}$$

$$\le \frac{b_{1i}}{p} \left| 1 - \frac{p}{p+1} b_{1i}^{-\frac{4}{p}} \right|. \tag{1.205}$$

Now we have the estimation for the error term $R(t,\kappa_1,\ldots,\kappa_m)^{(1)}$ for the first portion

$$\left|R(t,\kappa_1,\ldots,\kappa_m)^{(1)}\right| = c(m)\left(\frac{1}{3}\right)^t \left(\sum_{i=2}^m \int_0^{b_{12}} \mathrm{d}x_2 \cdots \int_0^{b_{1m}} \mathrm{d}x_m \frac{1}{\left(1+b_{1i}^2+\sum_{j=2,j\neq i}^m x_j^2\right)^{m/2}} \right.$$

$$\left. - \sum_{i=2}^m \int_0^{a_2} \mathrm{d}x_2 \cdots \int_0^{a_n} \mathrm{d}x_n \frac{1}{\left(1+\sum_{j=2}^m x_j^2\right)^{m/2}} \right) \tag{1.206}$$

$$\le c(m)\left(\frac{1}{3}\right)^t \left(\sum_{i=2}^m \int_0^{b_{12}} \mathrm{d}x_2 \cdots \int_0^{b_{1m}} \mathrm{d}x_m \frac{1}{\left(1+b_{1i}^2+\sum_{j=2,j\neq i}^m x_j^2\right)^{m/2}} \right.$$

$$\left. - \sum_{i=2}^m \int_0^{a_2} \mathrm{d}x_2 \cdots \int_0^{a_m} \mathrm{d}x_m \frac{1}{\left(1+b_{1i}^2+\sum_{j=2,j\neq i}^m x_j^2\right)^{m/2}} \right) \tag{1.207}$$

$$
\begin{aligned}
&+ c(m)\left(\frac{1}{3}\right)^{t}\left(\sum_{i=2}^{m}\int_{0}^{b_2}\mathrm{d}x_2\cdots\int_{0}^{b_{1m}}\mathrm{d}x_m\frac{1}{\left(1+b_{1i}^2+\sum_{j=2,j\neq i}^{m}x_j^2\right)^{m/2}}\right.\\
&\left.-\sum_{i=2}^{m}\int_{0}^{a_2}\mathrm{d}x_2\cdots\int_{0}^{a_m}\mathrm{d}x_m\frac{1}{\left(1+\sum_{j=2}^{m}x_j^2\right)^{m/2}}\right) \qquad (1.208)\\
&\le c(m)\left(\frac{1}{3}\right)^{t}(m-1)\left|\prod_{i=2}^{m}b_{1i}-\prod_{i=2}^{m}a_i\right|\\
&+ c(m)\left(\frac{1}{3}\right)^{t}\sum_{i=2}^{m}\left(\left(1-\frac{1}{\sqrt{p}}\right)^{p+1}a_i+\frac{b_{1i}}{p}\left|1-\frac{p}{p+1}b_{1i}^{-\frac{4}{p}}\right|\right)\prod_{i=2}^{m}a_i \qquad (1.209)\\
&= c(m)\left(\frac{1}{3}\right)^{t}\left[(m-1)\prod_{i=2}^{m}b_{1i}\left(1-\prod_{i=2}^{m}b_{1i}^{-\frac{2}{p}}\right)\right.\\
&\left.+\sum_{i=2}^{m}\left(\left(1-\frac{1}{\sqrt{p}}\right)^{p+1}b_{1i}^{1-\frac{2}{p}}+\frac{b_{1i}}{p}\left|1-\frac{p}{p+1}b_{1i}^{-\frac{4}{p}}\right|\right)\prod_{i=2}^{m}b_{1i}^{1-\frac{2}{p}}\right], \qquad (1.210)
\end{aligned}
$$

where

$$
c(m)=\begin{cases}\dfrac{2^{(m-3)/2}}{(\pi)^{m/2}}(m-2)!!\sqrt{\pi} & m \text{ is odd}\\[2ex] \dfrac{2^{m-2}}{(\pi)^{m/2}}\left[\frac{1}{2}(m-2)\right]! & m \text{ is even.}\end{cases}
$$

Therefore,

$$
\begin{aligned}
|R(t,\kappa_1,\ldots,\kappa_m)| &\le \sum_{n=1}^{m}\left|R(t,\kappa_1,\ldots,\kappa_m)^{(n)}\right| \qquad (1.211)\\
&\le c(m)\left(\frac{1}{3}\right)^{t}\sum_{n=1}^{m}\left[(m-1)\prod_{i=2}^{m}b_{ni}\left(1-\prod_{i=2}^{m}b_{ni}^{-\frac{2}{p}}\right)\right.\\
&\left.+\sum_{i=2}^{m}\left(\left(1-\frac{1}{\sqrt{p}}\right)^{p+1}b_{ni}^{1-\frac{2}{p}}+\frac{b_{ni}}{p}\left|1-\frac{p}{p+1}b_{ni}^{-\frac{4}{p}}\right|\right)\prod_{i=2}^{m}b_{ni}^{1-\frac{2}{p}}\right]. \qquad (1.212)
\end{aligned}
$$

Since $1-\prod_{i=2}^{m}b_{ni}^{-\frac{2}{p}}\to 0$ and $(1-\frac{1}{\sqrt{p}})^{p+1}\to 0$ as $t\to\infty$, it is easy to see that

$$
\lim_{t\to\infty}\frac{R(t,\kappa_1,\ldots,\kappa_m)}{\left(\frac{1}{3}\right)^{t}}=0. \qquad (1.213)
$$

□

PROOF OF THEOREM 12

Proof. The proof of this result is best seen by considering the average ICI for different problem orders m.

Case $m = 2$: Then, we have

$$E\left\{\mathrm{ICI}_t^{(2)}\right\} = 2\int_0^1\int_0^x \frac{y^{2(3^t)}}{x^{2(3^t)}}\mathrm{d}y\,\mathrm{d}x \tag{1.214}$$

$$= 2\int_0^1 \frac{1}{x^{2(3^t)}}\left[\frac{x^{2(3^t)+1}}{2(3^t)+1}\right]\mathrm{d}x \tag{1.215}$$

$$= \frac{2}{2(3^t)+1}\int_0^1 x\,\mathrm{d}x \tag{1.216}$$

$$= \frac{1}{2(3^t)+1}. \tag{1.217}$$

Case $m = 3$: Then, we have

$$E\left\{\mathrm{ICI}_t^{(3)}\right\} = 6\int_0^1\int_0^x\int_0^y \frac{y^{2(3^t)}+z^{2(3^t)}}{x^{2(3^t)}}\mathrm{d}z\,\mathrm{d}y\,\mathrm{d}x \tag{1.218}$$

$$= 6\int_0^1 \frac{1}{x^{2(3^t)}}\int_0^x \frac{2(3^t)+2}{2(3^t)+1}y^{2(3^t)+1}\mathrm{d}y\,\mathrm{d}x \tag{1.219}$$

$$= \frac{6}{2(3^t)+1}\int_0^1 x^2\mathrm{d}x \tag{1.220}$$

$$= \frac{2}{2(3^t)+1}. \tag{1.221}$$

Case $m = 4$: Then, we have

$$E\left\{\mathrm{ICI}_t^{(4)}\right\} = 24\int_0^1\int_0^x\int_0^y\int_0^z \frac{y^{2(3^t)}+z^{2(3^t)}+w^{2(3^t)}}{x^{2(3^t)}}\mathrm{d}w\,\mathrm{d}z\,\mathrm{d}y\,\mathrm{d}x \tag{1.222}$$

$$= 24\int_0^1 \frac{1}{x^{2(3^t)}}\int_0^x\int_0^y \left[y^{2(3^t)}z + \frac{2(3^t)+2}{2(3^t)+1}z^{2(3^t)+1}\right]\mathrm{d}z\,\mathrm{d}y\,\mathrm{d}x \tag{1.223}$$

$$= 24\int_0^1 \frac{1}{x^{2(3^t)}}\frac{1}{2}\int_0^x \frac{2(3^t)+3}{2(3^t)+1}y^{2(3^t)+2}\mathrm{d}y\,\mathrm{d}x \tag{1.224}$$

$$= \frac{12}{2(3^t)+1}\int_0^1 x^3\mathrm{d}x \tag{1.225}$$

$$= \frac{3}{2(3^t)+1}. \tag{1.226}$$

General case: For any $m \geq 2$ and $1 \leq i \leq m-2$, we have

$$E\left\{\mathrm{ICI}_t^{(m)}\right\} = m!\int_0^1\int_0^{c_1}\int_0^{c_2}\cdots\int_0^{c_{m-1}} \frac{\sum_{j=2}^m c_j^{2(3^t)}}{c_1^{2(3^t)}}\mathrm{d}c_m\,\mathrm{d}c_{m-1}\,\ldots\,\mathrm{d}c_1 \tag{1.227}$$

$$= \frac{m!}{(i-1)!} \frac{1}{2(3^t)+1} \int_0^1 \frac{1}{c_1^{2(3^t)}} \int_0^{c_1} \int_0^{c_2} \cdots \int_0^{c_{m-i}} [2(3^t)+1] c_{m-i+1}^{i-1}$$

$$\times \sum_{j=2}^{m-i} c_j^{2(3^t)} + [2(3^t)+i] c_{m-i+1}^{2(3^t)+i-1} \mathrm{d}c_{m-i+1} \ldots \mathrm{d}c_1 \tag{1.228}$$

$$= \frac{m(m-1)}{2(3^t)+1} \int_0^1 \frac{1}{c_1^{2(3^t)}} \int_0^{c_1} [2(3^t)+m-1] c_2^{2(3^t)+m-2} \mathrm{d}c_2 \, \mathrm{d}c_1 \tag{1.229}$$

$$= \frac{m(m-1)}{2(3^t)+1} \int_0^1 c_1^{m-1} \mathrm{d}c_1 \tag{1.230}$$

$$= \frac{m-1}{2(3^t)+1}, \tag{1.231}$$

which proves the result. □

ACKNOWLEDGMENTS

Portions of this work were completed while the author was on sabbatical leave at Helsinki University of Technology (now Aalto University), Espoo, Finland, in 2005. The author is grateful to Prof. Visa Koivunen for support during that time. The author would like to thank Prof. Errki Oja and Dr. Zhijian Yuan for numerous interactions during that time on the subject matter within this chapter.

NOTES

a. The system measure in Eq. (1.12) does not enforce uniqueness on the separated system outputs; however, such uniqueness is assured through signal prewhitening and coefficient orthogonality and thus will not be considered in what follows.

REFERENCES

[1] A. Hyvärinen, E. Oja, A fast fixed-point algorithm for independent component analysis, Neural Comput. 9 (7) (1997) 1483-1492.
[2] A. Hyvärinen, J. Karhunen, E. Oja, Independent Component Analysis, Wiley, New York, 2001.
[3] P.A. Regalia, E. Kofidis, Monotonic convergence of fixed-point algorithms for ICA, IEEE Trans. Neural Netw. 14 (2003) 943-949.
[4] E. Oja, Convergence of the symmetrical FastICA algorithm, in: Proc. 9th Int. Conf. Neural Inform. Processing, Singapore, vol. 3, November 2002, pp. 1368-1372.
[5] E. Oja, Z. Yuan, The FastICA algorithm revisited: convergence analysis, IEEE Trans. Neural Netw. 17 (6) (2006) 1370-1381.

[6] P. Tichavsky, Z. Koldovsky, E. Oja, Performance analysis of the FastICA algorithm and Cramer-Rao bounds for linear independent component analysis, IEEE Trans. Signal Process. 54 (4) (2006) 1189-1203.
[7] P. Tichavsky, Z. Koldovsky, E. Oja, Corrections to "Performance analysis of the FastICA algorithm and Cramer-Rao bounds for linear independent component analysis", IEEE Trans. Signal Process. 56 (4) (2008) 1715-1716.
[8] H. Shen, M. Kleinsteuber, K. Huper, Local convergence analysis of FastICA and related algorithms, IEEE Trans. Neural Netw. 19 (6) (2008) 1022-1032.
[9] E. Ollila, The deflation-based FastICA estimator: statistical analysis revisited, IEEE Trans. Signal Process. 58 (3) (2010) 1527-1541.
[10] S.C. Douglas, On the convergence behavior of the FastICA algorithm, in: Proc. Fourth Symp. Indep. Compon. Anal. Blind Signal Separation, Kyoto, Japan, April 2003, pp. 409-414.
[11] W.A. Gardner, Learning characteristics of stochastic-gradient-descent algorithms: a general study, analysis, and critique, Signal Process. 6 (2) (1984) 113-133.
[12] K.I. Diamantaras, S.-Y. Kung, Principal Component Neural Networks: Theory and Applications, Wiley, New York, 1996.
[13] N. Delfosse, P. Loubaton, Adaptive blind separation of independent sources: a deflation approach, Signal Process. 45 (1) (1995) 59-83.
[14] I.S. Gradshteyn, I.M. Ryzhik, Table of Integrals, Series, and Products, Academic Press, San Diego, CA, 1980.
[15] H.A. David, Order Statistics, second ed., Wiley, New York, 1980.
[16] S.C. Douglas, A statistical convergence analysis of the FastICA algorithm for two-source mixtures, in: Proc. 39th Asilomar Conf. Signals, Syst., Comput., Pacific Grove, CA, October 2005, pp. 335-339.
[17] S.C. Douglas, Z. Yuan, E. Oja, Average convergence behavior of the FastICA algorithm for blind source separation, in: Proc. 6th Int. Conf. Indep. Compon. Anal. Blind Source Separation, Charleston, SC, March 2006, pp. 790-798.

CHAPTER

Improved variants of the FastICA algorithm

2

Zbyněk Koldovský and Petr Tichavský
The Institute of Information Theory and Automation of the Czech Academy of Sciences, Czech Republic

2.1 INTRODUCTION

Blind Source Separation (BSS) represents a wide class of models and algorithmsthat have one goal in common: to retrieve unknown original signals from their mixtures [1]. In the instantaneous linear mixture model, the relation between unobserved original signals and observed measured signals is given by

$$\mathbf{X} = \mathbf{AS}, \tag{2.1}$$

where $\mathbf{X}$ and $\mathbf{S}$ are, respectively, matrices containing samples of the measured and the original signals. Their *ij*th element corresponds to the *j*th sample of the *i*th signal. We will consider the regular case where the numbers of rows in $\mathbf{X}$ and $\mathbf{S}$ are the same and are equal to d. $\mathbf{A}$ is a $d \times d$ regular *mixing* matrix representing the mixing system. The instantaneous model says that the *j*th original signal contributes to the *i*th measured signal with an attenuation factor of $\mathbf{A}_{ij}$, which is the *ij*th element of $\mathbf{A}$.

Independent component analysis (ICA) solves the BSS task on the basis of an assumption that the original signals $\mathbf{S}$ are statistically *independent*. Since the original signals are mixed through $\mathbf{A}$, the observed signals $\mathbf{X}$ are, in general, dependent. The ICA task thus can be formulated as the one to estimate the mixing matrix $\mathbf{A}$ or, equivalently, $\mathbf{W} \triangleq \mathbf{A}^{-1}$, called the *de-mixing matrix*, so that signals $\mathbf{Y} = \mathbf{WX}$ are as independent as possible.[a]

The solution of the ICA task is not uniquely determined. Any matrix $\mathbf{W}$ of the form

$$\mathbf{W} = \mathbf{\Lambda} \mathbf{P} \mathbf{A}^{-1}, \tag{2.2}$$

where $\mathbf{\Lambda}$ is a diagonal matrix with nonzero diagonal entries and $\mathbf{P}$ is a permutation matrix, separates the original signals from $\mathbf{X}$ up to their original order, scales, and signs. Therefore, we can later assume, without any loss of generality, that the variance of the source signals is equal to 1. Furthermore, the mean of the signals is irrelevant for purposes of the signals' independence and can be assumed equal to 0, or may be removed from the data in case it is nonzero.

Statistical (in)dependence can be measured in various ways depending on the assumptions applied to the model of the original signals. There are three basic

E. Bingham, S. Kaski, J. Laaksonen and J. Lampinen (Eds) Advances in Independent Component Analysis and Learning Machines.
http://dx.doi.org/10.1016/B978-0-12-802806-3.00002-6

models used in ICA/BSS.[b] The first one assumes that the signal is a sequence of identically and independently distributed (i.i.d.) random variables. As the condition of separability of such signals requires that no more than one signal is Gaussian, the approach is called *non-Gaussianity*-based [8]. The second approach takes the *nonstationarity* of signals into account by modeling them as independently distributed Gaussian variables whose variances are changing in time. The third basic model considers weakly stationary Gaussian processes. These signals are separable if their spectra are distinct; therefore, it is said to be based on the *spectral diversity* or nonwhiteness.

2.1.1 NON-GAUSSIANITY-BASED MODEL

In this model, each original signal is modeled as an i.i.d. sequence. Therefore, the nth sample of the ith original signal, which is the nth element of the ith row of $\mathbf{S}$, also denoted as $s_i(n)$, has the probability density function (p.d.f.) f_{s_i}. Since the signals are assumed to be independent, the joint density of $s_1(n), \ldots, s_d(n)$ is equal to the product of the corresponding marginals,

$$f_{s_1,\ldots,s_d} = \prod_{i=1}^{d} f_{s_i}. \tag{2.3}$$

The corresponding notation of distributions and p.d.f.s will also be used for the measured signals $\mathbf{X}$ and the separated signals $\mathbf{Y}$.

A common criterion for measuring independence of separated signals is the *Kullback-Leibler divergence* between their joint density and the product of marginal densities, which is indeed their *mutual information* defined as

$$I(\mathbf{Y}) = \int_{\mathcal{R}^d} f_{y_1,\ldots,y_d}(\xi_1,\ldots,\xi_d) \ln \frac{f_{y_1,\ldots,y_d}(\xi_1,\ldots,\xi_d)}{\prod_{i=1}^{d} f_{y_i}(\xi_i)} d\xi_1,\ldots,d\xi_d. \tag{2.4}$$

Assume for now that the components of $\mathbf{Y}$ are not correlated and are normalized to have variances equal to 1. Then, it holds that

$$I(\mathbf{Y}) = \sum_{i=1}^{d} H(y_i) + \text{const.}, \tag{2.5}$$

where $H(y_i)$ is the entropy of the ith separated signal defined as

$$H(y_i) = -\int_{\mathcal{R}} f_{y_i}(\xi) \ln f_{y_i}(\xi) d\xi. \tag{2.6}$$

Hence, the minimization of Eq. (2.4) is equivalent to the minimization of the entropies of all signals, which is the principle also used by FastICA.

2.1.2 THE FastICA ALGORITHM

FastICA is one of the most widely used ICA algorithms for the linear mixing model, a fixed-point algorithm first proposed by Hyvärinen and Oja [9,10]. Following

Eq. (2.5), it is based on the optimization of a contrast function measuring the non-Gaussianity of the separated source. We will show later that an optimal measure of the non-Gaussianity requires knowledge of the density function. In FastICA, a nonlinear contrast function is chosen so that it can be appropriate for large-scale densities.

2.1.2.1 Preprocessing

The first step of many ICA algorithms, including FastICA, consists of removing the sample mean (the mean is irrelevant for the signals' dependence), scaling the signals to have unit variances (the original scale is also irrelevant as it cannot be retrieved due to the indeterminacy of ICA), and de-correlating them. The de-correlation is a necessary condition for independence. Such a transformation is appropriately expressed as

$$\mathbf{Z} = \hat{\mathbf{C}}^{-1/2}\,(\mathbf{X} - \overline{\mathbf{X}}), \tag{2.7}$$

where

$$\hat{\mathbf{C}} = (\mathbf{X} - \overline{\mathbf{X}})(\mathbf{X} - \overline{\mathbf{X}})^T/N \tag{2.8}$$

is the sample covariance matrix; $\overline{\mathbf{X}}$ is the sample mean, $\overline{\mathbf{X}} = \mathbf{X} \cdot \mathbf{1}_N \mathbf{1}_N^T/N$ and $\mathbf{1}_N$ denotes the $N \times 1$ vector of ones. Another popular solution is to apply the principal component analysis to rows of $\mathbf{X} - \overline{\mathbf{X}}$.

Now, the output $\mathbf{Z}$ contains de-correlated and unit variance data in the sense that $\mathbf{Z}\mathbf{Z}^T/N = \mathbf{I}$ (the identity matrix) and their mutual information can be written as in Eq. (2.5). This property remains valid if and only if $\mathbf{Z}$ is multiplied by a unitary matrix $\mathbf{U}$. Therefore, the separating transform can be searched through finding an appropriate $\mathbf{U}$ such that $\mathbf{U}\mathbf{U}^T = \mathbf{I}$ and $\mathbf{UZ}$ are as independent as possible. The constraint on $\mathbf{U}$ is called the *orthogonal constraint*.

2.1.2.2 The FastICA algorithm for one unit

The algorithm estimates one row of the de-mixing matrix $\mathbf{U}$ as a vector $\mathbf{u}^T$ that is a stationary point (minimum or maximum) of

$$\hat{\mathrm{E}}[G(\mathbf{u}^T\mathbf{Z})] \stackrel{\text{def}}{=} G(\mathbf{u}^T\mathbf{Z})\mathbf{1}_N/N$$

subject to $\|\mathbf{u}\| = 1$, where $G(\cdot)$ is a suitable nonlinear and nonquadratic function, is applied element-wise to vector arguments. The latter expression is indeed the sample mean of $G(\cdot)$ over samples of $\mathbf{u}^T\mathbf{Z}$; and $\hat{\mathrm{E}}[\cdot]$ denotes the sample mean operator.

Finding $\mathbf{u}^T$ proceeds iteratively. Starting with a random initial unit norm vector $\mathbf{u}$, the algorithm iterates

$$\mathbf{u}^+ \leftarrow \mathbf{Z}g(\mathbf{Z}^T\mathbf{u}) - \mathbf{u}\,g'(\mathbf{u}^T\mathbf{Z})\mathbf{1}_N \tag{2.9}$$

$$\mathbf{u} \leftarrow \mathbf{u}^+/\|\mathbf{u}^+\| \tag{2.10}$$

until convergence is achieved. Here, $g(\cdot)$ and $g'(\cdot)$ denote the first and second derivatives of the function $G(\cdot)$. The application of $g(\cdot)$ and $g'(\cdot)$ to the vector $\mathbf{u}^T\mathbf{Z}$ is also element-wise. Classical widely used functions $g(\cdot)$ include "pow3," that is, $g(x) = x^3$ (then the algorithm performs kurtosis minimization); "tanh," that is, $g(x) = \tanh(x)$; and "gauss," $g(x) = x\exp(-x^2/2)$.

It is not known in advance which column of $\mathbf{U}$ is being estimated: it largely depends on the initialization. If all independent components were estimated in parallel, the algorithm could be written as

$$\mathbf{U}^+ \leftarrow g(\mathbf{UZ})\mathbf{Z}^T - \mathtt{diag}[g'(\mathbf{UZ})\mathbf{1}_N]\,\mathbf{U} \tag{2.11}$$

$$\mathbf{u}_k \leftarrow \mathbf{u}_k^+/\|\mathbf{u}_k^+\|, \quad k = 1, \ldots, d, \tag{2.12}$$

where $\mathbf{u}_k^+$ stands for the kth row of $\mathbf{U}^+$, $\mathbf{u}_k$ stands for the kth row of $\mathbf{U}$, and $\mathtt{diag}[\mathbf{v}]$ stands for a diagonal matrix with diagonal elements taken from the vector $\mathbf{v}$. A result of this iterative process will be denoted as $\mathbf{U}^{1\mathrm{U}}$.

2.1.2.3 The symmetric FastICA algorithm

The symmetric FastICA is designed to estimate all separated signals simultaneously. One step of the parallel estimation proceeds through Eq. (2.11) and each is completed by a symmetric orthonormalization. Specifically, starting with a random unitary matrix $\mathbf{U}$, the method iterates

$$\mathbf{U}^+ \leftarrow g(\mathbf{UZ})\mathbf{Z}^T - \mathtt{diag}[g'(\mathbf{UZ})\mathbf{1}_N]\,\mathbf{U} \tag{2.13}$$

$$\mathbf{U} \leftarrow (\mathbf{U}^+\mathbf{U}^{+T})^{-1/2}\mathbf{U}^+ \tag{2.14}$$

until convergence is achieved. Note that $\mathbf{U}$ is orthogonal due to Eq. (2.14). The resulting matrix of the symmetric algorithm will be denoted as $\mathbf{U}^{\mathrm{SYM}}$.

The stopping criterion is typically

$$1 - \min(|\mathtt{diag}(\mathbf{U}^T\mathbf{U}_{\mathrm{old}})|) < \epsilon \tag{2.15}$$

for a suitable positive constant ϵ; here $\mathbf{U}_{\mathrm{old}}$ denotes the resulting matrix of the previous iteration.

Besides the symmetric algorithm, there also exists Deflation FastICA that estimates all signals. The deflation approach, which is common for many other ICA algorithms [11], estimates the components successively under orthogonality conditions. The accuracy of Deflation FastICA depends on the order of components as they were separated by the algorithm. The order is determined by the initialization.

2.1.2.4 Summary

The separated signals (independent components) are finally equal to

$$\hat{\mathbf{S}} = \mathbf{UZ} = \mathbf{UD}(\mathbf{X} - \overline{\mathbf{X}}), \tag{2.16}$$

where $\mathbf{D}$ stands for the preprocessing transform; for example, $\mathbf{D} = \mathbf{C}^{-1/2}$ as in Eq. (2.7). The whole separating (de-mixing) matrix is thus equal to

$$\mathbf{W} = \mathbf{UD}. \tag{2.17}$$

Note also that the mean $\mathbf{W}\overline{\mathbf{X}}$ could be added back to $\hat{\mathbf{S}}$.

Since $\mathbf{U}$ is orthogonal, sample correlations of the separated signals $\hat{\mathbf{S}}$ are exactly equal to 0. This is the consequence of the orthogonality constraint.

2.1.3 LATER DEVELOPMENTS OF FastICA

Since the first papers on the FastICA algorithm were published, the algorithm has become one of the most successful and most frequently used methods for ICA. It was subject to intensive interest of many researchers, which gave rise to many theoretical and practical analyses of its behavior and modifications. Here, we refer to some of them.

Statistical properties of the algorithm, especially its accuracy when finite data are available, were studied in [12–14] and later in [15–17]. The results were often compared with the corresponding Cramér-Rao bound derived, for example, in [5,14, 18,19].

The algorithm was also adapted for operation with complex-valued signals [20–22]; the Cramér-Rao bound for such a case was studied in [23], and identifiability issues were studied in [24].

Speed of the algorithm was improved for FastICA with "pow3" nonlinearity in a novel method called RobustICA [25]. Another way of speed enhancement for FastICA with general contrast functions was achieved by replacing the nonlinear contrast functions such as "tanh" or "gauss" by suitable rational functions [26]. With these functions, the statistical properties of FastICA remain nearly the same but the evaluation of rational function is faster on most processors.

Stability issues were studied in [27] where it was shown that in the case where separated independent components have multimodal distributions (the p.d.f. has two or more peaks), it happens with nonzero probability that deflation FastICA gets stuck in a false solution which does not correspond to the separation of all sources. Only the algorithm with the "pow3" nonlinearity (kurtosis) can guarantee a zero probability of this phenomenon. For the deflation FastICA, the order of the separated components appeared to be crucial for stability of the algorithm. An improved algorithm which optimizes the order of the separated components in FastICA was proposed in [28]. For symmetric FastICA there was a simple test of saddle points proposed to improve the success rate of the algorithm in [14].

An improved FastICA with adaptive choice of nonlinearity is the subject of the EFICA algorithm [29]. The fact that the nonlinearity influences the algorithm's statistical accuracy was already known, and other FastICA variants endowed with an adaptive choice had previously been proposed; see, for example, [28,30–32].

FastICA properties were also studied in the presence of additive noise. In [33], an unbiased variant was proposed based on the assumption of known covariances of the noise. Later, it was shown in [34] that One-unit FastICA tends to estimating the minimum mean square solution rather than to identifying the inversion of the mixing matrix.

2.2 ACCURACY OF ONE-UNIT AND SYMMETRIC FastICA

2.2.1 PERFORMANCE EVALUATION

The accuracy of separation can be evaluated through a comparison of the estimated mixing matrix with the original one or of the separated signals with the original ones. The original quantities must be known, which happens only in simulated experiments: some independent signals are mixed by a generated mixing matrix, an ICA algorithm is applied to the mixed signals, and the resulting separating matrices or separated signals are evaluated. The evaluation method must take into account the ICA indeterminacy, especially the random order of separated signals.

Let $\mathbf{G}$ be the so-called *gain matrix* defined as

$$\mathbf{G} = \mathbf{WA}. \tag{2.18}$$

Ideally, $\mathbf{G}$ is equal to $\mathbf{\Lambda P}$ as follows from Eq. (2.2). Here, $\mathbf{\Lambda}$ is the diagonal matrix representing the signals' scale indeterminacy, while $\mathbf{P}$ is the permutation matrix determining their order. In practice, $\mathbf{G} \approx \mathbf{\Lambda P}$ due to estimation errors in $\mathbf{W}$.

The Amari's index evaluates the separation accuracy as a whole with the aid of a nonnegative value

$$I = \sum_{i=1}^{d} \left(\frac{\sum_{j=1}^{d} |\mathbf{G}_{ij}|}{\max_k |\mathbf{G}_{ik}|} - 1 \right) + \sum_{j=1}^{d} \left(\frac{\sum_{i=1}^{d} |\mathbf{G}_{ij}|}{\max_k |\mathbf{G}_{kj}|} - 1 \right). \tag{2.19}$$

The criterion reflects the fact that $\mathbf{G}$ should contain one and only one dominant element per row and column.

To evaluate each separated signal individually, it is popular to use standard measures such as Signal-to-Interference ratio (SIR). However, before the computation of SIR, the separated signals must be correctly assigned to the original ones. A straightforward way is to match the separated and original signals based on dominant elements of $\mathbf{G}$ under the condition that the matched pairs of signals are disjoint. The most common approach, called *greedy*, finds the maximal (in absolute value) element of $\mathbf{G}$, assigns the corresponding signals, and repeats the process until all signals are paired. A more sophisticated non-greedy pairing based on the Kuhn-Munkres algorithm was proposed in [35].

Once the permutation matrix $\mathbf{P}$ is found, and the separated signals are re-ordered, the kth separated signal, denoted as $\hat{s}_k(n)$, is equal to

$$\hat{s}_k(n) = \mathbf{G}_{k1}s_1(n) + \cdots + \mathbf{G}_{kk}s_k(n) + \cdots + \mathbf{G}_{kd}s_d(n).$$

The SIR of the kth separated signal equals

$$\mathrm{SIR}_k = \frac{|\mathbf{G}_{kk}|^2\sigma_k^2}{\sum_{i=1,i\neq k}^{d}|\mathbf{G}_{ki}|^2\sigma_i^2}, \tag{2.20}$$

where σ_i^2 is the variance of the ith original signal. Henceforth, the variances will be assumed equal to 1; that is, $\sigma_i^2 = 1$, $i = 1, \ldots, d$. This assumption can be used without any loss of generality because of the indeterminacy in signals' scales.

The reciprocal value of SIR is named the Interference-to-Signal Ratio (ISR)

$$\mathrm{ISR}_k = \frac{\sum_{i=1,i\neq k}^{d}|\mathbf{G}_{ki}|^2}{|\mathbf{G}_{kk}|^2}. \tag{2.21}$$

2.2.2 CRAMÉR-RAO LOWER BOUND

Cramér-Rao lower bound (CRLB) is a general bound for the variance of an unbiased estimator [36]. Consider a vector of parameters $\boldsymbol{\theta}$ being estimated from a data vector $\mathbf{x}$, where the latter has probability density $f_{\mathbf{x}|\boldsymbol{\theta}}(\mathbf{x}|\boldsymbol{\theta})$. Let $\hat{\boldsymbol{\theta}}$ be an unbiased estimator of $\boldsymbol{\theta}$. If the following *Fisher information matrix* (FIM) exists

$$\mathbf{F}_{\boldsymbol{\theta}} = \mathrm{E}_{\boldsymbol{\theta}}\left[\frac{1}{f_{\mathbf{x}|\boldsymbol{\theta}}^2}\frac{\partial f_{\mathbf{x}|\boldsymbol{\theta}}(\mathbf{x}|\boldsymbol{\theta})}{\partial\boldsymbol{\theta}}\left(\frac{\partial f_{\mathbf{x}|\boldsymbol{\theta}}(\mathbf{x}|\boldsymbol{\theta})}{\partial\boldsymbol{\theta}}\right)^T\right], \tag{2.22}$$

then, under mild regularity conditions, it holds that

$$\operatorname{cov}\hat{\boldsymbol{\theta}} \geq \mathrm{CRLB}_{\boldsymbol{\theta}} = \mathbf{F}_{\boldsymbol{\theta}}^{-1},$$

where $\operatorname{cov}\hat{\boldsymbol{\theta}}$ is the covariance matrix of $\hat{\boldsymbol{\theta}}$.

Now we apply the idea of the Cramér-Rao theory to the elements of an ISR matrix whose ijth element is defined as

$$\mathrm{ISR}_{ij} = \mathrm{E}\left[\frac{|\mathbf{G}_{ij}|^2}{|\mathbf{G}_{ii}|^2}\right], \tag{2.23}$$

to derive an algorithm-independent lower bound on values of its elements.

Let the separated signals be already re-ordered and scaled so that $\mathbf{G} = \mathbf{I} + \boldsymbol{\epsilon}$ where $\boldsymbol{\epsilon}$ is a "small" matrix of errors. Then the elements of the ISR matrix can be approximated as

$$\mathrm{ISR}_{ij} \approx \mathrm{E}[|\epsilon_{ij}|^2], \tag{2.24}$$

and the lower bound can be defined as the CRLB for $\boldsymbol{\epsilon}$; see also [37]. Note that we are only interested in the nondiagonal elements of Eq. (2.24), because the asymptotic behavior of the ISR is independent of the diagonal terms (assuming "small" errors).

2.2.2.1 Non-Gaussian i.i.d. signals

Details of the computation of the CRLB is given in [19] with a small correction in [38]. The bound says that

$$\mathrm{ISR}_{ij} \geq \frac{1}{N}\frac{\kappa_j}{\kappa_i\kappa_j - 1}, \quad i \neq j, \tag{2.25}$$

where

$$\kappa_i = \mathrm{E}\left[(\psi_i(x))^2\right] \tag{2.26}$$

and

$$\psi_i(x) = -\frac{f_i'(x)}{f_i(x)} \tag{2.27}$$

is the so-called score function of f_i. The same result was also observed elsewhere in the literature; see, for example, [5,18,39]; for the complex-domain case see [23].

It can be shown that $\kappa_i \geq 1$ where the equality holds if and only if f_i is Gaussian; see Appendix E in [14]. Hence, the denominator of Eq. (2.25) becomes equal to 0 only if both κ_i and κ_j are equal to 1, which means that both the ith and jth signals have Gaussian distributions. This is in accordance with the primary requirement that only one original signal can have the Gaussian p.d.f. It can also be seen that the bound is minimized when $\kappa_i \to +\infty$ and $\kappa_j \to +\infty$, which can be interpreted as the signals being non-Gaussian as much as possible.

2.2.2.2 Piecewise stationary non-Gaussian signals

The above CRLB, indeed, follows from a more general bound that was derived for a piecewise stationary non-Gaussian model of signals in [40,41]. In that model, signals are assumed to obey the i.i.d. model separately within M blocks. Let the blocks have, for simplicity, the same length. Then, the bound says that

$$\mathrm{ISR}_{ij} \geq \frac{1}{N} \cdot \frac{A_{ij}}{A_{ij}A_{ji} - 1} \cdot \frac{\overline{\sigma}_j^2}{\overline{\sigma}_i^2}, \quad i \neq j, \tag{2.28}$$

where

$$A_{ij} = \frac{1}{M} \sum_{\ell=1}^{M} \frac{\sigma_i^{2(\ell)}}{\sigma_j^{2(\ell)}} \kappa_j^{(\ell)} \tag{2.29}$$

$$\overline{\sigma}_i^2 = \frac{1}{M} \sum_{\ell=1}^{M} \sigma_i^{2(\ell)}. \tag{2.30}$$

$\sigma_i^{2(\ell)}$ denotes the variance of the ith signal within the ℓth block, and $\kappa_i^{(\ell)}$ is defined as

$$\kappa_i^{(\ell)} = \mathrm{E}\left[\left(\psi_i^{(\ell)}(x)\right)^2\right], \tag{2.31}$$

where $\psi_i^{(\ell)} = -(\bar{f}_i^{(\ell)})'/\bar{f}_i^{(\ell)}$. $\bar{f}_i^{(\ell)}$ denotes the p.d.f. of the ith original signal on the ℓth block, that is, $f_i^{(\ell)}$, but normalized to the unit variance (the variance of $f_i^{(\ell)}$ is involved in $\sigma_i^{2(\ell)}$). Recall the simplifying assumption that the original signals have unit scales, which means that $\overline{\sigma}_i^2 = 1, i = 1, \ldots, d$.

The shapes of the expressions on the right-hand sides of Eqs. (2.25) and (2.28) are analogous to those of other (more general) Cramér-Rao bounds derived for ICA/BSS or related disciplines, such as independent vector analysis (IVA); an interested reader is referred to [42,43].

2.2.3 ASYMPTOTIC BEHAVIOR OF FastICA

Let $\mathbf{G}^{1U}$ and $\mathbf{G}^{SYM}$, respectively, be the gain matrices obtained by One-unit and Symmetric FastICA using the nonlinear function $g(\cdot)$. Let the function be even, which means that the corresponding $G(\cdot)$ is symmetric, and also let the p.d.f.s of signals be symmetric. It was shown in [14] that, for $i \neq j$, elements of $N^{1/2}\mathbf{G}_{ij}^{1U}$ and $N^{1/2}\mathbf{G}_{ij}^{SYM}$ have asymptotically Gaussian distribution $\mathcal{N}(0, V_{ij}^{1U})$ and $\mathcal{N}(0, V_{ij}^{SYM})$, where

$$V_{ij}^{1U} = \frac{\beta_i - \mu_i^2}{(\mu_i - \rho_i)^2} \tag{2.32}$$

$$V_{ij}^{SYM} = \frac{\beta_i - \mu_i^2 + \beta_j - \mu_j^2 + (\mu_j - \rho_j)^2}{(|\mu_i - \rho_i| + |\mu_j - \rho_j|)^2} \tag{2.33}$$

with $\mu_i = \mathrm{E}[s_i g(s_i)]$, $\rho_i = \mathrm{E}[g'(s_i)]$, $\beta_i = \mathrm{E}[g^2(s_i)]$, and $g'(\cdot)$ being the first derivative of $g(\cdot)$. It is sufficient to assume that the above derivative and expectations exist. The expressions for nonsymmetric distributions were derived in [17].

Next, it can be shown that Eq. (2.32) achieves its minimum for $g(\cdot)$ being equal to the score function of the distribution f_i, that is, for

$$g(x) = \psi_i(x) = -\frac{f_i'(x)}{f_i(x)}.$$

In that case, it is easy to compute that $\mu_i = 1$ and $\rho_i = \beta_i = \kappa_i$.

Assume for now that the distributions of all signals are the same, which means that the above quantities are independent of the index i. That is, $g(x) = \psi(x)$ and $\rho_i = \beta_i = \kappa$, and then, according to Eqs. (2.32) and (2.33),

$$\mathrm{var}[\mathbf{G}_{ij}^{1U}] \approx \frac{1}{N} V_{ij}^{1U} = \frac{1}{N}\frac{1}{\kappa - 1} \tag{2.34}$$

$$\mathrm{var}[\mathbf{G}_{ij}^{SYM}] \approx \frac{1}{N} V_{ij}^{SYM} = \frac{1}{N}\left(\frac{1}{4} + \frac{1}{2}\frac{1}{\kappa - 1}\right). \tag{2.35}$$

For the same case, the CRLB from Eq. (2.25) takes the form

$$\mathrm{ISR}_{ij} \geq \frac{1}{N}\frac{\kappa}{\kappa^2 - 1}, \quad i \neq j. \tag{2.36}$$

Comparisons of Eqs. (2.34) and (2.35) with Eq. (2.36) for $\kappa \geq 1$ are shown in Figure 2.1. One-unit FastICA for the optimum case approaches the CRLB when $\kappa \to \infty$, while Symmetric FastICA is nearly efficient for κ lying in a neighborhood of 1. The latter case, however, means that the distributions of signals are close to the Gaussian distribution, so the signals are hard to separate, and the CRLB itself goes to infinity. For $\kappa \to \infty$, the performance of Symmetric FastICA (Eq. 2.35) is limited by a constant, which is due to the orthogonal constraint (the sample covariance matrix of the separated signals is exactly equal to the identity matrix).

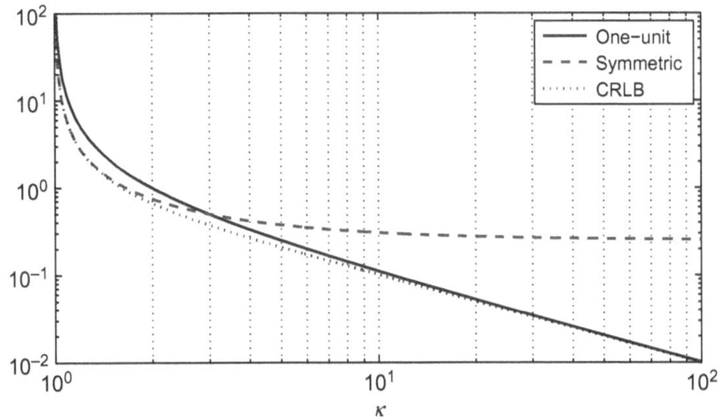

FIGURE 2.1

A comparison of One-unit FastICA, Symmetric FastICA, and the corresponding CRLB for the case when all signals have the same distributions and $g(x) = \psi(x)$. The expressions are plotted as functions of $\kappa \geq 1$ (here $N = 1$).

2.2.4 CHOICE OF THE NONLINEARITY

From the previous analysis it follows that it is not possible to suggest a nonlinearity that would be optimum for all signals, because they need not have the same distribution. The distributions need not be known as we face a blind problem; moreover, score functions of the distributions need not be smooth as required by FastICA. Improved variants of FastICA therefore endow the algorithm by an adaptive choice of the nonlinearity [28,30–32].

It is also not possible to choose a nonlinearity that would enable FastICA to separate all non-Gaussian distributions. An example was shown in [26]: consider signals having the same p.d.f. as $s = \beta b + \sqrt{1-\beta^2} q$ where b and q stand for binary (BPSK) and Laplacean random variables, respectively, and $\beta \in [0, 1]$. For many nonlinearities (e.g., "tanh") it holds that $\mu_i - \rho_i > 0$ for $\beta = 0$ while $\mu_i - \rho_i < 0$ for $\beta = 1$ (if not, other distributions of b and q can be chosen). It then follows that there exists $\beta \in (0, 1)$ such that $\tau_i = \mu_i - \rho_i = 0$. From Eqs. (2.32) and (2.33), it follows that FastICA cannot separate such a distribution (although being non-Gaussian) using the given nonlinearity.

The original variants of FastICA use general-purpose nonlinearities such as "tanh," because it is useful for many signals' distributions that are met in practice. In [26], it was suggested to replace "tanh" by rational functions that are similarly appropriate for separating long-tailed distributions. One such nonlinearity, henceforth referred to as "rati," is

$$g(x) = \frac{x}{1 + x^2/4}. \tag{2.37}$$

The advantage of using the rational function is that it requires a significantly lower computational burden than "tanh" due to its evaluation on most CPUs. As a result, FastICA using the rational function is typically twice as fast as the algorithm with "tanh."

The suitability of any rational function to separate a given distribution with FastICA can be easily inspected and compared with "tanh" using the analytical expressions on the right-hand side of Eq. (2.32) or (2.33).

2.3 GLOBAL CONVERGENCE

The global convergence of FastICA was theoretically proven in special cases only. For example, if the nonlinearity is "pow3," the global convergence of Symmetric FastICA was proven in [44] but only for the theoretical case in which an infinite amount of samples is available. In practice, the behavior of FastICA is also known to be quite good when "tanh" or other nonlinearities are used.

Nevertheless, if it is run, for instance, 10,000 times from random initial de-mixing matrices, the algorithm gets stuck in an unwanted solution in 1-100 cases. These cases are recognized by an exceptionally low value of SIR achieved. The rate of false solutions depends on the dimension of the model, on the stopping rule, and on the length of the data. But it never vanishes completely. For example, when separating d signals all having uniform distribution, the failure rates of Symmetric FastICA using the stopping rule (Eq. 2.15), respectively, with $\epsilon = 10^{-4}$ and $\epsilon = 10^{-5}$ are shown in Table 2.1.

Table 2.1 Number of Failures of Symmetric FastICA with the "tanh" Nonlinearity among 10,000 Trails

	$N = 200$	$N = 500$	$N = 1000$	$N = 10,000$
$d = 2$ and $\varepsilon = 10^{-4}$	85	57	59	46
$d = 2$ and $\varepsilon = 10^{-5}$	49	16	15	12
$d = 2$ and s.p.check	**00**	**0**	**0**	**0**
$d = 3$ and $\varepsilon = 10^{-4}$	49	5	4	6
$d = 3$ and $\varepsilon = 10^{-5}$	43	0	1	0
$d = 3$ and s.p.check	**00**	**0**	**0**	**0**
$d = 4$ and $\varepsilon = 10^{-4}$	95	9	4	11
$d = 4$ and $\varepsilon = 10^{-5}$	85	2	0	5
$d = 4$ and s.p.check	**05**	**0**	**0**	**0**
$d = 5$ and $\varepsilon = 10^{-4}$	166	2	4	11
$d = 5$ and $\varepsilon = 10^{-5}$	151	1	2	2
$d = 5$ and s.p.check	**17**	**0**	**0**	**0**

Notes: *d is the dimension of the signal mixture; ε is the stopping parameter in Eq. (2.15); the acronym "s.p.check" denotes the algorithm endowed by the test of saddle points. Bold values commonly emphasize the best results in a given comparison. We propose to boldify also the name of the best method, which is the method in the same row. So the whole row will be written in bold.*

2.3.1 TEST OF SADDLE POINTS

A detailed investigation of the false solutions showed that they lie approximately halfway (in the angular sense) between a pair of original signals, thus, in saddle points. Although these points are not stable, the algorithm can stop when getting to their close neighborhood as the following iteration step is too small.

Specifically, the false solutions typically contain two components $u_1(n)$ and $u_2(n)$ that are close to $(s_k(n) + s_\ell(n))/\sqrt{2}$ and $(s_k(n) - s_\ell(n))/\sqrt{2}$, for certain $k, \ell \in \{1, \ldots, d\}$. Thus, they should be transformed into

$$u_1'(n) = (u_1(n) + u_2(n))/\sqrt{2} \quad \text{and} \quad u_2'(n) = (u_1(n) - u_2(n))/\sqrt{2}. \tag{2.38}$$

It was suggested in [14] to complete the algorithm by checking all $\binom{d}{2}$ pairs of the estimated independent components for a possible improvement via the saddle points. If the test for a saddle point is positive, it is suggested to perform several additional iterations of the original algorithm, starting from the improved estimate (Eq. 2.38).

The selection between given candidates (u_k, u_ℓ) and (u_k', u_ℓ') can be done by maximizing the criterion (a measure of total non-Gaussianity of the components),

$$c(u_k, u_\ell) = (\hat{\mathrm{E}}[G(u_k(n))] - G_0)^2 + (\hat{\mathrm{E}}[G(u_\ell(n))] - G_0)^2$$

where $G_0 = \mathrm{E}[G(\xi)]$ and ξ is a standard Gaussian variable; $\hat{\mathrm{E}}[\cdot]$ stands for the sample mean operator. For example, in the case of the nonlinearity "tanh," $G(x) = \log\cosh(x)$ and $G_0 \approx 0.3746$.

The number of failures after this test of saddle points is compared in Table 2.1. This table shows zero rate after the test except for the most difficult case when the data length is $N = 200$. Nevertheless, even in this case the rate of failures has significantly dropped compared to the original FastICA.

2.4 APPROACHING CRAMÉR-RAO BOUND

The analysis of the FastICA variants and the comparison with the corresponding CRLB showed that there is room for improvements in terms of accuracy. Highly non-Gaussian signals can be accurately separated using an appropriate nonlinearity that is close to the score function of the distribution. Since distributions of signals can be different, the nonlinearity should be chosen different for each signal. However, the accuracy of Symmetric FastICA is limited by the orthogonal constraint, which mainly limits the separation of highly non-Gaussian signals. By contrast, One-unit FastICA is less effective when separating signals having distributions that are close to the Gaussian. Only the symmetric version can guarantee global convergence, that is, the separation of all signals.

These conclusions gave rise to a new, more sophisticated, algorithm named EFICA [29]. EFICA is initialized by the outcome of Symmetric FastICA endowed by the test of saddle points described in the previous section. The partly separated signals are used to select optimal nonlinearities g_i, $i = 1, \ldots, d$, for each separated signal, and used in fine-tuning of rows in $\mathbf{W}$. Finally, the whole $\mathbf{W}$ is refined using weighted symmetric orthogonalizations in a way that the orthogonal constraint is avoided. This is done with optimal weights derived from an analysis of a weighted symmetric algorithm.

2.4.1 WEIGHTED SYMMETRIC FastICA

Consider a variant of the symmetric algorithm where different nonlinear functions $g_k(\cdot)$, $k = 1, \ldots, d$ are used in Eq. (2.13) to estimate each row of $\mathbf{U}^+$. Then, before the symmetric orthogonalization step (Eq. 2.14), the rows of $\mathbf{U}^+$ are re-weighted by positive weights. One iteration of such algorithm is thus

$$\mathbf{U}^+ \leftarrow g(\mathbf{UZ})\mathbf{Z}^T - \text{diag}[g'(\mathbf{UZ})\mathbf{1}_N]\,\mathbf{U} \tag{2.39}$$

$$\mathbf{U}^+ \leftarrow \text{diag}[c_1, \ldots, c_d] \cdot \mathbf{U}^+ \tag{2.40}$$

$$\mathbf{U} \leftarrow (\mathbf{U}^+\mathbf{U}^{+T})^{-1/2}\mathbf{U}^+, \tag{2.41}$$

where $g(\cdots)$ is an element-wise function applying $g_k(\cdot)$, $k = 1, \ldots, d$, to the corresponding rows of the argument.

The key step in deriving EFICA is to analyze this algorithm, which was done in [29] in the same way as in [14]. The result is that the nondiagonal normalized gain matrix elements for this method, $N^{1/2}\mathbf{G}_{ij}^{\text{WS}}$, have asymptotically Gaussian distribution $\mathcal{N}(0, V_{ij}^{\text{WS}})$, where

$$V_{ij}^{\text{WS}} = \frac{c_i^2\gamma_i + c_j^2(\gamma_j + \tau_j^2)}{(c_i\tau_i + c_j\tau_j)^2}, \quad i \neq j, \tag{2.42}$$

where $\gamma_i = \beta_i - \mu_i^2$ and $\tau_i = |\mu_i - \rho_i|$.

2.4.2 EFICA

EFICA utilizes the weighted symmetric orthogonalization within its last refinement stage, that is, after the initialization, choice of nonlinearities, and fine-tuning. One such orthogonalization is performed for each separated signal. The weights in Eq. (2.40) are chosen such that V_{ij}^{WS} is minimized, specifically, for the ith separated signal, c_i is put equal to 1, and

$$c_j^{\text{OPT}} = \arg\min_{c_j, c_i = 1} V_{ij}^{\text{WS}} = \frac{\tau_j\gamma_i}{\tau_i(\gamma_j + \tau_j^2)}, \quad j \neq i. \tag{2.43}$$

Since c_j^{OPT} also depends on i, the weights must be selected different for each signal. Only the ith row of $\mathbf{U}$ after Eq. (2.41) is then used as the ith row for the final de-mixing transform. Consequently, the rows of the final transform are no more orthogonal in general, which means that the algorithm is not constrained to produce exactly orthogonal components.

By putting Eq. (2.43) into Eq. (2.42), we arrive at the asymptotic variance of the nondiagonal normalized gain matrix elements by EFICA, which is

$$V_{ij}^{\mathrm{EF}} \approx \frac{1}{N} \frac{\gamma_i(\gamma_j + \tau_j^2)}{\tau_j^2 \gamma_i + \tau_i^2(\gamma_j + \tau_j^2)}, \quad i \neq j. \tag{2.44}$$

Comparing Eq. (2.44) with Eqs. (2.32) and (2.33), the former can always be shown to be smaller than the latter two, provided that the same nonlinearity is used for all signals.

If the nonlinearities $g_i, i = 1, \ldots, d$, match the score functions of the signals, then

$$\tau_i = \gamma_i = \kappa_i - 1$$

and Eq. (2.44) becomes equal to the CRLB (Eq. 2.25). It means that EFICA is asymptotically efficient in that special case.[c]

EFICA can be implemented to work efficiently only with a class of distributions for which it is possible to choose appropriate nonlinearities supplying the score functions. The original EFICA implementation from [45] assumes signals having a generalized Gaussian distribution; see Appendix for the definition of this distribution family. A more general implementation using Pham's parametric score function least-square estimator [46] was proposed in [32].

In principle, EFICA does not differ much from FastICA in terms of computational complexity, so it retains its popular property, which is high speed. On the other hand, it outperforms FastICA in terms of accuracy and global convergence (stability), which was demonstrated by various experiments even with real-world signals. Some further improvements of EFICA in terms of speed and accuracy were proposed in [26] and [32].

2.4.2.1 Example

A simulated example was conducted where 13 signals of the generalized Gaussian distribution, each with a different value of the parameter α, respectively, equal to 0.1, 0.3, 0.5, 0.8, 1, 1.5, 1.9, 2, 2.1, 2.5, 4, 8, and 10, were mixed by a random mixing matrix and separated. The experiment was repeated 100 times with a fixed length of data $N = 5000$. The achieved average SIR of the signals separated by EFICA and by other ICA methods (Symmetric FastICA with the "tanh" nonlinearity, JADE by Cardoso and Souloumiac [47], and NPICA by Boscolo et al. [48]) was computed and is shown in Figure 2.2 as a function of α (one value per separated signal). Likewise, the CRLB computed using Eqs. (2.25) and (2.62) is shown in Figure 2.2.

The CRLB exists only for $\alpha > 1$. EFICA approaches the bound, which confirms its efficiency for the generalized Gaussian family. FastICA and JADE do not approach the CRLB, which is mainly caused by the orthogonal constraint. NPICA is close to the CRLB up to some failures that deteriorate the average SIR. However, NPICA requires a much higher computational load than EFICA as it utilizes a nonparametric modeling of the signals' distributions.

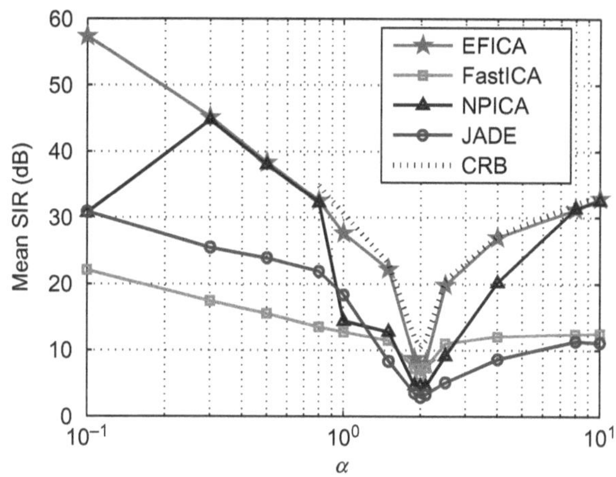

FIGURE 2.2

The average SIR of 13 components having the generalized Gaussian distribution with α, respectively, equal to 0.1, 0.3, 0.5, 0.8, 1, 1.5, 1.9, 2, 2.1, 2.5, 4, 8, and 10.

2.4.3 BLOCK EFICA

Block EFICA is a generalization of the EFICA algorithm for piecewise stationary non-Gaussian signals proposed in [41]. The model, first mentioned in Section 2.2.2.2, assumes that the original signal can be partitioned into a set of M blocks, so that the signals are i.i.d. within each block. The distributions may have different variances and even different distributions on distinct blocks.

Block EFICA searches for appropriate nonlinearities similarly to EFICA, but separately for each block of the preseparated signals. Assuming that the selected nonlinearities match true score functions and that variance of the signals is constant over the blocks, the asymptotic variance of the nondiagonal normalized gain matrix elements by Block EFICA was shown to be

$$V_{ij}^{\mathrm{BEF}} = \frac{\overline{\kappa}_j}{\overline{\kappa}_i \overline{\kappa}_j - 1}, \quad i \neq j, \tag{2.45}$$

where $\overline{\kappa}_i = \frac{1}{M}\sum_{\ell=1}^{M} \kappa_i^{(\ell)}$. This result corresponds with the CRLB in Eq. (2.28) when taking $(\sigma^2)_i^{(\ell)} = 1$ for all i and ℓ.

2.5 FastICA IN PRESENCE OF ADDITIVE NOISE

In this section, we will assume that the mixed signals also contain additive noise, so the mixing model is

$$\mathbf{X} = \mathbf{AS} + \mathbf{N}, \tag{2.46}$$

where $\mathbf{N}$ has the same size as $\mathbf{X}$ and its rows contain samples of noise. The noise signals are assumed to be Gaussian i.i.d. and uncorrelated[d] with the covariance matrix equal to $\sigma^2\mathbf{I}$. It is worth noting that when $\sigma^2 > 0$, the tasks to identify $\mathbf{A}$ and to separate $\mathbf{S}$ are no longer equivalent. We will henceforth focus on the separation of $\mathbf{S}$; an unbiased estimation of $\mathbf{A}$ through FastICA assuming known σ^2 was studied in [33].

2.5.1 SIGNAL-TO-INTERFERENCE-PLUS-NOISE RATIO

An appropriate criterion for the evaluation of separated signals by $\mathbf{W}$ is now the Signal-to-Interference-plus-Noise Ratio (SINR).[e] For the kth separated signal, the SINR is equal to [49]

$$\mathrm{SINR}_k = \frac{|\mathbf{G}_{kk}|^2}{\sum_{i=1, i\neq k}^{d} |\mathbf{G}_{ki}|^2 + \sigma^2 \sum_{i=1}^{d} |\mathbf{W}_{ki}|^2}. \tag{2.47}$$

The values of SINR are bounded unless $\sigma^2 = 0$. The maximum SINR is achieved for

$$\mathbf{W}^{\mathrm{MMSE}} = \mathbf{A}^T(\mathbf{A}\mathbf{A}^T + \sigma^2\mathbf{I})^{-1}, \tag{2.48}$$

which simultaneously minimizes the mean square distance between the original and separated signals, that is,

$$\mathbf{W}^{\mathrm{MMSE}} = \arg\min_{\mathbf{W}} \mathrm{E}[\|\mathbf{S} - \mathbf{W}\mathbf{X}\|_F^2]. \tag{2.49}$$

By putting $\mathbf{W}^{\mathrm{MMSE}}$ into Eq. (2.47), the ultimate bound for the SINR of the kth signal is

$$\frac{\mathbf{V}_{kk}^2}{\sum_{i\neq k}^{d} \mathbf{V}_{ki}^2 + \sigma^2 \sum_{i=1}^{d} (\mathbf{V}\mathbf{A}^{-1})_{ki}^2}, \tag{2.50}$$

where $\mathbf{V} = (\mathbf{I} + \sigma^2(\mathbf{A}^T\mathbf{A})^{-1})^{-1}$. It is worth noting that the latter bound depends on $\mathbf{A}$ unlike the Cramér-Rao bounds for the noise-free cases (Section 2.2.2).

The asymptotic expansion of Eq. (2.50) for "small" σ^2 was derived in [49] and gives

$$\min \mathrm{SINR}_k = \frac{1}{\sigma^2\|\mathbf{b}_k\|^2} - B_k + \mathcal{O}(\sigma^2), \tag{2.51}$$

where

$$B_k = 2 + \frac{1}{\|\mathbf{b}_k\|^4}\left(\sum_{i\neq k}^{d} (\mathbf{B}\mathbf{B}^T)_{ki}^2 - 2\sum_{i=1}^{d} \mathbf{B}_{ki}(\mathbf{B}\mathbf{B}^T\mathbf{B})_{ki}\right),$$

$\mathbf{B} = \mathbf{A}^{-1}$, and $\mathbf{b}_k^T$ denotes the kth row of $\mathbf{B}$.

The first term in Eq. (2.51) reveals that if the rows of $\mathbf{A}^{-1}$ have the same norm, the ultimate bound (Eq. 2.50) is approximately the same for each signal (provided that $\mathbf{A}$ is well conditioned).

2.5.2 BIAS FROM THE MINIMUM MEAN-SQUARED ERROR SOLUTION

Without analysis, it is not clear whether FastICA aims to approach the de-mixing transform $\mathbf{W}^{\text{MMSE}}$ or $\mathbf{A}^{-1}$ when noise is present. A more practical method seems to be the former transform as it yields the optimum signals in terms of SINR, that is, the minimum mean-squared error solution

$$\mathbf{S}^{\text{MMSE}} = \mathbf{W}^{\text{MMSE}}\mathbf{X}. \tag{2.52}$$

We therefore define the bias of an estimated separating matrix $\mathbf{W}$ as

$$\text{E}[\mathbf{W}](\mathbf{W}^{\text{MMSE}})^{-1} - \mathbf{D}, \tag{2.53}$$

where $\mathbf{D}$ is the diagonal matrix that normalizes $\mathbf{S}^{\text{MMSE}}$ to unit scales. The definition of $\mathbf{D}$ comes from the fact that an optimum blind algorithm is expected to yield normalized $\mathbf{S}^{\text{MMSE}}$ since their original scales are unknown to it.

It was shown in [34] that, for "small" σ^2, $\mathbf{D}$ satisfies

$$\mathbf{D} = \mathbf{I} + \frac{1}{2}\sigma^2 \mathtt{diag}[\mathbf{H}_{11}, \ldots, \mathbf{H}_{dd}] + \mathcal{O}(\sigma^3), \tag{2.54}$$

where $\mathbf{H} = (\mathbf{A}^T\mathbf{A})^{-1}$.

2.5.2.1 Bias of algorithms using the orthogonal constraint

The orthogonal constraint requires that

$$\text{E}[\mathbf{WX}(\mathbf{WX})^T] = \mathbf{W}(\mathbf{AA}^T + \sigma^2\mathbf{I})\mathbf{W}^T = \mathbf{I}, \tag{2.55}$$

so the bias of all constrained algorithms is lower bounded by

$$\min_{\mathbf{W}} \|\mathbf{W}(\mathbf{W}^{\text{MMSE}})^{-1} - \mathbf{D}\|_F \quad \text{w.r.t.} \quad \mathbf{W}(\mathbf{AA}^T + \sigma^2\mathbf{I})\mathbf{W}^T = \mathbf{I}. \tag{2.56}$$

It was shown in [50] that $\mathbf{W}$ solving the minimization problem (Eq. 2.56) has the property that

$$\mathbf{W}(\mathbf{W}^{\text{MMSE}})^{-1} = \mathbf{I} + \sigma^2\Gamma + \mathcal{O}(\sigma^3)$$

where $\mathbf{\Gamma}$ is a nonzero matrix obeying $\mathbf{\Gamma} + \mathbf{\Gamma}^T = \mathbf{H}$.

It follows that the bias (Eq. 2.53) of ICA algorithms which use the orthogonal constraint has the asymptotic order $\mathcal{O}(\sigma^2)$.

2.5.2.2 Bias of One-unit FastICA

Consider the situation that FastICA is applied to $\mathbf{S}^{\text{MMSE}}$. An optimum unbiased solution in the sense of Eq. (2.53) is the diagonal matrix $\mathbf{D}$. It was shown in [50] that

$$\text{E}[\mathbf{w}_k^{1\text{U}}] \propto \mathbf{e}_k + O(\sigma^3), \tag{2.57}$$

where $\mathbf{w}_k^{1\text{U}}$ denotes the kth row of the de-mixing transform by One-unit FastICA (when initialized by $\mathbf{D}$ and then applied to $\mathbf{S}^{\text{MMSE}}$); $\mathbf{e}_k$ denotes the kth row of the identity matrix.

It follows that the asymptotic bias of the one-unit approach has the order $O(\sigma^3)$, that is, lower than $O(\sigma^2)$.

2.5.2.3 Bias of Symmetric FastICA and EFICA

The biases of FastICA and EFICA derived in the same way satisfy [50]

$$\mathrm{E}[\mathbf{W}^{\mathrm{alg}}](\mathbf{W}^{\mathrm{MMSE}})^{-1} - \mathbf{D} = \frac{1}{2}\sigma^2 \mathbf{H} \odot (\mathbf{1}_{d\times d} - \mathbf{I} + \mathbf{M}^{\mathrm{alg}}) + O(\sigma^3), \tag{2.58}$$

where the superscript ^{alg} signifies the algorithm (either Symmetric FastICA or EFICA). In both cases, $\mathbf{M}$ is not diagonal; $\mathbf{1}_{d\times d}$ is the $d \times d$ matrix of ones. It follows that the bias of both algorithms has the order $O(\sigma^2)$; hence, the bias is asymptotically higher than that of One-unit FastICA.

2.5.3 1FICA

EFICA is an optimal estimator of the separating matrix in terms of the estimation variance when the mixed signals do not contain any noise. However, if the noise is present, the estimate by EFICA is biased and need not be optimal in terms of SINR. By contrast, the above results show that the bias of One-unit FastICA has at least the order $\mathcal{O}(\sigma^3)$.

The only problem is to modify One-unit FastICA to guarantee the estimation of all components. The 1FICA algorithm derived in [34] (also for complex-valued signals) was designed to meet this requirement. It proceeds in three steps.

1. Because of a good global convergence behavior, the initialization is taken from Symmetric FastICA using nonlinearity "tanh" or "rati" followed by the test of saddle points.
2. Each row of the de-mixing transform is fine-tuned through performing few one-unit iterations using an adaptively chosen nonlinearity.
3. To restrain the global solution, the resulting row is accepted if not being too distant from the initialization; otherwise, the solution will be the outcome of the first step.

Under mild assumptions, it follows that 1FICA has the same asymptotic bias as One-unit FastICA.

APPENDIX: GENERALIZED GAUSSIAN DISTRIBUTIONS

The normalized random variable distributed according to the generalized Gaussian law has the density function with a shape parameter $\alpha > 0$ defined as

$$f_\alpha(x) = \frac{\alpha\beta_\alpha}{2\Gamma(1/\alpha)} \exp\left\{-(\beta_\alpha|x|)^\alpha\right\}, \tag{2.59}$$

where $\Gamma(\cdot)$ is the Gamma function, and

$$\beta_\alpha = \sqrt{\frac{\Gamma(3/\alpha)}{\Gamma(1/\alpha)}}. \tag{2.60}$$

This generalized Gaussian family encompasses the ordinary standard normal distribution for $\alpha = 2$, the Laplacean distribution for $\alpha = 1$, and the uniform distribution in the limit $\alpha \to \infty$.

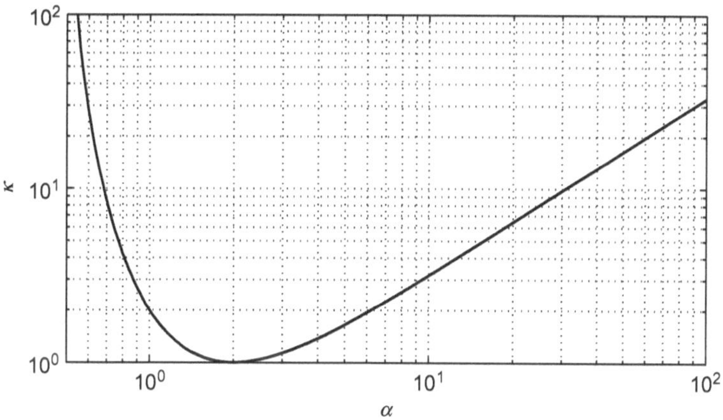

FIGURE 2.3

The moment κ of the generalized Gaussian p.d.f. as a function of the shape parameter α according to Eq. (2.62).

The score function of the distribution is

$$\psi_\alpha(x) = -\frac{\frac{\partial f_\alpha(x)}{\partial x}}{f_\alpha(x)} = \frac{|x|^{\alpha-1}\,\mathrm{sign}(x)}{\mathrm{E}_\alpha[|x|^\alpha]}, \tag{2.61}$$

which is continuous only for $\alpha > 1$. It can be shown that κ defined similar to Eq. (2.26) depends on α as

$$\kappa_\alpha = \mathrm{E}_\alpha[\psi_\alpha^2(x)] = \{\mathrm{E}_\alpha[|x|^\alpha]\}^2 = \begin{cases} \frac{\Gamma\left(2-\frac{1}{\alpha}\right)\Gamma\left(\frac{3}{\alpha}\right)}{\left[\Gamma\left(1+\frac{1}{\alpha}\right)\right]^2} & \text{for } \alpha > 1/2 \\ +\infty & \text{otherwise.} \end{cases} \tag{2.62}$$

The dependence of κ_α on $\alpha \in [0.5, 100]$ is displayed in Figure 2.3. For $\alpha < 0.5$, κ_α goes to infinity and the CRLB does not exist. It may follow that, for $\alpha < 0.5$, there might be estimators whose variances decrease faster than N^{-1} as $N \to +\infty$.

ACKNOWLEDGMENTS

We thank Prof. Erkki Oja for his hospitality and fruitful discussions that we had during our visit at HUT in 2006 and later while writing joint papers on statistical analysis of FastICA and on EFICA. This work was supported by the Czech Science Foundation through Project No. 14-13713S.

NOTES

a. The beginnings of ICA can be dated to 1986 when Herault and Jutten published their paper [2] on a learning algorithm that was able to separate independent signals. Later, the concept of ICA was most clearly stated by Comon in [3], which is one of the most cited papers on ICA. Presently, there are several books and proceedings devoted to this important topic of signal processing [1,4–7].

b. Some authors associate the non-Gaussianity-based model with ICA only. They classify the methods using other models as belonging under the general flag of BSS.

c. It should be noted that the analysis of FastICA as well as EFICA is local. Therefore, to be more precise, we should say that the asymptotic efficiency of EFICA is ensured when its global convergence is guaranteed.

d. The case when noise signals have general covariance matrix $\mathbf{C_N}$ can be transformed into the mixing model with uncorrelated noise where the unknown mixing matrix is $\sigma \mathbf{C_N}^{-1/2}\mathbf{A}$.

e. Note that SIR does not take into account the presence of the residual noise in separated signals.

REFERENCES

[1] A. Cichocki, S.-I. Amari, Adaptive Signal and Image Processing: Learning Algorithms and Applications, Wiley, New York, 2002.

[2] J. Herault, C. Jutten, Space or time adaptive signal processing by neural network models, AIP Conf. Proc. 151 (1986) 206-211.

[3] P. Comon, Independent component analysis: a new concept?, Signal Process. 36 (3) (1994) 287-314.

[4] A. Cichocki, R. Zdunek, A.H. Phan, S.I. Amari, Nonnegative Matrix and Tensor Factorizations: Applications to Exploratory Multi-Way Data Analysis and Blind Source Separation, Wiley, Chichester, 2009.

[5] P. Comon, C. Jutten, Handbook of Blind Source Separation: Independent Component Analysis and Applications, Academic Press, Oxford, 2010, 859 pp.

[6] A. Hyvärinen, J. Karhunen, E. Oja, Independent Component Analysis, Wiley-Interscience, New York, 2001.

[7] T.-W. Lee, Independent Component Analysis: Theory and Applications, Kluwer, Boston, MA, 1998, 237 pp.

[8] J. Eriksson, V. Koivunen, Identifiability, separability, and uniqueness of linear ICA models, IEEE Signal Process. Lett. 11 (7) (2004) 601-604.

[9] A. Hyvärinen, E. Oja, A fast fixed-point algorithm for independent component analysis, Neural Comput. 9 (1997) 1483-1492.

[10] A. Hyvärinen, Fast and robust fixed-point algorithms for independent component analysis, IEEE Trans. Neural Netw. 10 (1999) 626-634.

[11] N. Delfosse, P. Loubaton, Adaptive blind separation of independent sources: a deflation approach, Signal Process. 45 (1995) 59-83.

[12] S.C. Douglas, A statistical convergence analysis of the FastICA algorithm for two-source mixtures, in: Conference Record of the Thirty-Ninth Asilomar Conference on Signals, Systems and Computers, November 2005, pp. 335-339.

[13] A. Hyvärinen, One-unit contrast functions for independent component analysis: a statistical analysis, in: Neural Networks for Signal Processing VII (Proc. IEEE NNSP Workshop 1997), Amelia Island, FL, 1997, pp. 388-397.

[14] P. Tichavský, Z. Koldovský, E. Oja, Performance analysis of the FastICA algorithm and Cramér-Rao bounds for linear independent component analysis, IEEE Trans. Signal Process. 54 (4) (2006) 1189-1203.
[15] E. Ollila, The deflation-based FastICA estimator: statistical analysis revisited, IEEE Trans. Signal Process. 58 (3) (2010) 1527-1541.
[16] H. Shen, M. Kleinsteuber, K. Huper, Local convergence analysis of FastICA and related algorithms, IEEE Trans. Neural Netw. 19 (6) (2008) 1022-1032.
[17] T. Wei, Asymptotic analysis of the generalized symmetric FastICA algorithm, in: IEEE Workshop on Statistical Signal Processing (SSP 2014), 2014, pp. 460-463.
[18] J.-F. Cardoso, Blind signal separation: statistical principles, Proc. IEEE 90 (8) (1998) 2009-2026.
[19] Z. Koldovský, P. Tichavský, E. Oja, Cramér-Rao lower bound for linear independent component analysis, in: IEEE International Conference on Acoustics, Speech and Signal Processing (ICASSP 2005), Philadelphia, vol. III, March 2005, pp. 581-584.
[20] E. Bingham, A. Hyvärinen, A fast fixed-point algorithm for independent component analysis of complex valued signals, Int. J. Neural Syst. 10 (I) (2000) 1-8.
[21] H.L. Li, T. Adali, Algorithms for complex ML ICA and their stability analysis using Wirtinger calculus, IEEE Trans. Signal Process. 58 (12) (2010) 6156-6167.
[22] Y. Zhang, S.A. Kassam, Optimum nonlinearity and approximation in complex FastICA, in: Proceedings of the 46th Conference on Information Sciences and Systems (CISS), 2012, pp. 1-6.
[23] B. Loesch, B. Yang, Cramér-Rao bound for circular and noncircular complex independent component analysis, IEEE Trans. Signal Process. 61 (2) (2013) 365-379.
[24] J. Eriksson, V. Koivunen, Complex random vectors and ICA models: identifiability, uniqueness, and separability, IEEE Trans. Inf. Theory 52 (3) (2006) 1017-1029.
[25] V. Zarzoso, P. Comon, M. Kallel, How fast is FastICA?, in: Proceedings of the 14th European Signal Processing Conference (EUSIPCO-2006), Florence, Italy, September 4-8, 2006.
[26] P. Tichavský, Z. Koldovský, E. Oja, Speed and accuracy enhancement of linear ICA techniques using rational nonlinear functions, in: Proceedings of the 7th International Conference on Independent Component Analysis (ICA2007), September 2007, pp. 285-292.
[27] T. Wei, On the spurious solutions of the FastICA algorithm, in: IEEE Statistical Signal Processing Workshop 2014, 2014, pp. 161-164.
[28] J. Miettinen, K. Nordhausen, H. Oja, S. Taskinen, Deflation-based FastICA with adaptive choices of nonlinearities, IEEE Trans. Signal Process. 62 (21) (2014) 5716-5724.
[29] Z. Koldovský, P. Tichavský, E. Oja, Efficient variant of algorithm FastICA for independent component analysis attaining the Cramér-Rao lower bound, IEEE Trans. Neural Netw. 17 (5) (2006) 1265-1277.
[30] J.-C. Chao, S.C. Douglas, Using piecewise linear nonlinearities in the natural gradient and FastICA algorithms for blind source separation, in: IEEE International Conference on Acoustics, Speech and Signal Processing, 2008, pp. 1813-1816.
[31] A. Dermoune, T. Wei, FastICA algorithm: five criteria for the optimal choice of the nonlinearity function, IEEE Trans. Signal Process. 61 (8) (2013) 2078-2087.
[32] J. Málek, Z. Koldovský, S. Hosseini, Y. Deville, A variant of EFICA algorithm with adaptive parametric density estimator, in: 8th International Workshop on Electronics, Control, Modelling, Measurement, and Signals (ECMS 2007), Liberec, Czech Republic, May 2007, pp. 79-84.

[33] A. Hyvärinen, Gaussian moments for noisy independent component analysis, IEEE Signal Process. Lett. 6 (6) (1999) 145-147.
[34] Z. Koldovský, P. Tichavský, Blind instantaneous noisy mixture separation with best interference-plus-noise rejection, in: Proceedings of the 7th International Conference on Independent Component Analysis (ICA2007), September 2007, pp. 730-737.
[35] P. Tichavský, Z. Koldovský, Optimal pairing of signal components separated by blind techniques, IEEE Signal Process. Lett. 11 (2) (2004) 119-122.
[36] R.C. Rao, Linear Statistical Inference and Its Applications, second ed., Wiley, New York, 1973.
[37] E. Doron, A. Yeredor, P. Tichavský, Cramér-Rao lower bound for blind separation of stationary parametric Gaussian sources, IEEE Signal Process. Lett. 14 (6) (2007) 417-420.
[38] P. Tichavský, Z. Koldovský, E. Oja, Corrections of the "Performance analysis of the FastICA algorithm and Cramér-Rao bounds for linear independent component analysis", IEEE Trans. Signal Process. 56 (4) (2008) 1715-1716.
[39] E. Ollila, K. Hyon-Jung, V. Koivunen, Compact Cramér-Rao bound expression for independent component analysis, IEEE Trans. Signal Process. 56 (4) (2008) 1421-1428.
[40] J.-F. Cardoso, D.T. Pham, Separation of non-stationary sources, algorithms and performance, in: S.J. Roberts, R.M. Everson (Eds.), Independent Components Analysis: Principles and Practice, Cambridge University Press, Cambridge, 2001, pp. 158-180.
[41] Z. Koldovský, J. Málek, P. Tichavský, Y. Deville, S. Hosseini, Blind separation of piecewise stationary non-Gaussian sources, Signal Process. 89 (12) (2009) 2570-2584.
[42] T. Adali, M. Anderson, G.-S. Fu, Diversity in independent component and vector analyses: identifiability, algorithms, and applications in medical imaging, IEEE Signal Process. Mag. 31 (3) (2014) 18-33.
[43] A. Yeredor, Blind separation of Gaussian sources with general covariance structures: bounds and optimal estimation, IEEE Trans. Signal Process. 58 (10) (2010) 5057-5068.
[44] E. Oja, Z. Yuan, The FastICA algorithm revisited: convergence analysis, IEEE Trans. Neural Netw. 17 (6) (2006) 1370-1381.
[45] Matlab codes [online]. Available from: http://itakura.ite.tul.cz/zbynek/downloads.htm.
[46] D.T. Pham, P. Garat, Blind separation of mixture of independent sources through a quasi-maximum likelihood approach, IEEE Trans. Signal Process. 45 (7) (1997) 1712-1725.
[47] J.-F. Cardoso, A. Souloumiac, Blind beamforming from non-Gaussian signals, radar and signal processing, IEE Proc. F 140 (6) (1993) 362-370.
[48] R. Boscolo, H. Pan, V.P. Roychowdhury, Independent component analysis based on nonparametric density estimation, IEEE Trans. Neural Netw. 15 (1) (2004) 55-65.
[49] Z. Koldovský, P. Tichavský, Methods of fair comparison of performance of linear ICA techniques in presence of additive noise, in: IEEE International Conference on Acoustics, Speech and Signal Processing (ICASSP 2006), Toulouse, vol. V, May 2006, pp. 873-876.
[50] Z. Koldovský, P. Tichavský, Asymptotic analysis of bias of FastICA-based algorithms in presence of additive noise, Technical report no. 2181, ÚTIA, AV ČR, 2007.

CHAPTER

A unified probabilistic model for independent and principal component analysis

3

Aapo Hyvärinen

Department of Computer Science, Helsinki Institute for Information Technology, University of Helsinki, Helsinki, Finland

3.1 INTRODUCTION

Principal component analysis (PCA) and independent component analysis (ICA) are two fundamental methods for unsupervised learning. In the machine learning and neural networks literature, they have a relatively long history. Neural learning for PCA started with Oja's rule [1] and its extensions [2]. An early connection between PCA and ICA was given by nonlinear versions of PCA criteria [3]. The general theory of ICA is explained in [4].

Let us consider a linear generative model

$$x_i = \sum_{j=1}^{n} a_{ij} s_j, \quad i = 1, \ldots, n, \tag{3.1}$$

where the x_i are observed random variables, the s_j are latent random variables (components) which are assumed mutually independent, and the a_{ij} are parameters. Depending on further assumptions, this framework can implement ICA or PCA. In particular, if we assume that the s_j are non-Gaussian, we obtain the basic version of ICA. The typical goal of ICA is to "separate sources" in the sense that we want to recover the original s_i.

On the other hand, if we assume that the components s_j are Gaussian and have different variances, we obtain a model which may be related to PCA, depending on what further assumptions, such as the orthogonality of the matrix $\mathbf{A}$ which collects the coefficients a_{ij}, are made. This approach to PCA is slightly unconventional, but we will see below that it is equivalent to the classic one. The goal in PCA is not so much to recover (all) the original s_i but to find the subspace spanned by a limited number of the s_i (and the corresponding columns of the matrix $\mathbf{A}$) which explains the largest amount of variance of the data.

In this chapter, our purpose is to develop a probabilistic model based on Eq. (3.1) which unifies PCA and ICA in the sense that maximization of the likelihood performs either PCA or ICA depending on the specific constraints and the data. The basic

E. Bingham, S. Kaski, J. Laaksonen and J. Lampinen (Eds) Advances in Independent Component Analysis and Learning Machines.
http://dx.doi.org/10.1016/B978-0-12-802806-3.00003-8

idea in our model is to modify the ICA assumptions so that we explicitly model the variances of the components, and then integrate them out in a Bayesian framework.

Using such a variant of the linear generative model, we show the following. First, if the components are assumed Gaussian in the model, and we constrain $\mathbf{A}$ to be orthogonal, maximization of the likelihood performs PCA. Second, if the components are assumed non-Gaussian in the model, maximization of the likelihood separates original non-Gaussian components, that is, recovers the mixing matrix up to trivial indeterminacies like ICA, again assuming that $\mathbf{A}$ is constrained orthogonal, and further that the data are prewhitened.

3.2 VARIANCE OF COMPONENTS AS SEPARATE PARAMETER

3.2.1 DEFINITION OF NEW MODEL

It is well known that in the linear model (Eq. 3.1), the variances of the components cannot be recovered. This is because we can always rescale a component s_j as $\gamma_j s_j$, redefine the mixing coefficients as a_{ij}/γ_j, and the model is equivalent in the sense that the observed data have the same distribution.

The conventional approach to ICA is to define that the variances of the components are equal to one. This approach simplifies the problem, but it seems to have the drawback that the connection to PCA is lost, because PCA is dependent on the principal components having distinct variances.

We propose here to consider the variances of the components as separate parameters. Further, we propose to integrate those parameters out in a Bayesian approach.

Thus, define σ_j^2 to be the variance of the jth independent component. Denote the vector collecting the σ_j as $\boldsymbol{\sigma}$, and denote $\mathbf{W} = (\mathbf{w}_1, \mathbf{w}_2, \ldots, \mathbf{w}_n)^T = \mathbf{A}^{-1}$. Then, we have, using the well-known derivation of the ICA likelihood [4]:

$$p(\mathbf{x}|\mathbf{A}, \boldsymbol{\sigma}) = |\det \mathbf{W}| \prod_j \frac{1}{\sigma_j} p_j \left(\frac{\mathbf{w}_j^T \mathbf{x}}{\sigma_j} \right), \tag{3.2}$$

where p_i denotes the probability density function (p.d.f.) of the s_i when it is standardized to unit variance.

In order to be able to integrate out the σ_j in closed form, we have to restrict ourselves to a special form of the p_i. We consider the generalized Laplacian distribution, also called the generalized Gaussian distribution. The p.d.f. is given by

$$p_j(s|\alpha_j) = \frac{1}{Z(\alpha_j)} \exp(-|s|^{\alpha} C(\alpha_j)), \tag{3.3}$$

where α is the parameter controlling the shape of the density. For $\alpha = 2$, we obtain the Gaussian density, for $\alpha < 2$, we obtain (highly) peaked, super-Gaussian densities, and for $\alpha > 2$, flat, sub-Gaussian densities. The constants Z and C are needed to normalize the p.d.f. and to make its variance equal to one, and they are well-known (although different conventions of parameterization exist), but irrelevant for our purposes.

3.2.2 INTEGRATING OUT THE VARIANCE PARAMETER

To handle the variances in a Bayesian framework, we first need to define a prior for the σ_j. We choose to use the noninformative (Jeffreys') prior:

$$p(\boldsymbol{\sigma}) = \prod_j \frac{1}{\sigma_j}, \tag{3.4}$$

where obviously σ_j are constrained positive.

Consider an identically and independently distributed (i.i.d.) sample of $\mathbf{x}$, denoted individually as $\mathbf{x}(t), t = 1, \ldots, T$ and as a whole in matrix form as $\mathbf{X} = (\mathbf{x}(1), \ldots, \mathbf{x}(T))$. We have

$$\begin{aligned} p(\mathbf{X}, \boldsymbol{\sigma} | \mathbf{A}, \boldsymbol{\alpha}) &= p(\mathbf{X} | \mathbf{A}, \boldsymbol{\sigma}, \boldsymbol{\alpha}) p(\boldsymbol{\sigma}) = p(\boldsymbol{\sigma}) \prod_t p(\mathbf{x}(t) | \mathbf{A}, \boldsymbol{\sigma}, \boldsymbol{\alpha}) \\ &= |\det \mathbf{W}|^T \prod_j \frac{1}{\sigma_j^{T+1} Z(\alpha_j)^T} \exp\left(-\sum_t \left| \frac{\mathbf{w}_j^T \mathbf{x}(t)}{\sigma_j} \right|^{\alpha_j} C(\alpha_j) \right). \end{aligned} \tag{3.5}$$

To integrate out $\boldsymbol{\sigma}$, make the following change of variables:

$$u_j = \sum_t \left| \frac{\mathbf{w}_j^T \mathbf{x}(t)}{\sigma_j} \right|^{\alpha_j} C(\alpha_j) \tag{3.6}$$

$$\Leftrightarrow \sigma_j = \left[\sum_t \left| \mathbf{w}_j^T \mathbf{x}(t) \right|^{\alpha_j} C(\alpha_j) \right]^{1/\alpha_j} u_j^{-1/\alpha_j} \tag{3.7}$$

$$\Rightarrow \frac{\mathrm{d}\sigma_j}{\mathrm{d}u_j} = -\frac{1}{\alpha_j} \left[\sum_t \left| \mathbf{w}_j^T \mathbf{x}(t) \right|^{\alpha_j} C(\alpha_j) \right]^{1/\alpha_j} u_j^{-1/\alpha_j - 1}, \tag{3.8}$$

which enables the integration out as

$$\begin{aligned} p(\mathbf{X} | \mathbf{A}, \boldsymbol{\alpha}) &= \int p(\mathbf{X}, \boldsymbol{\sigma} | \mathbf{A}, \boldsymbol{\alpha}) \mathrm{d}\boldsymbol{\sigma} \\ &= \int |\det \mathbf{W}|^T \prod_j \frac{1}{Z(\alpha_j)^T} \left[\sum_t \left| \mathbf{w}_j^T \mathbf{x}(t) \right|^{\alpha_j} C(\alpha_j) \right]^{-(T+1)/\alpha_j} \\ &\quad \times u_j^{(T+1)/\alpha_j} \exp(-u_j) \frac{1}{\alpha_j} \left[\sum_t \left| \mathbf{w}_j^T \mathbf{x}(t) \right|^{\alpha_j} C(\alpha_j) \right]^{1/\alpha_j} u_j^{-1/\alpha_j - 1} \mathbf{du} \\ &= |\det \mathbf{W}|^T \prod_j \frac{C(\alpha_j)^{-T/\alpha_j}}{Z(\alpha_j)^T} \left[\sum_t \left| \mathbf{w}_j^T \mathbf{x}(t) \right|^{\alpha_j} \right]^{-T/\alpha_j} \\ &\quad \times \int u_j^{T/\alpha_j - 1} \exp(-u_j) \mathrm{d}u_j \\ &= |\det \mathbf{W}|^T \prod_j \frac{C(\alpha_j)^{-T/\alpha_j}}{Z(\alpha_j)^T} \left[\sum_t \left| \mathbf{w}_j^T \mathbf{x}(t) \right|^{\alpha_j} \right]^{-T/\alpha_j} \Gamma(T/\alpha_j), \end{aligned} \tag{3.9}$$

where Γ denotes the conventional gamma function. Note that the integrals here are in the positive quadrant since u_j as well as σ_j are by definition positive.

Thus, we have the following log-likelihood:

$$\frac{1}{T}\log p(\mathbf{X}|\mathbf{A},\boldsymbol{\alpha}) = \log|\det \mathbf{W}| - \sum_j \frac{1}{\alpha_j}\log\left[\sum_t \left|\mathbf{w}_j^T\mathbf{x}(t)\right|^{\alpha_j}\right] + f(\alpha_j, T), \tag{3.10}$$

where f denotes a function depending on α_j and T alone:

$$f(\alpha_j, T) = \frac{1}{T}\log\Gamma(T/\alpha_j) - \frac{1}{\alpha_j}\log C(\alpha_j) - \log Z(\alpha_j). \tag{3.11}$$

3.2.3 ALTERNATIVE APPROACH MAXIMIZING JOINT LIKELIHOOD

An alternative, non-Bayesian approach is possible by considering the joint likelihood of $\mathbf{A}$ and $\boldsymbol{\sigma}$, directly given in Eq. (3.2) when evaluated for the whole sample. Again, define the p_i as in Eq. (3.3). Thus, we have the joint log-likelihood

$$\frac{1}{T}\log(\mathbf{X}|\mathbf{A},\boldsymbol{\sigma}) = \log|\det \mathbf{W}| + \frac{1}{T}\sum_j\sum_t - \left|\frac{\mathbf{w}_j^T\mathbf{x}(t)}{\sigma_j}\right|^{\alpha_j} C(\alpha_j) - \log\sigma_j - \log Z(\alpha_j). \tag{3.12}$$

Now, for a fixed $\mathbf{W}$, we can find the maxima of this likelihood with respect to $\boldsymbol{\sigma}$ in closed form as

$$\hat{\sigma}_j(\mathbf{w}_j) = \left[\frac{\alpha_j C(\alpha_j)}{T}\sum_t \left|\mathbf{w}_j^T\mathbf{x}(t)\right|^{\alpha}\right]^{1/\alpha} \tag{3.13}$$

and we can plug this in the joint likelihood to obtain after some manipulations

$$\frac{1}{T}\log(\mathbf{X}|\mathbf{A},\hat{\boldsymbol{\sigma}}(\mathbf{W})\boldsymbol{\alpha}) = \log|\det \mathbf{W}| - \sum_j \frac{1}{\alpha_j}\log\left[\sum_t \left|\mathbf{w}_j^T\mathbf{x}(t)\right|^{\alpha_j}\right] + \tilde{f}(\alpha_j, T) \tag{3.14}$$

with

$$\tilde{f}(\alpha_j, T) = \frac{1}{\alpha_j}\left[\log\frac{T}{\alpha_j} - 1\right] - \frac{1}{\alpha_j}\log C(\alpha_j) - \log Z(\alpha_j). \tag{3.15}$$

We see that Eq. (3.14) is equal to Eq. (3.10) except for the additive functions f and $\tilde{f}$, which do not depend on the sample or $\mathbf{W}$, although they do depend on the parameters α_j. Simple numerical simulations show that in fact f and $\tilde{f}$ are practically equal for any reasonable α and $T \geq 100$.

3.2.4 COMPARISON WITH CONVENTIONAL LIKELIHOOD

The likelihood in Eq. (3.10) is formally rather similar to the conventional log-likelihood of ICA, which in the case of the generalized Gaussian density can be written as

$$\frac{1}{T}\log \tilde{p}(\mathbf{X}|\mathbf{A},\boldsymbol{\alpha}) = \log|\det \mathbf{W}| - \frac{1}{T}\sum_{j}\sum_{t}\left|\mathbf{w}_j^T\mathbf{x}(t)\right|^{\alpha_j} C(\alpha_j) + \bar{f}(\alpha_j, T), \tag{3.16}$$

where the function $\bar{f}$ is defined as

$$\bar{f}(\alpha_j, T) = -\log Z(\alpha_j). \tag{3.17}$$

Thus, we see the interesting phenomenon that our new likelihood in Eq. (3.10) contains the logarithmic function between the two summations. If the α_j are fixed, this logarithm is the main difference between the two likelihoods, in addition to the different "weighting" factors $C(\alpha_j)$ and $1/\alpha_j$.

The new likelihood in Eq. (3.10) has the interesting property that it is homogeneous with respect to the rows' norms of **W**. That is, if we multiply the rows of **W** by any scalar factors, the likelihood is constant. This seems to be an interesting reflection of the fact that the scales of the rows of **W** cannot be determined in the generative model.

To recapitulate, we have derived an alternative likelihood, given in Eq. (3.10), for the linear generative model in Eq. (3.1). The likelihood was obtained in closed form for the case of the generalized Gaussian density with parameters α_j controlling the shape of the densities.

3.3 ANALYSIS OF MAXIMUM LIKELIHOOD ESTIMATION

Next we show how maximization of the new likelihood in Eq. (3.10) can perform PCA or ICA in special circumstances.

3.3.1 CONSTRAINT ON SEPARATING MATRIX

Since the objective function is constant with respect to the norms of the rows $\mathbf{w}_i$, we can constrain them, purely for reasons of numerical stability, to be equal to unity. In fact, we decide to constrain the matrix **W** to be orthogonal in what follows:

$$\mathbf{W}\mathbf{W}^T = \mathbf{I}. \tag{3.18}$$

This is justified to the extent that such a constraint is often used in PCA and in ICA, assuming whitened data in ICA. Furthermore, the constraint allows for stronger theoretical results below. The constraint is equivalent to constraining **A** to be orthogonal.

3.3.2 ESTIMATION FOR DATA MODELED AS GAUSSIAN

First, we show that if the data are modeled as (or assumed to be) Gaussian, maximization of the new likelihood in Eq. (3.10) performs PCA. This is given in the following theorem:

Theorem 13. *Fix $\alpha_j = 2$ for all j, which means modeling the latent components as Gaussian. Assume that the eigenvalues of the data (sample) covariance matrix* *are distinct. When the likelihood in Eq. (3.10) is maximized under the constraint of orthogonality of* **A***, the global maximum is attained when* **A** *contains the eigenvectors of the covariance matrix of the data* **X** *as its columns.*

Proof. Denote by $\mathbf{C}$ the covariance matrix of the sample. Ignoring the additive constant f for notational simplicity, the log-likelihood in Eq. (3.10) then becomes

$$L = -\frac{1}{2}\sum_j \log\left[\mathbf{w}_j^T \mathbf{C}\mathbf{w}_j\right], \tag{3.19}$$

since the log-determinant of $\mathbf{W}$ is 0. Denote by $\mathbf{C} = \mathbf{U}\mathrm{diag}(\lambda_i)\mathbf{U}^T$ the eigenvalue decomposition of the covariance matrix, and make the change of variables $\mathbf{Q} = \mathbf{W}\mathbf{U}$. Then, we have

$$\begin{aligned} L &= -\frac{1}{2}\sum_j \log\left[\mathbf{q}_j^T \mathrm{diag}(\lambda_i)\mathbf{q}_j\right] = -\frac{1}{2}\sum_j \log\left[\sum_i q_{ij}^2\lambda_i\right] \\ &= -\frac{1}{2}\sum_j \log\left[\sum_i b_{ij}\lambda_i\right], \end{aligned} \tag{3.20}$$

where we denote $b_{ij} = q_{ij}^2$. Due to orthogonality of $\mathbf{W}$ and $\mathbf{U}$, the matrix $\mathbf{B}$ is doubly stochastic, which means its rows and columns have sum equal to one. Let us write $f(u) = -\frac{1}{2}\log(u)$, so we have

$$L(\mathbf{B}) = \sum_{j=1}^{n} f\left(\sum_i b_{ij}\lambda_i\right). \tag{3.21}$$

The function f is strictly convex. For any strictly convex function f, for any j, and for any set of distinct λ_i, we have

$$f\left(\sum_i b_{ij}\lambda_i\right) \leq \sum_i b_{ij} f(\lambda_i) \tag{3.22}$$

with equality if and only if exactly one of the b_{ij} is nonzero. Thus, we have

$$L(\mathbf{B}) \leq \sum_j \sum_i b_{ij} f(\lambda_i) = \sum_i f(\lambda_i) \tag{3.23}$$

with equality in the $\leq$ only if the b_{ij} has exactly one nonzero element for each j, which implies that $\mathbf{B}$ is a permutation matrix. Thus, we see that L is maximized when $\mathbf{B}$ is a permutation matrix. This corresponds to $\mathbf{Q}$ being a signed permutation matrix, and $\mathbf{A} = \mathbf{W}^T = \mathbf{U}\mathbf{Q}$ thus contains the eigenvectors in $\mathbf{U}$ as its columns, and the theorem is proven. □

Note that the theorem does not apply to the conventional ICA likelihood in Eq. (3.16), because in that case we would have the log-likelihood without the additional logarithm as

$$\tilde{L}(\mathbf{W}) = -\sum_j \frac{1}{2}\left(\mathbf{w}_j^T \mathbf{C}\mathbf{w}_j\right) = -\frac{1}{2}\mathrm{tr}(\mathbf{W}\mathbf{C}\mathbf{W}^T) = -\frac{1}{2}\mathrm{tr}(\mathbf{C}). \tag{3.24}$$

Thus, the conventional likelihood is constant under the assumption of Gaussianity and the constraint of orthogonality, as is well known.

Further note that for the theorem to hold, we do not need to assume that the data actually are Gaussian, we only need to assume that we model the data as Gaussian in the sense of setting $\alpha_j = 2$ in the estimation procedure. Of course, if the α_j are estimated from the data instead of being fixed *a priori*, the assumption could presumably be replaced by assuming that the data actually are Gaussian.

3.3.3 ESTIMATION FOR DATA ASSUMED TO BE NON-GAUSSIAN

Next, we analyze the behavior of the new likelihood when the data follow the conventional ICA model, with non-Gaussian components; this is equivalent to our model with non-Gaussian components. Furthermore, we assume that in the estimation, we use a non-Gaussian version of the likelihood, that is, $\alpha_j \neq 2$.

Importantly, we assume that the data are whitened in contrast to the preceding section. Like in the preceding section, we constrain $\mathbf{W}$ to be orthogonal. For simplicity, we consider the case where the α_j (non-Gaussianity models) are fixed *a priori*, although the result is unlikely to change essentially if we estimate the α_j.

A complication in the analysis is that our log-likelihood is not smooth, while most related analysis [3,5] assumes smooth functions. We restrict ourselves here to an approximative analysis, where we apply the existing smoothness-based analysis to our method, essentially assuming that we use a smooth approximation of our new likelihood.

Using such a smoothness approximation, the consistency of our new likelihood can be easily shown. Consider each summand in the log-likelihood, of the form $\log\left[\sum_t |\mathbf{w}_j^T\mathbf{x}(t)|^{\alpha_j}\right]$. This is a logarithm of an objective of the form $\sum_t G(\mathbf{w}_j^T\mathbf{x}(t))$. Such an objective (without logarithm) was analyzed, for example, in [5], where it was shown that it reaches the maximum at the independent components, assuming the data are white, the norm of $\mathbf{w}_j$ is constrained to unity, and crucially, that a certain "nonpolynomial cumulant" is positive. The nonpolynomial cumulant is positive if the model of non-Gaussianity is reasonable for the data; in our case, the generalized Laplacian distribution with the given α must be a reasonable approximation of the distribution of the component. Typically, this is not very restrictive: if the data are sparse (super-Gaussian), taking $\alpha < 2$ usually makes this condition hold.

Assuming that the condition on the reasonable non-Gaussianity model holds, we can easily see that our likelihood enables estimation of the model. The likelihood is simply a sum of logarithms of functions which are each maximized at the independent components. The situation only differs from the ordinary case of the ordinary ICA likelihood in the existence of the logarithm. Since logarithm is a monotonic function, the maxima are the same, and we have thus shown that maximization of the new likelihood estimates the ICA model (under the reservations and approximations given above).

3.4 CONCLUSION

We proposed to consider the variances of components in a linear-mixing model as independent parameters. This enabled a unification of PCA and ICA in the form of the likelihood in Eq. (3.10). In fact, it is intuitively clear that the conventional assumption of unit variance of the components in ICA makes it impossible to analyze the variances of the components. As we have shown here, the conventional assumption can be removed by considering the variances as additional parameters to be estimated, and eventually integrated out.

The unified model is primarily proposed here as an interesting theoretical framework. Future research is needed to see if it is useful in practice. The two-stage approach of first doing PCA and then ICA has been quite successful in practice, so it remains to be seen if a unified approach could be better for any practical applications.

REFERENCES

[1] E. Oja, A simplified neuron model as a principal component analyzer, J. Math. Biol. 15 (1982) 267-273.
[2] E. Oja, J. Karhunen, On stochastic approximation of the eigenvectors and eigenvalues of the expectation of a random matrix, J. Math. Anal. Appl. 106 (1985) 69-84.
[3] E. Oja, The nonlinear PCA learning rule in independent component analysis, Neurocomputing 17 (1) (1997) 25-46.
[4] A. Hyvärinen, J. Karhunen, E. Oja, Independent Component Analysis. Wiley Interscience, New York, 2001.
[5] A. Hyvärinen, E. Oja, Independent component analysis by general nonlinear Hebbian-like learning rules, Signal Process. 64 (3) (1998) 301-313.

CHAPTER

Riemannian optimization in complex-valued ICA

4

Visa Koivunen[1] and Traian Abrudan[2]

[1]*Department of Signal Processing and Acoustics, School of Electrical Engineering, Aalto University, Espoo, Finland,* [2]*Department of Computer Science, University of Oxford, Oxford, UK*

4.1 INTRODUCTION

Complex random vectors are commonly used in applications such as wireless communications, harmonic analysis, biomedical sensors (e.g., fMRI), sensor array signal processing, and radar. Many spectrally efficient modulation schemes as well as some of the recent radio transceiver developments are prime examples of this, all being based on complex-valued signal models. Furthermore, the observed data in these applications are typically multivariate by nature. The observation contains in most cases a mixture of multiple signals, for example a few spatially multiplexed data streams or many returns from multiple radar targets and other scatterers with potentially different Doppler and clutter. An important signal processing task is then to separate the original source signals from the observed mixture signals [1,14]. The multichannel k-variate received signal $\mathbf{z} = (z_1, \ldots, z_k)^T$ (sensor outputs) is modeled in terms of the *source signals* $s_1, \ldots, s_d$ possibly corrupted by additive *noise vector* $\mathbf{n}$, that is

$$\begin{aligned} \mathbf{z} &= \mathbf{A}\mathbf{s} + \mathbf{n} \\ &= \mathbf{a}_1 s_1 + \cdots + \mathbf{a}_d s_d + \mathbf{n}, \end{aligned} \tag{4.1}$$

where $\mathbf{A} = \begin{pmatrix}\mathbf{a}_1 & \cdots & \mathbf{a}_d\end{pmatrix}$ is the $k \times d$ *system matrix* and $\mathbf{s} = (s_1, \ldots, s_d)^T$ contains the source signals. It is assumed that $d \leq k$. If $d \leq k$, then a low rank structure where the signals lie in a d-dimensional subspace is commonly assumed. In practice, the system matrix is used to describe the sensor array geometry in array processing applications, frequency flat MIMO channel in wireless multiantenna communication systems [17], and mixing systems in the case of source separation problems, for example. All the components above are assumed to be complex-valued, and $\mathbf{s}$ and $\mathbf{n}$ are assumed to be mutually statistically independent with zero mean. Moreover, it is commonly assumed that noise $\mathbf{n}$ and/or sources $\mathbf{s}$ possess circularly symmetric distributions, most commonly circular complex Gaussian distribution [16]. However, for blind source separation problems, the sources may not be Gaussian distributed in order to have an identifiable independent component analysis (ICA) model, see [1] for detailed identifiability conditions in complex-valued cases.

E. Bingham, S. Kaski, J. Laaksonen and J. Lampinen (Eds) Advances in Independent Component Analysis and Learning Machines.
http://dx.doi.org/10.1016/B978-0-12-802806-3.00004-X

The amount of prior information in signal separation varies based on the signal separation task at hand. In wireless communications and radar, the transmitted waveforms may be known or orthogonal, they may arrive at the sensors from different directions and experience independent channels. This type of side information on source signals or system matrix facilitates highly reliable signal separation, and consequently resolving multiple targets in radar and demodulating multiple datastreams in communications.

In blind source separation (BSS) that is often based on ICA, both the mixing system **A** and the sources **s** are unknown. Hence, it is a particularly demanding signal processing task. The goal in ICA is to solve the mixing matrix and consequently to separate the sources from their mixtures exploiting only the assumption that sources are mutually statistically independent. ICA is a powerful technique of multichannel data analysis and signal processing (see [2] and its bibliography). It has found several applications including audio and speech signal separation, machine learning, biomedical signal processing, image processing, and data mining. We are particularly interested in the complex-valued ICA model considered in [3,4] to name only a couple.

In the following, we assume that sensor outputs follow the noiseless complex-valued ICA model, that is

$$\mathbf{z} = \mathbf{A}\mathbf{s},$$

where $\mathbf{s} \in \mathbf{C}^d$ has mutually *statistically independent* components $s_1, \ldots, s_d$, and without any loss of generality, assume that $E[\mathbf{s}] = \mathbf{0}$. As is common in ICA, we assume that the number of sources is equal to the number of sensors, so $k = d$ and that the mixing matrix $\mathbf{A} \in \mathbf{C}^{k \times k}$ is of full rank. Due to fundamental ambiguities of the separation problem exploiting the independence assumption [1], ICA should be understood as the determination of a matrix $\mathcal{B}$, called the *separating matrix*, that satisfies

$$\hat{\mathbf{s}} = \mathcal{B}\mathbf{z} = \mathbf{D}\mathbf{s},$$

where $\mathbf{D}$ is a $k \times k$ scaled permutation matrix, that is, $\hat{\mathbf{s}}$ contains permuted and scaled components of $\mathbf{s}$. For the separation to be possible (up to above ambiguities), at most one of the sources can obey circular Complex Normal (CN) distribution, but noncircular complex sources can have CN distribution with distinct circularity coefficient [1]. The whitening transform and fundamental results on source separation for noncircular complex-valued random vectors are discussed in detail in [1].

Many widely used ICA methods perform the signal separation in two stages [13,14]. First, the signals are prewhitened so that the signals will become uncorrelated. For whitened mixtures **y** the separating matrix will be a unitary matrix, so

$$\mathbf{W}^H\mathbf{y} = \hat{\mathbf{s}}$$

for some unitary matrix $\mathbf{W} \in \mathbf{C}^{k \times k}$, and thus

$$\mathbf{G} = \mathbf{W}^H\mathbf{B}$$

is a separating matrix for the original observed mixtures. Matrix **B** above is the whitening matrix, that is

$$\mathbf{B}^H\mathbf{B} = C(\mathbf{z})^{-1},$$

where $C(\mathbf{z})$ is the positive definite covariance matrix of the observed data $\mathbf{z}$. Consequently, the search for the separating matrix may be performed in the space of unitary matrices. The objective is then to make the estimated sources statistically as independent as possible under the constraint that matrix $\mathbf{W}$ is unitary. This leads to a real-valued objective function with a constraint on unitary matrix structure. In general, such problem may be formulated as follows:

$$\text{minimize } \mathcal{J}(\mathbf{W}) \text{ subject to} \tag{4.2}$$

$$\mathbf{W}^H\mathbf{W} = \mathbf{I}_n. \tag{4.3}$$

More detailed versions of the geodesic algorithms are presented in this chapter, their theoretical properties and additional simulation examples may be found in [5,6]. Furthermore, Matlab codes for the steepest descent (SD) [5] and related conjugate gradient algorithms [6] are publicly available at Matlab Central.

4.2 OVERVIEW OF OPTIMIZATION UNDER UNITARY MATRIX CONSTRAINT

Constrained optimization problems arise frequently in many sensor array and multichannel signal processing applications. Most of the adaptive algorithms minimize an error criterion in an iterative manner, subject to some matrix equality constraint. Solving this type of problem is typically done using numerical optimization because a closed-form solution may not exist, or if it exists it is very tedious to find.

Two main approaches for solving this type optimization problem have been proposed in the literature. The first one requires solving the *constrained* optimization problem on the Euclidean space by using classical gradient-based algorithms. The approach does not take advantage of the group property of unitary matrices, that is, the fact that unitary matrices are closed under the multiplication operation. This means that the product of two unitary matrices is again a unitary matrix. A typical method in this class is the unconstrained gradient-based method where the constraint is enforced in a separate step by projecting $\tilde{\mathbf{W}}_{k+1}$ onto the space of unitary matrices, see [2] for an example. This is not the case under additive updates used in most adaptive algorithms. The departure from the unitary property may be significant and a significant effort will be spent on projecting back the constraint space, instead of moving toward the optimum. Consequently, this type of algorithm achieves a lower convergence speed, as demonstrated in [5]. Furthermore, the Euclidean gradient algorithms combined with a projection step needed to satisfy the constraint does not take into account the curvature of the constrained surface. Hence, only linear convergence is achieved [7].

Another widely used constrained optimization approach is the method of Lagrange multipliers. The method of Lagrange multipliers is widely used for optimizing a function of multiple variables subject to one or more scalar constraints.

The method introduces a set of real scalar parameters λ_i called *Lagrange multipliers*. A new cost function $\mathcal{L}(\mathbf{W})$ called *Lagrangian* is constructed by combining the original cost function $\mathcal{J}(\mathbf{W})$ and an additional term containing the constraints weighted by the Lagrange multipliers. This objective function is optimized w.r.t. both the elements of $\mathbf{W}$ and the λ_i's. By finding stationary points of the constrained cost function of $\mathcal{J}(\mathbf{W})$ one finds the stationary points of the unconstrained cost function $\mathcal{L}(\mathbf{W})$. Using this approach for solving the ICA problem may be unfeasible in large dimensional problems because of the large number of constraints. An example of such ICA method is the Joint Approximate Diagonalization of Eigenmatrices (JADE) algorithm [3] where the number of constraints would become intolerably large, see [5].

The second class of methods requires solving an *unconstrained* optimization problem on the differentiable manifold determined by the constrained set. This manifold can be considered as a constrained surface that is embedded on a higher-dimensional Euclidean space. The idea of optimization over manifolds is illustrated in Figure 4.1. The optimization works as follows. Optimizing a differentiable cost function $\mathcal{J}(\mathbf{W})$ on a differentiable manifold chooses a point $\mathbf{W}$ on the manifold and a search direction $\mathbf{H}$ tangent to the manifold, at $\mathbf{W}$ (for detail, see Figure 4.1). In the next iteration, step one moves along a curve $\mathcal{C}(t)$ which emanates from $\mathbf{W}$ in the direction of $\mathbf{H}$. This curve is a *local parametrization* and it is used to describe a neighborhood around a given point on the manifold. Therefore, $\mathcal{C}(t)$ is contained on the manifold and satisfies the condition $\mathcal{C}(0) = \mathbf{W}$. Moreover, its derivative at $\mathbf{W}$ coincides with the search direction $\mathbf{H}$, that is, $\mathcal{C}'(0) = \mathbf{H}$. Optimization of the cost function in one dimension along the curve $\mathcal{C}(t)$ is performed at each iteration. By moving along these curves, the constraint on unitary matrices is always satisfied at each iteration.

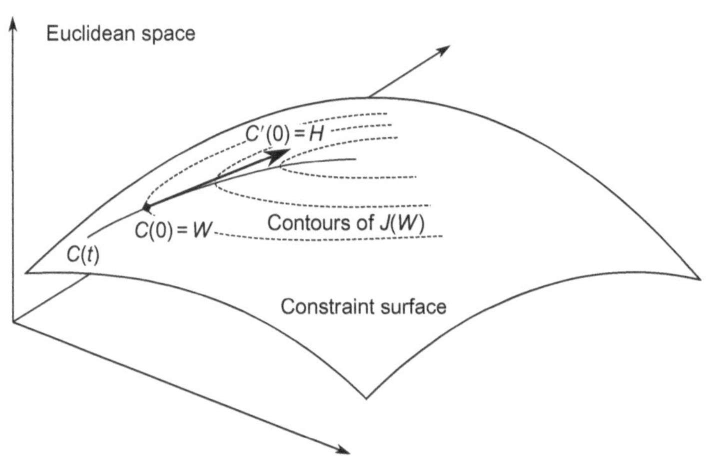

FIGURE 4.1

SD optimization on manifolds.

Differential geometry-based optimization algorithms have the property that they move along directions which are tangent to the manifold. Depending on how the local parametrization is defined, the algorithm can be classified to *projection-based algorithms* and *geodesic algorithms*. Geodesic algorithms move toward the optimum along the locally shortest paths [15].

An important aspect to be considered is that the unitary matrices are closed under the standard matrix multiplication operation, that is, they form the Lie group of $n \times n$ unitary matrices $U(n)$. This property may be exploited in optimization. A multiplicative update by a unitary matrix guarantees that the result will also be unitary matrix. The most important benefit is that geodesics are described by simple formulas, hence they may be computed very efficiently. Furthermore, differential geometry-based algorithms can exploit the reduced dimension of the manifold, unlike the method of Lagrange multipliers.

Nongeodesic algorithms such as [8] move along straight lines tangent to the manifold and deviate from the unitary constraint at every iteration. This is due to the fact that the manifold is a "curved space." Therefore, this algorithm has the same drawback as its Euclidean counterpart; that is, the constraint restoration procedure needs to be applied after every iteration. In the following section, geodesic methods for optimization under unitary matrix constraints are presented.

4.3 GEODESIC METHOD FOR OPTIMIZING UNDER UNITARY CONSTRAINT

An appealing approach to optimize a cost function on a Riemannian manifold is to move along geodesics. Geodesics are locally the shortest paths on a Riemannian manifold [9]. They correspond to straight lines on Euclidean space. Riemannian algorithms for optimizing a cost function under unitary matrix constraint use the *exponential map* as a local parametrization. Consequently, the algorithms employ a multiplicative update rule. Intuitively this means that a rotation is applied to the previous estimate to obtain the new one. Each iteration k of the algorithm consists of the following step:

$$\mathbf{W}_{k+1} = \exp(-\mu_k \mathbf{H}_k^{\mathrm{R}})\mathbf{W}_k = \mathbf{R}_k \mathbf{W}_k \tag{4.4}$$

where $-\mathbf{H}_k^{\mathrm{R}}$ is the directional vector of the geodesic. It is a skew-Hermitian matrix. Consequently, its matrix exponential

$$\mathbf{R}_k = \exp(-\mu_k \mathbf{H}_k^{\mathrm{R}})$$

is a unitary matrix. Since $\mathbf{W}_k$ is a unitary matrix, $\mathbf{W}_{k+1}$ will remain unitary at every iteration because of the group property. In this way, the constraint is satisfied automatically and no projection operation is needed.

The key advantages are the convenient expressions for the geodesics and parallel transport. Geodesics are expressed in terms of matrix exponential of skew-Hermitian matrices. There exists highly efficient numerical methods for computing the matrix exponential, see [10] and reference therein. In comparison, the projection-based method [8] requires the computation of SVD of arbitrary matrices. Another important property of the Lie group is that transporting vectors from a tangent space to another may be done in a very simple manner.

4.3.1 STEEPEST DESCENT ALGORITHM

The main benefit of the Riemannian SD algorithm for the optimization under unitary matrix constraint is that it is very simple to implement. Each iteration of the Riemannian SD algorithm is comprised of two subsequent stages. The first one is to find the Riemannian gradient which points to the steepest ascent direction on the manifold. In the second stage one takes a step along the geodesic emanating in the direction of the negative gradient. The Riemannian gradient of the smooth objective function $\mathcal{J}$ at a point $\mathbf{W}_k \in U(n)$ is given by:

$$\nabla^{\mathrm{R}}\mathcal{J}(\mathbf{W}_k) = \frac{\partial \mathcal{J}}{\partial \mathbf{W}^*}(\mathbf{W}_k) - \mathbf{W}_k \left[\frac{\partial \mathcal{J}}{\partial \mathbf{W}^*}(\mathbf{W}_k)\right]^H \mathbf{W}_k. \tag{4.5}$$

where $\frac{\partial \mathcal{J}}{\partial \mathbf{W}^*}(\mathbf{W}_k)$ is defined as

$$\frac{\partial \mathcal{J}}{\partial \mathbf{A}^*} \triangleq \frac{1}{2}\left(\frac{\partial \mathcal{J}}{\partial \mathbf{A}_R} + \jmath \frac{\partial \mathcal{J}}{\partial \mathbf{A}_I}\right). \tag{4.6}$$

and represents the gradient of the cost function $\mathcal{J}$ on the Euclidean space at a given $\mathbf{W}$ [11]. See [5] for a detailed derivation. The geodesic emanating from $\mathbf{W}_k$ along the SD direction

$$-\nabla^{\mathrm{R}}\mathcal{J}(\mathbf{W}_k)$$

on $U(n)$ is given by:

$$\mathcal{W}(\mu) = \exp(-\mu \mathbf{G}_k)\mathbf{W}_k, \quad \text{where} \tag{4.7}$$

$$\mathbf{G}_k \triangleq \nabla^{\mathrm{R}}\mathcal{J}(\mathbf{W}_k)\mathbf{W}_k^H \in \mathfrak{u}(n). \tag{4.8}$$

$\mathbf{G}_k$ is the gradient of $\mathcal{J}$ at $\mathbf{W}_k$ after translation into the tangent space at the identity element. Consequently, the matrix $\mathbf{G}_k$ is skew-Hermitian, that is, $\mathbf{G}_k = -\mathbf{G}_k^H$. The skew-Hermitian structure of $\mathbf{G}_k$ brings important computational benefits when computing the matrix exponential [5], as well as when performing the line search. The Riemannian SD algorithm on $U(n)$ has been derived in [5] and it is summarized in Table 4.1.

In the above algorithm, we choose the step size using the Armijo rule. In this way, the algorithm takes an initial step along the geodesic. Then, two other choices are checked by evaluating the cost function for the cases of doubling or halving the step size. Since the step size evolves in a dyadic basis, the geodesic methods are very suitable for the Armijo step. This is due to the fact that doubling the step size does not require any expensive computation, just squaring the rotation matrix as in the scaling and squaring procedure. For normal matrices, the computation of the matrix exponential via matrix squaring *prevents the round-off error accumulation* [12]. An Armijo type of geodesic SD algorithm has this advantage, since the argument of the matrix exponential is skew-Hermitian. Moreover, when the step size is halved, the

Table 4.1 SD Algorithm Along Geodesics on $U(n)$

1	Initialization: $k = 0$, $\mathbf{W}_k = \mathbf{I}$
2	Compute the Riemannian gradient direction $\mathbf{G}_k$: $\boldsymbol{\Gamma}_k = \frac{\partial \mathcal{J}}{\partial \mathbf{W}^*}(\mathbf{W}_k)$, $\mathbf{G}_k = \boldsymbol{\Gamma}_k \mathbf{W}_k^H - \mathbf{W}_k \boldsymbol{\Gamma}_k^H$
3	Evaluate $\langle \mathbf{G}_k, \mathbf{G}_k \rangle_{\mathbf{I}} = (1/2)\text{trace}\{\mathbf{G}_k^H \mathbf{G}_k\}$. If it is sufficiently small, then stop
4	Determine $\mu_k = \arg\min_\mu \mathcal{J}(\exp(-\mu \mathbf{G}_k)\mathbf{W}_k)$
5	Update: $\mathbf{W}_{k+1} = \exp(-\mu_k \mathbf{G}_k)\mathbf{W}_k$
6	$k := k + 1$ and go to step **2**

appropriate rotation matrix may be available from the scaling and squaring procedure which is often combined with other methods for approximating the matrix exponential. This allows reducing the complexity because an expensive numerical operation may be avoided. It is known that in a stationary case where the matrices involved in the cost function do not vary in time, the above SD algorithm employed with the Armijo step size rule almost always converges to a local minimum, assuming that it is not initialized at a stationary point. One can similarly come up with a conjugate gradient algorithm that typically converges faster to the optimum solution, see [6].

4.4 EXAMPLE ON SIGNAL SEPARATION IN MIMO SYSTEM

In this section, will provide an example of how blind source separation using ICA and the presented optimization method may be applied in separating multiple independent datastreams in a multiantenna (MIMO) communication system [17]. All the transmissions share the same frequency resources. Hence, the receivers observe mixtures of the transmitted streams. This type of transmission scheme is called *spatial multiplexing* and it is used in 4G wireless systems as well as wireless local area networks to give significant increase of the spectral efficiency of the transmission. Spectral efficiency may in fact linearly increase as a function of the minimum of the number of transmit or receive antennas whereas increasing the transmit power will only provide logarithmic increase. An example of a MIMO communication system is depicted in Figure 4.2.

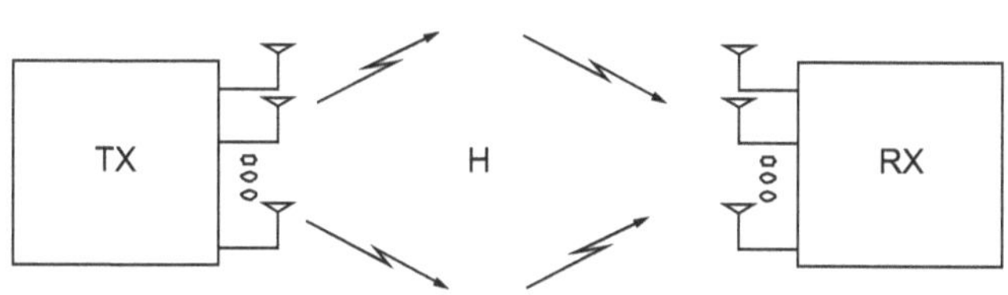

FIGURE 4.2

In spatial multiplexing MIMO systems, multiple data streams are transmitted using the same frequency resources. In rich scattering environments, significant increase in spectral efficiency is achieved. The receiver observes mixture signals that need to be separated in order to demodulate them reliably.

Separating signals blindly in an MIMO communication systems may be done by exploiting the statistical independence assumption of the transmitted signals. The JADE algorithm [3] is a widely used method for solving signal separation problems. The JADE algorithm is comprised of two stages. First, a *prewhitening* transform of the received signal is performed. The second-stage employees *unitary rotations* to make the signals as independent as possible. This stage may be formulated as an optimization problem under unitary matrix constraint. No closed-form solution is available except for the simplest cases, that is 2-by-2 unitary matrices. The second stage may be efficiently implemented by using the proposed SD on the unitary group.

A number of m independent zero-mean signals are sent through each of the m transmit antennas and they are received by r receive antennas. The frequency flat MIMO channel matrix $\mathbf{H}$ now defines an $r \times m$ mixing matrix ($r \geq m$). We use the most common signal model used in BSS. The $r \times N$ matrix $\mathbf{Z}$ corresponding to the received signal may be written as $\mathbf{Z} = \mathbf{HS} + \mathbf{V}$, where $\mathbf{S}$ is an $m \times N$ matrix corresponding to the m transmitted signals and $\mathbf{V}$ is the additive white noise. In the prewhitening stage, the received signal is decorrelated based on the eigendecomposition of the correlation matrix. The prewhitened received signal is given by

$$\mathbf{Y} = \mathbf{\Lambda}_m^{-1/2}\mathbf{U}_m^H\mathbf{Z},$$

where $\mathbf{U}_m$ and $\mathbf{\Lambda}_m$ contain the m eigenvectors and the m eigenvalues corresponding to the signal subspace, respectively.

In the second stage, the goal is to determine a unitary matrix $\mathbf{W}$ such that an estimate of the transmitted signals

$$\hat{\mathbf{S}} = \mathbf{WY}$$

is obtained up to phase and permutation ambiguities, which are inherent to any blind methods. The unitary matrix may be found by exploiting the information provided by the fourth-order cumulants of the whitened signals. The JADE algorithm looks for the minimum of the following criterion

$$\mathcal{J}_{\text{JADE}}(\mathbf{W}) = \sum_{i=1}^{m} \text{off}\{\mathbf{W}^H\hat{\mathbf{M}}_i\mathbf{W}\} \tag{4.9}$$

w.r.t. $\mathbf{W}$, under the unitarity constraint on $\mathbf{W}$. Hence, we have a minimization problem on $U(m)$ considered in this chapter. The eigenmatrices $\hat{\mathbf{M}}_i$ are estimated from the fourth-order cumulants. They need to be diagonalized as well. The operator off$\{\cdot\}$ sums up the squared magnitudes of the off-diagonal elements of a matrix. Consequently, this criterion penalizes the departure of all eigenmatrices from the diagonal structure. The Euclidean gradient of the JADE cost function is

$$\mathbf{\Gamma}_{\mathbf{W}} = 2\sum_{i=1}^{m} \hat{\mathbf{M}}_i\mathbf{W}\big[\mathbf{W}^H\hat{\mathbf{M}}_i\mathbf{W} - \mathbf{I} \odot (\mathbf{W}^H\hat{\mathbf{M}}_i\mathbf{W})\big],$$

where $\odot$ denotes the element-wise matrix multiplication.

The performance is studied quantitatively in terms of convergence speed of the *JADE criterion* and as well as using the *Amari distance* (performance index) [2]. Let us define a matrix capturing the cross-talk between the separated independent sources:

$$\hat{\mathbf{F}} = \hat{\mathcal{B}}\mathbf{A}.$$

Amari distance is then defined as follows:

$$d(\hat{\mathbf{F}}) = \frac{1}{2k(k-1)}\left\{\sum_{i=1}^{k}\left(\sum_{j=1}^{k}\frac{|\hat{f}_{ij}|}{\max_{\ell}|\hat{f}_{i\ell}|}-1\right)+\sum_{j=1}^{k}\left(\sum_{i=1}^{k}\frac{|\hat{f}_{ij}|}{\max_{\ell}|\hat{f}_{\ell j}|}-1\right)\right\}$$

where $\hat{f}_{ij} = [\hat{\mathbf{F}}]_{ij}$. Under perfect separation $d(\hat{\mathbf{F}}) = 0$. When the estimator fails to separate the sources, the value of $d()$ increases. Amari distance $d()$ is scaled so that the maximum value is 1.

This JADE criterion (Eq. 4.9) on the other hand quantifies how well the eigenmatrices $\hat{\mathbf{M}}_i$ are jointly diagonalized. This indicates how good an optimization solution is found, that is, the unitary rotation stage of the BSS. The Amari distance d_{A} is measured between the true channel matrix (known as ground truth in simulations) $\mathbf{H}$ and the obtained estimate of the channel matrix $\hat{\mathbf{H}}$. It can be used as a performance measure for the entire BSS technique as well. In terms of deviation from the unitary constraint, the performance is measured by using a *unitarity criterion*:

$$\Delta_{k+1} = \left\|\tilde{\mathbf{W}}_{k+1}^{H}\tilde{\mathbf{W}}_{k+1} - \mathbf{I}\right\|_F^2 \tag{4.10}$$

in a logarithmic scale.

A number of $m = 4$ independent data streams are transmitted, three QPSK signals (complex-valued modulation scheme) and one BPSK signal. The signal-to-noise ratio is SNR = 20 dB and the channel taps are assumed to be independent and random with power distributed according to a Rayleigh distribution. The results are averaged over 100 random realizations of the (4×6) MIMO matrix $\mathbf{H}$ and (4×1000) signal matrix.

In the first simulation, we study the performance of three optimization algorithms: the classical Euclidean SD algorithm which enforces the unitarity property of $\mathbf{W}$ by a projection operation after each iteration, the Euclidean SD with the Lagrange multipliers method emphasizing the unitarity property, and the presented Riemannian SD algorithm. Figure 4.3 shows observed mixtures of data streams at the receivers, and separated signals by using JADE with the developed Riemannian SD algorithm. The performance of the three algorithms in terms of convergence speed and accuracy of satisfying the unitary constraint are presented in Figure 4.4. It can be observed that the datastreams are well-separated and reliable demodulation of the data can be easily achieved. The phase ambiguities (rotation) can be easily solved by taking into account how I- and Q-components are related in each modulation scheme. Permutation ambiguity can be solved in a higher level of the protocol stack since the frame structure of the data conveys information about the source and destination of the data stream.

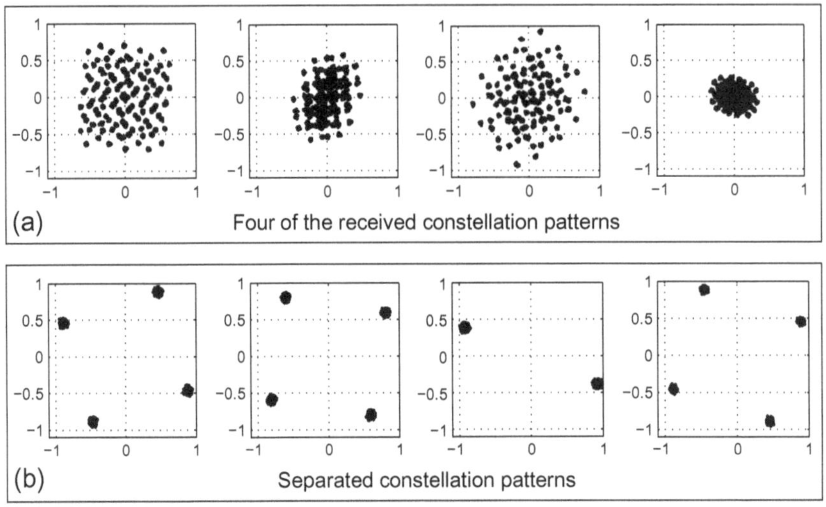

FIGURE 4.3

The constellation patterns corresponding to (a) four of the six received signals and (b) the four recovered signals by using JADE with the developed SD algorithm. In any ICA solution, there are inherent permutation and phase ambiguities which may be noticed as a rotation of the constellation.

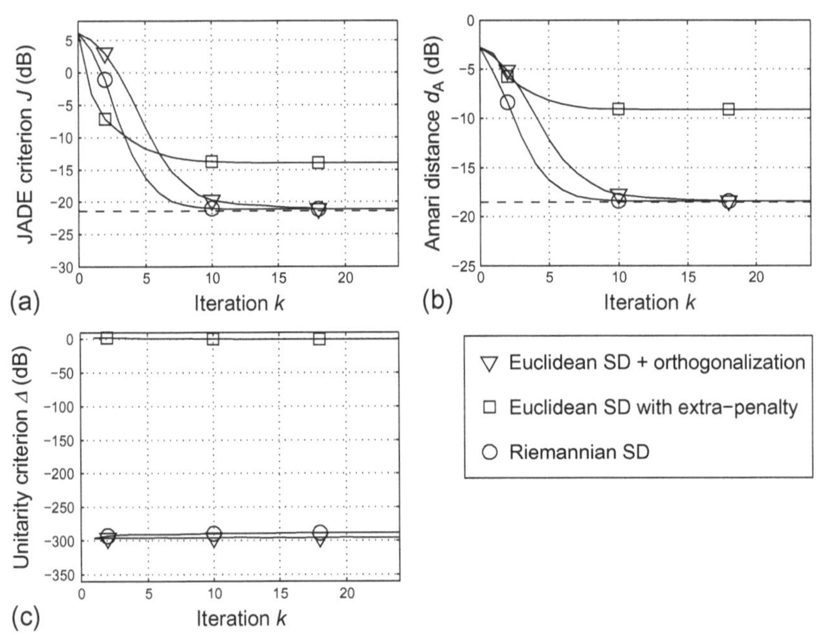

FIGURE 4.4

A comparison between the conventional optimization methods operating on the Euclidean space (the classical SD algorithm with enforcing unitarity, the extra-penalty SD method) and the Riemannian SD algorithm from Table 4.1. The horizontal thick dotted line in subplots (a) and (b) represents the solution of the original JADE algorithm [3]. The performance measures are the JADE criterion $\mathcal{J}_{\mathrm{JADE}}(\mathbf{W}_k)$ (Eq. 4.9), Amari distance d_{A} and the unitarity criterion Δ_k (4.10) vs. the iteration step. The developed Riemannian SD algorithm outperforms the conventional methods.

The developed SD converges to the JADE solution the fastest. Moreover, the Amari index converges the fastest for the SD algorithm. The unitarity property is also satisfied with a high fidelity.

4.5 CONCLUSION

In this chapter, we presented an optimization method for solving complex-valued ICA problems. This approach is suitable for ICA techniques that perform the separation in two stages, that is, first decorrelate the observed mixture signals and then look for unitary transforms that make the data as independent as possible. We focused on the second stage by presenting a Riemannian geometry approach for optimization of a real-valued cost function of complex-valued matrix argument W, such that the constraint that W is an unitary matrix is always satisfied. We presented an SD algorithm that operates on the Lie group of unitary matrices [5]. The algorithm moves toward the optimum along the geodesics, that is, locally shortest path in the constraint surface. The developed method fully exploits the recent advances in computing matrix exponential when finding the rotation matrices for the iterative update. The iterative update is multiplicative, hence exploiting the group property that products of unitary matrices are unitary matrices. This allows for automatically satisfying the constraint at each step. The developed algorithm is applied to BSS in multiantenna (MIMO) wireless systems where multiple datastreams are transmitted simultaneously using the same frequency resources. Spatially multiplexed complex-valued data streams were recovered with high fidelity while satisfying the constraint on unitary matrices all the time.

REFERENCES

[1] J. Eriksson, V. Koivunen, Complex random vectors and ICA models: identifiability, uniqueness and separability, IEEE Trans. Inform. Theory 52 (3) (2006) 1017-1029.

[2] A. Hyvärinen, J. Karhunen, E. Oja, Independent Component Analysis, John Wiley & Sons, New York, 2001.

[3] J. Cardoso, A. Souloumiac, Blind beamforming for non-Gaussian signals, IEE Proc. F 140 (6) (1993) 362-370.

[4] E. Bingham, A. Hyvarinen, A fast fixed-point algorithm for independent component analysis of complex-valued signals, Int. J. Neural Syst. 10 (01) (2000) 1-8.

[5] T. Abrudan, J. Eriksson, V. Koivunen, Steepest descent algorithms for optimization under unitary matrix constraint, IEEE Trans. Signal Process. 56 (3) (2008) 1134-1147.

[6] T. Abrudan, J. Eriksson, V. Koivunen, Conjugate gradient algorithm for optimization under unitary matrix constraint, Signal Process. 89 (9) (2009) 1704-1714.

[7] S.T. Smith, Optimization techniques on Riemannian manifolds, Fields Inst. Commun. 3 (1994) 113-136.

[8] J.H. Manton, Optimization algorithms exploiting unitary constraints, IEEE Trans. Signal Process. 50 (2002) 635-650.

[9] M.P. do Carmo, Riemannian Geometry. Mathematics: Theory and Applications, Birkhauser, Boston, 1992.
[10] A. Iserles, A. Zanna, Efficient computation of the matrix exponential by general polar decomposition, SIAM J. Numer. Anal. 42 (2005) 2218-2256.
[11] D.H. Brandwood, A complex gradient operator and its applications in adaptive array theory, IEE Proc. F H 130 (1983) 11-16.
[12] C. Moler, C. van Loan, Nineteen dubious ways to compute the exponential of a matrix, twenty-five years later, SIAM Rev. 45 (1) (2003) 3-49.
[13] S.C. Douglas, Self-stabilized gradient algorithms for blind source separation with orthogonality constraints, IEEE Trans. Neural Netw. 11 (2000) 1490-1497.
[14] A. Hyvärinen, E. Oja, A fast fixed-point algorithm for independent component analysis, Neural Comput. 9 (1997) 1483-1492.
[15] Y. Nishimori, Learning algorithm for independent component analysis by geodesic flows on orthogonal group, Int. Joint Conf. Neural Netw. 2 (1999) 933-938.
[16] E. Ollila, D.E. Tyler, V. Koivunen, H.V. Poor, Complex elliptically symmetric distributions: survey, new results and applications, IEEE Trans. Signal Process. 60 (11) (2012) 5597-5625.
[17] C.B. Papadias, Globally convergent blind source separation based on a multiuser kurtosis maximization criterion, IEEE Trans. Signal Process. 48 (2000) 3508-3519.

CHAPTER

Nonadditive optimization 5

Zhirong Yang[1] and Irwin King[2]

[1]*Department of Computer Science, Aalto University, Espoo, Finland,* [2]*The Chinese University of Hong Kong, Shatin, Hong Kong, China*

5.1 INTRODUCTION

Many machine learning and signal processing tasks can be formulated as an optimization problem, where a cost function $\mathcal{J}(\boldsymbol{\theta})$ is minimized over the model parameters $\boldsymbol{\theta}$. Conventional optimization methods employ the following procedure: first find a direction $\mathbf{u}$ which often comes from the derivatives and then search a suitable step size η along the line in that direction. We call such updating method *additive optimization* because the new estimate is obtained by adding the direction vector times the step size to the current estimate: $\boldsymbol{\theta}^{\text{new}} = \boldsymbol{\theta} + \eta \cdot \mathbf{u}$.

Despite its popularity, the additive optimization can be problematic or difficult to use. In some tasks, the objective is accompanied with constraints such as nonnegativity, orthogonality, or stochasticity. The new estimate $\boldsymbol{\theta}^{\text{new}}$ may violate the constraints, and additional projection is thus needed to obtain valid solutions. Moreover, tuning the updating step size η is often not easy: overly small sizes can lead to slow convergence and overly big step sizes can cause divergence. Comprehensive line search methods exist, but they often rely on additional assumptions on the curvature and introduce even more hyper-parameters.

In this chapter, we review several nonadditive optimization methods in our research with Prof. Oja, which overcome the drawbacks in the additive approach. After a brief introduction of additive optimization in the next section, we recapitulate in Section 5.3 the essential steps in the FastICA algorithm, which is an elegant and efficient approximation to the Newton method. Section 5.4 presents the algorithms derived from the fixed points of the cost function, with an example in visualization. Section 5.5 gives the geodesic update rule in the presence of orthogonality constraint. Section 5.6 presents the multiplicative updates, including its theory and applications to both nonnegative data and unconstrained optimization problems. Discussions and potential future work are given in Section 5.7.

E. Bingham, S. Kaski, J. Laaksonen and J. Lampinen (Eds) Advances in Independent Component Analysis and Learning Machines.
http://dx.doi.org/10.1016/B978-0-12-802806-3.00005-1

5.2 ADDITIVE OPTIMIZATION

Let the target optimization problem be

$$\underset{\boldsymbol{\theta}}{\text{minimize}} \quad \mathcal{J}(\boldsymbol{\theta}) \tag{5.1}$$

$$\text{subject to} \quad \boldsymbol{\theta} \in \mathcal{S}, \tag{5.2}$$

where $\mathcal{J}$ is the objective function and $\mathcal{S}$ is the subset specified by the constraints. We do not assume the convexity of $\mathcal{J}$.

Additive optimization attempts to solve the above problem by iteratively applying $\boldsymbol{\theta}^{\text{new}} = \boldsymbol{\theta} + \eta \cdot \mathbf{u}$, where $\mathbf{u}$ is a descent direction obtained from, for example, negative gradient or Quasi-Newton (e.g., [1]). However, $\boldsymbol{\theta}^{\text{new}}$ is not necessarily in $\mathcal{S}$ any more. Therefore, an additional projection step is needed: $\boldsymbol{\theta}^{\text{new}}_{\text{projected}} = \min_{\boldsymbol{\zeta} \in \mathcal{S}} D(\boldsymbol{\zeta}||\boldsymbol{\theta}^{\text{new}})$, where $D(\cdot)$ is a certain divergence measure such as the Euclidean distance.

Selecting the best step size η in general remains tricky. This requires an inner loop for searching a suitable step along the line specified by $\mathbf{u}$. In addition to extra computation cost in evaluating the objective function several times, the line search often makes additional assumptions on the curvature, for instance, the Wolfe conditions [2].

Below we show several ways to develop nonadditive optimization algorithms without tuning η or without the extra projection step.

5.3 FAST FIXED-POINT APPROXIMATED NEWTON ALGORITHMS

The Newton's method in optimization is free of tuning η:

$$\boldsymbol{\theta}^{\text{new}} = \boldsymbol{\theta} - \mathbf{H}^{-1}\nabla, \tag{5.3}$$

where ∇ and $\mathbf{H}$ are the gradient and Hessian of $\mathcal{J}$ to $\boldsymbol{\theta}$, respectively. In general, $\mathbf{H}$ is difficult to obtain due to its complex form or inhibitive computation cost. However, in certain problems, we can make efficient approximation to the Hessian, with the known example FastICA [3].

The derivation is given below. Assume $\mathbf{x}$ is whitened, that is $\mathbb{E}\{\mathbf{x}\mathbf{x}^T\} = \mathbf{I}$. We can develop a fast fixed-point approximated Newton algorithm (FFAN) for the following problem:

$$\underset{\mathbf{w}}{\text{minimize}} \quad \mathcal{J}(\mathbf{w}) = \mathbb{E}\left\{G\left(\mathbf{w}^T\mathbf{x}\right)\right\} \tag{5.4}$$

$$\text{subject to} \quad \|\mathbf{w}\| = 1, \tag{5.5}$$

where G is a smooth function.

Denote g and g' the first and second derivatives of G, respectively. The gradient and Hessian of $\mathcal{J}$ to $\mathbf{w}$ are, respectively, $\nabla = \mathbb{E}\{g(\mathbf{w}^T\mathbf{x})\mathbf{x}\}$ and $H = \mathbb{E}\{g'(\mathbf{w}^T\mathbf{x})\mathbf{x}\mathbf{x}^T\}$. FFAN makes the approximation

$$H \approx \mathbb{E}\left\{g'\left(\mathbf{w}^T\mathbf{x}\right)\right\}\mathbb{E}\left\{\mathbf{x}\mathbf{x}^T\right\} = \mathbb{E}\left\{g'\left(\mathbf{w}^T\mathbf{x}\right)\right\} \cdot \mathbf{I}. \tag{5.6}$$

Inserting this approximated Hessian to the Newton's method, we have

$$\mathbf{w} \leftarrow \mathbf{w} - \mathbb{E}\left\{g'\left(\mathbf{w}^T\mathbf{x}\right)\right\}^{-1} \nabla. \tag{5.7}$$

Due to the subsequent normalization $\mathbf{w} \leftarrow \mathbf{w}/\|\mathbf{w}\|$, we can multiply the scalar $\mathbb{E}\left\{g'\left(\mathbf{w}^T\mathbf{x}\right)\right\}$ to the right-hand side of Eq. (5.7), which yields

$$\mathbf{w} \leftarrow \mathbb{E}\left\{g'\left(\mathbf{w}^T\mathbf{x}\right)\right\}\mathbf{w} - \nabla, \tag{5.8}$$

$$\mathbf{w} \leftarrow \mathbf{w}/\|\mathbf{w}\|. \tag{5.9}$$

In addition to independent component analysis, we have applied the same technique in finding discriminative direction in classification [4].

5.4 FIXED-POINT ALGORITHMS FOR KERNEL LEARNING

Directly analyzing the fixed-point condition $\frac{\partial \mathcal{J}}{\partial \boldsymbol{\theta}} = 0$ in the form $\boldsymbol{\theta} = f(\boldsymbol{\theta})$ can also lead to update rules that are free of tuning step sizes. In this section, we illustrate this idea first by a simple setting in mode seeking and then by a more comprehensive setting in nonlinear dimensionality reduction, both objectives are a function over pairwise squared Euclidean distances.

In Parzen density estimation with Gaussian kernels (width σ) and N data points in $\mathbb{R}^d$, a density function $p(\mathbf{x})$ can be asymptotically estimated by

$$p(\mathbf{x}) \approx \hat{p}(\mathbf{x}) = \frac{1}{N\sigma^d(2\pi)^{d/2}} \sum_{i=1}^{N} \exp\left(-\frac{\|\mathbf{x}-\mathbf{x}_i\|^2}{2\sigma^2}\right). \tag{5.10}$$

Thus minimizing $\mathcal{J}(\mathbf{x}) = -\hat{p}(\mathbf{x})$ over $\mathbf{x}$ approximately seeks the modes in the empirical distribution, and has found its application in, for example, image segmentation [5]. The fixed points of $\mathcal{J}(\mathbf{x})$ appear when

$$\frac{\partial \mathcal{J}(\mathbf{x})}{\partial \mathbf{x}} = -\frac{1}{N\sigma^{d+2}(2\pi)^{d/2}} \sum_{i=1}^{N} \exp\left(-\frac{\|\mathbf{x}-\mathbf{x}_i\|^2}{2\sigma^2}\right)(\mathbf{x}-\mathbf{x}_i) = 0, \tag{5.11}$$

or

$$\mathbf{x} = \frac{\sum_{i=1}^{N} \exp\left(-\frac{\|\mathbf{x}-\mathbf{x}_i\|^2}{2\sigma^2}\right)\mathbf{x}_i}{\sum_{i=1}^{N} \exp\left(-\frac{\|\mathbf{x}-\mathbf{x}_i\|^2}{2\sigma^2}\right)}. \tag{5.12}$$

This suggests the update rule

$$\mathbf{x}^{\text{new}} = \frac{\sum_{i=1}^{N} \exp\left(-\frac{\|\mathbf{x}-\mathbf{x}_i\|^2}{2\sigma^2}\right)\mathbf{x}_i}{\sum_{i=1}^{N} \exp\left(-\frac{\|\mathbf{x}-\mathbf{x}_i\|^2}{2\sigma^2}\right)}, \tag{5.13}$$

which is known as the mean-shift algorithm and is essentially a realization of the EM algorithm [6].

Next, we show the application of the same technique to Heavy-tailed Symmetric Stochastic Neighbor Embedding (HSSNE; [7]) which is a nonlinear dimensionality reduction method for visualization. HSSNE minimizes the objective

$$\mathcal{J}(\mathbf{Y}) = \sum_{ij,i\neq j} P_{ij} \ln \frac{P_{ij}}{Q_{ij}}, \tag{5.14}$$

where $P_{ij} \geq 0$ are given pairwise proximities between data objects, $\sum_{ij} P_{ij} = 1$, $P_{ii} = 0$, $\mathbf{Y} = [\mathbf{y}_1, \ldots, \mathbf{y}_N]^T$, $q_{ij} = (\alpha \|\mathbf{y}_i - \mathbf{y}_j\|^2 + c)^{-1/\alpha}$, with $\alpha \geq 0$ and $c \geq 0$, and $Q_{ij} = q_{ij} / \sum_{ab} q_{ab}$.

Denote $S_{ij} = -\frac{\partial q_{ij}}{\partial \|\mathbf{y}_i - \mathbf{y}_j\|^2} = q_{ij}^{\alpha}$. We have the fixed-point equation

$$\frac{\partial \mathcal{J}(\mathbf{Y})}{\partial Y_{ik}} = \sum_j \left(P_{ij} - Q_{ij}\right) S_{ij} \left(Y_{ik} - Y_{jk}\right) = 0. \tag{5.15}$$

Reorganizing a bit

$$Y_{ik} \sum_j P_{ij} S_{ij} = Y_{ik} \sum_j Q_{ij} S_{ij} + \sum_j \left(P_{ij} - Q_{ij}\right) S_{ij} Y_{jk}, \tag{5.16}$$

we get the following fixed-point update rule [7]:

$$Y_{ik}^{\text{new}} = \frac{Y_{ik} \sum_j B_{ij} + \sum_j \left(A_{ij} - B_{ij}\right) Y_{jk}}{\sum_j A_{ij}}, \tag{5.17}$$

where $A_{ij} = P_{ij} S_{ij}$ and $B_{ij} = Q_{ij} S_{ij}$. It has been shown that iteratively applying the above update rule monotonically decreases the first-order Taylor expansion of $\mathcal{J}(Y)$ [7]. It also approximates the Newton's method by using a single-step estimate of the second-order update [7]. Such an approximation technique was also used in the mean-shift algorithm as a generalized expectation-maximization solution [6].

5.5 GEODESIC UPDATES IN STIEFEL MANIFOLDS

Optimization over orthogonal matrices is often needed in latent variable analysis such as ICA. A general form of optimization problem is

$$\underset{\mathbf{W}}{\text{minimize}} \quad \mathcal{J}(\mathbf{W}) \tag{5.18}$$

$$\text{subject to } \mathbf{W} \in \left\{ \mathbf{W} \in \mathbb{R}^{m \times r} | m \geq r, \mathbf{W}^T \mathbf{W} = \mathbf{I} \right\}. \tag{5.19}$$

The gradient $\nabla = \partial \mathcal{J} / \partial \mathbf{W}$ can be decomposed into two components, one in the tangent space of Stiefel manifold and the other perpendicular to the tangent space. By removing the second component, we obtain the gradient in the Stiefel manifold [8,9]: $\mathcal{G} = \nabla - \mathbf{W} \nabla^T \mathbf{W}$. A similar rectified gradient form was given in [10,11].

However, the additive optimization updates in general do not follow the geodesics in the Stiefel manifold. The new estimate after each update

$$\mathbf{W}^{\text{new}} = \mathbf{W} + \eta \left(\nabla - \mathbf{W} \nabla^T \mathbf{W} \right) \tag{5.20}$$

usually goes out of the Stiefel manifold. To overcome this problem, we have found that the exponential map in Stiefel manifolds [9] works well in dimensionality reduction [12]. The resulting update rule is nonadditive:

$$\mathbf{W}^{\text{new}} = \text{expm}\left(\eta\left(\nabla\mathbf{W}^T - \mathbf{W}\nabla^T\right)\right)\mathbf{W}, \tag{5.21}$$

where expm represents the matrix exponential [13].

5.6 MULTIPLICATIVE UPDATES

Nonnegativity has shown to be a powerful constraint in learning. When used in matrix factorization problems, it often leads to part-based representations and finds applications in signal processing and cluster analysis (see, e.g., [14]).

The research direction has attracted much effort since Lee and Seung's work on nonnegative matrix factorization (NMF; [15]). Given a nonnegative data matrix $\mathbf{X}$, NMF seeks two low-rank and nonnegative matrices $\mathbf{W}$ and $\mathbf{H}$ such that $\mathbf{X} \approx \mathbf{WH}$. The approximation error can be measured by $D(\mathbf{X}||\mathbf{WH})$, where $D()$ is Euclidean distance or nonnormalized Kullback-Leibler divergence.

Conventional additive optimization methods such as steepest descent require line search of the step size and they cannot guarantee the nonnegativity of $\mathbf{W}$ and $\mathbf{H}$ after each update. Lee and Seung propose a very different optimization method called multiplicative updates (e.g., for $\mathbf{W}$):

1. the gradient is first decomposed into two nonnegative parts

$$\nabla = \frac{\partial D(\mathbf{X}||\mathbf{WH})}{\partial \mathbf{W}} = \nabla^+ - \nabla^-; \tag{5.22}$$

   the decomposition often naturally comes from the structure of the cost function; then,
2. iteratively apply a simple update rule

$$W_{ik}^{\text{new}} = W_{ik}\frac{\nabla_{ik}^-}{\nabla_{ik}^+}. \tag{5.23}$$

For example, when $D(\mathbf{X}||\mathbf{WH}) = \frac{1}{2}\sum_{ij}[X_{ij} - (\mathbf{WH})_{ij}]^2$, $\nabla^+ = \mathbf{WHH}^T$, $\nabla^- = \mathbf{XH}^T$, and $W_{ik}^{\text{new}} = W_{ik}\frac{(\mathbf{XH}^T)_{ik}}{(\mathbf{WHH}^T)_{ik}}$.

When $\mathbf{W}$ is nonnegatively initialized, everything in the right-hand side of Eq. (5.23) is nonnegative and so remains the new estimate.[a] More attractively, the multiplicative updates do not require a user-supplied learning step size. The iterations have been shown to monotonically decrease the approximation error [16].

Due to its simplicity, the multiplicative updates have gained great popularity in recent years. The monotonicity guarantee of the original NMF algorithms is proven with an EM-like approach: first construct an auxiliary upper-bounding function of objective function at the current estimate, of which the minimization gives the multiplicative update rule. This approach is also known as majorization-minimization (see, e.g., [17]). However, constructing the auxiliary function was a challenging work in general for other divergences and for other matrix decomposition forms.

In [18], we have proposed a unified method for developing NMF algorithms with monotonicity guarantee for a wide range of divergences. The principle of deriving a multiplicative update rule and the corresponding auxiliary function of a given separable objective is summarized as the following procedure.

1. Transform the objective function into the form of finite generalized polynomials. Use the logarithm limit wherever needed.
2. Upper bound each monomial according to their concavity or convexity by using their first-order Taylor expansion or the Jensen inequality, respectively.
3. If there are three or more individual upper-bounds, combine them into two monomials according to their exponents. Form the auxiliary function.
4. Take the derivative of the auxiliary function with respect to the factorizing matrix variable.
5. Apply the logarithm limit if needed. Employ L'Hôpital's rule when the limit has the form $\frac{0}{0}$.
6. Obtain the multiplicative update rule by setting the derivative to zero.

The above procedure has been extended to nonseparable cases such as the γ- and Rényi-divergences [18], to the quadratic factorization case where one or more factorizing matrices may appear twice in the approximation [18,19], to NMF problems with orthogonality or stochasticity constraint, and to structural decomposition beyond matrix factorization [20–22]. The resulting algorithms have been successfully applied to various learning tasks such as part-based representation learning, clustering, bi-clustering, and graph matching [18–30].

Interestingly, the multiplicative update rule Eq. (5.23) also empirically works for problems beyond nonnegativity and matrix factorization/decomposition [31]. We again consider the nonlinear dimensionality reduction problem in Eq. (5.14), where we fix $\alpha = c = 1$, that is, $q_{ij} = \left(1 + \|\mathbf{y}_i - \mathbf{y}_j\|^2\right)^{-1}$, to avoid notational clutter. This learning objective is called t-distributed stochastic neighbor embedding (t-SNE; [32]), which is an unconstrained optimization problem.

We write $Z_{ik} = \exp(Y_{ik})$ for the t-SNE objective. The gradient of $\mathcal{J}$ with respect to $\mathbf{Z}$ thus is

$$\nabla_{ik} = \frac{\partial \mathcal{J}}{\partial Z_{ik}} = \frac{\partial \mathcal{J}}{\partial Y_{ik}} \frac{1}{Z_{ik}}. \tag{5.24}$$

The gradient to Y can be decomposed into two nonnegative parts:

$$\begin{aligned} \frac{1}{4}\frac{\partial \mathcal{J}}{\partial Y_{ik}} = & \left[(\mathbf{D}^{\mathbf{A}} + \mathbf{B})\mathbf{Y}^{+} + (\mathbf{D}^{B} + \mathbf{A})\mathbf{Y}^{-}\right]_{ik} \\ & - \left[(\mathbf{D}^{\mathbf{A}} + \mathbf{B})\mathbf{Y}^{-} + (\mathbf{D}^{\mathbf{B}} + \mathbf{A})\mathbf{Y}^{+}\right]_{ik}, \end{aligned} \tag{5.25}$$

where $Y^{+}_{ik} = \frac{|Y_{ik}| + Y_{ik}}{2}$, $Y^{-}_{ik} = \frac{|Y_{ik}| - Y_{ik}}{2}$, $A_{ij} = P_{ij} q_{ij}$, and $B_{ij} = Q_{ij} q_{ij}$; $\mathbf{D}^{\mathbf{A}}$ and $\mathbf{D}^{\mathbf{B}}$ are diagonal matrices with $D^{A}_{ii} = \sum_j A_{ij}$ and $D^{B}_{ii} = \sum_j B_{ij}$. Therefore, according to the reformulating principle Eq. (5.23) for multiplicative updates, we obtain

$$Z_{ik}^{\text{new}} = Z_{ik} \frac{\left[(\mathbf{D^A} + \mathbf{B})\mathbf{Y}^- + (\mathbf{D^B} + \mathbf{A})\mathbf{Y}^+\right]_{ik}}{\left[(\mathbf{D^A} + \mathbf{B})\mathbf{Y}^+ + (\mathbf{D^B} + \mathbf{A})\mathbf{Y}^-\right]_{ik}} \tag{5.26}$$

because the term $\frac{1}{Z_{ik}} > 0$ appears in both numerator and denominator and thus cancels out. After each multiplicative update, Y_{ik} can be easily obtained by taking the logarithm of Z_{ik}.

The t-SNE objective function value empirically decreases by using the above multiplicative update rule. The experiment results can be found in [31]. Recently Peltonen and Lin [33] have also successfully applied the same technique to information retrieval-based neighbor embedding.

5.7 DISCUSSION

In this chapter, we have reviewed several nonadditive optimization techniques for machine learning and signal processing. Our practice has shown that additive updates are not the only paradigm for optimization. Nonadditive methods that make use of the problem structure can bring better convergence and convenience.

Our review focused on the methods developed through recent collaboration with Prof. Oja. There are many other nonadditive optimization approaches not covered in this chapter. Research in this direction will continue to study more efficient and robust optimization methods.

NOTES

a. To avoid numerical problems one can positively initialize $\mathbf{W}$ plus a small constant ϵ to both ∇^- and ∇^+.

REFERENCES

[1] J. Nocedal, Updating quasi-Newton matrices with limited storage, Math. Comput. 35 (151) (1980) 773-782.

[2] J. Nocedal, S. Wright, Numerical Optimization, Chapter 3. Line Search Methods. Springer, New York, 1999.

[3] A. Hyvärinen, J. Karhunen, E. Oja, Independent Component Analysis, John Wiley & Sons, New York, 2001.

[4] Z. Yang, J. Laaksonen, A fast fixed-point algorithm for two-class discriminative feature extraction, in: Proceedings of 16th International Conference on Artificial Neural Networks (ICANN), Part II, 2006, pp. 330-339.

[5] D. Comaniciu, M. Peter, Mean shift: a robust approach toward feature space analysis, IEEE Trans. Pattern Anal. Mach. Intell. 24 (5) (2002) 603-619.

[6] M. Carreira-Perpiñán, Gaussian mean-shift is an EM algorithm, IEEE Trans. Pattern Anal. Mach. Intell. 29 (5) (2007) 767-776.

[7] Z. Yang, I. King, Z. Xu, E. Oja, Heavy-tailed symmetric stochastic neighbor embedding, in: Advances in Neural Information Processing Systems, 2009.
[8] A. Edelman, The geometry of algorithms with orthogonality constraints, SIAM J. Matrix Appl. 20 (2) (1998) 303-353.
[9] Y. Nishimori, S. Akaho, Learning algorithms utilizing quasi-geodesic flows on the Stiefel manifold, Neurocomputing 67 (2005) 106-135.
[10] E. Oja, A simplified neuron model as a principal component analyzer, J. Math. Biol. 15 (1982) 267-273.
[11] E. Oja, Principal components, minor components, and linear neural networks, Neural Netw. 5 (1992) 927-935.
[12] Z. Yang, Discriminative learning with application to interactive facial image retrieval (Doctoral dissertation), TKK Dissertations in Information and Computer Science TKK-ICS-D9, Helsinki University of Technology, Faculty of Information and Natural Sciences, Department of Information and Computer Science, Espoo, Finland, 2008.
[13] R. Horn, C. Johnson, Topics in Matrix Analysis, Cambridge University Press, Cambridge, 1991.
[14] A. Cichocki, R. Zdunek, A.H. Phan, S. Amari, Nonnegative Matrix and Tensor Factorization, John Wiley & Sons, New York, 2009.
[15] D. Lee, H. Seung, Learning the parts of objects by non-negative matrix factorization, Nature 401 (1999) 788-791.
[16] D. Lee, H. Seung, Algorithms for non-negative matrix factorization, Adv. Neural Inform. Process. Syst. 13 (2001) 556-562.
[17] D.R. Hunter, K. Lange, A tutorial on MM algorithms, Am. Stat. 58 (1) (2004) 30-37.
[18] Z. Yang, E. Oja, Unified development of multiplicative algorithms for linear and quadratic nonnegative matrix factorization, IEEE Trans. Neural Netw. 22 (12) (2011) 1878-1891.
[19] Z. Yang, E. Oja, Quadratic nonnegative matrix factorization, Pattern Recogn. 45 (4) (2012) 1500-1510.
[20] Z. Yang, E. Oja, Clustering by low-rank doubly stochastic matrix decomposition, in: ICML, 2012.
[21] Z. Yang, T. Hao, O. Dikmen, X. Chen, E. Oja, Clustering by nonnegative matrix factorization using graph random walk, in: Advances in Neural Information Processing Systems, 2012.
[22] Z. Zhu, Z. Yang, E. Oja, Multiplicative updates for learning with stochastic matrices, in: Proceedings of the 18th conference Scandinavian Conferences on Image Analysis (SCIA), 2013, pp. 143-152.
[23] Z. Yang, E. Oja, Linear and nonlinear projective nonnegative matrix factorization, IEEE Trans. Neural Netw. 21 (5) (2010) 734-749.
[24] Z. Yang, H. Zhang, Z. Yuan, E. Oja, Kullback-Leibler divergence for nonnegative for nonnegative matrix factorization, in: Proceedings of 21st International Conference on Artificial Neural Networks (ICANN), 2011, pp. 14-17.
[25] H. Zhang, T. Hao, Z. Yang, E. Oja, Pairwise clustering with t-PLSI, in: Proceedings of 22nd International Conference on Artificial Neural Networks (ICANN), 2012, pp. 411-418.
[26] Z. Lu, Z. Yang, E. Oja, Selecting β-divergence for nonnegative matrix factorization by score matching, in: Proceedings of 22nd International Conference on Artificial Neural Networks (ICANN), 2012, pp. 419-426.

[27] H. Zhang, Z. Yang, E. Oja, Adaptive multiplicative updates for projective nonnegative matrix factorization, in: Proceedings of 19th International Conference on Neural Information Processing (ICONIP), 2012, pp. 277-284.
[28] Z. Yang, H. Zhang, E. Oja, Online projective nonnegative matrix factorization for large datasets, in: Proceedings of 19th International Conference on Neural Information Processing (ICONIP), 2012, pp. 285-290.
[29] H. Zhang, Z. Yang, E. Oja, Improving cluster analysis by co-initializations, Pattern Recogn. Lett. 45 (2014) 71-77.
[30] H. Zhang, Z. Yang, E. Oja, Adaptive multiplicative updates for quadratic nonnegative matrix factorization, Neurocomputing 134 (2014) 206-213.
[31] Z. Yang, C. Wang, E. Oja, Multiplicative updates for t-SNE, in: IEEE International Workshop on Machine Learning for Signal Processing (MLSP), 2010.
[32] L. van der Maaten, G. Hinton, Visualizing data using t-SNE, J. Mach. Learn. Res. 9 (2008) 2579-2605.
[33] J. Peltonen, Z. Lin, Multiplicative update for fast optimization of information retrieval based neighbor embedding, in: IEEE International Workshop on Machine Learning for Signal Processing (MLSP), 2013, pp. 1-6.

CHAPTER

Image denoising, local factor analysis, Bayesian Ying-Yang harmony learning[1]

6

Guangyong Chen, Fengyuan Zhu, Pheng Ann Heng and Lei Xu
Department of Computer Science and Engineering, The Chinese University of Hong Kong, Shatin, N.T., Hong Kong, China

6.1 A BRIEF OVERVIEW ON DENOISING STUDIES

Distortion is often introduced into images through capturing instruments, data transmission media, image quantization, and discrete source of radiation. There are two typical image distortions [1]. One is called blur, which is intrinsic to image acquisition systems. The second distortion comes from environmental disturbances, which is usually called noises. Typically, an observed image X is regarded as generated from an additive model $X = \hat{X} + E$ with $\hat{X}$ representing the clean image and E denoting the noise that is independent from $\hat{X}$, as shown in Figure 6.1. This chapter focuses on removing this additive noise E, which is a challenging ill-posed problem. In the past decades, many efforts have been made and a brief overview is provided as follows.

6.1.1 SPATIAL FILTERING

A spatial filter is an image operator where each pixel x_t is changed by a function of the intensities of pixels in a neighbor $\mathcal{N}(x_t) \in \mathbb{R}^{d\times d}$. Based on the assumption that statistical property of images keep piecewise constant, numerous methods have been proposed for different noise types, such as mean filter, Gaussian filter, Wiener filter, weighted mean filter, anisotropic filter, bilateral filter for Gaussian noise, and median filter for Laplace noise [2].

Under the assumption of piecewise flat, the mean filter estimate

$$\bar{x}_t = \frac{1}{\#\mathcal{N}(x_t)} \sum_{x_j \in \mathcal{N}(x_t)} x_j, \tag{6.1}$$

where $\#S$ denotes the cardinality of the set S. When a pixel value x in $\mathcal{N}(x_t)$ is affected by a noise e from $G(e|0, \sigma_e^2)$, it follows from $x = \hat{x}_t + e$ that we have $G(x|\hat{x}_t, \sigma_e^2)$,

[1]All related codes and data set are available on the website: http://appsrv.cse.cuhk.edu.hk/~gychen/.

E. Bingham, S. Kaski, J. Laaksonen and J. Lampinen (Eds) Advances in Independent Component Analysis and Learning Machines.
http://dx.doi.org/10.1016/B978-0-12-802806-3.00006-3

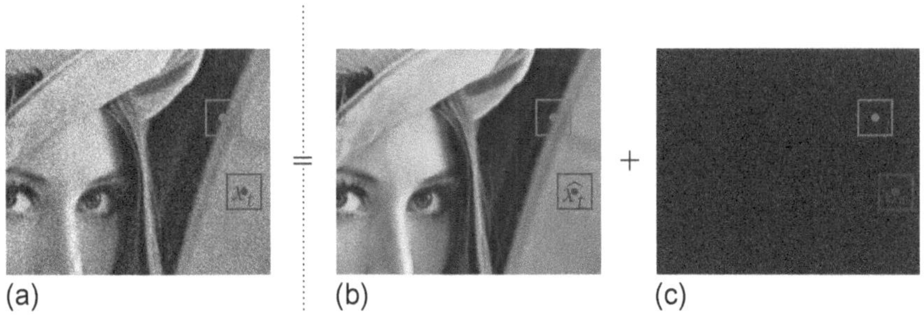

FIGURE 6.1

The additive noise model. (a) An observed image, (b) the clean image, and (c) the additive noise. An example of flat patch is marked in the deep-color box, while the light color box contains a sharp edge.

where $G(u|\mu, \sigma^2)$ denotes a Gaussian distribution with the mean μ and the variance σ^2. By Eq. (6.1), we obtain

$$\bar{x}_t \sim G\left(\bar{x}_t|\hat{x}_t, \frac{\sigma_e^2}{d^2}\right), \tag{6.2}$$

which indicates that the reconstructed pixel $\bar{x}_t$ is a rough approximation of the original mean $\hat{x}_t$, whose accuracy heavily depends on the size $d \times d$ of $\mathcal{N}(x_t)$. A big d seemly increases the accuracy, but requires the property of local flat in a large neighbor. Usually, it is difficult to choose a suitable d.

Assume that the noise variance σ_e^2 is known and also $\hat{x}_t$ comes from a Gaussian with the mean $\bar{x}_t$ and the variance σ_x^2, the Wiener filter further improves the mean filter by

$$\bar{x}_{t|\text{Wiener}} = \bar{x}_t + \frac{\sigma_x^2}{\sigma_x^2 + \sigma_e^2}(x_t - \bar{x}_t). \tag{6.3}$$

It involves $\bar{x}_t$ and thus still faces the difficulty of choosing a suitable d. Also, it is not easy to estimate σ_x^2.

Moreover, the assumption of piecewise constant does not hold anywhere on the images, such as the sharp edges shown in the blue box in Figure 6.1. Consequently, removing noise may blur sharp edges and destroy other subtle image information. Actually, all the spatial filtering methods suffer this same difficulty and limitation.

6.1.2 TRANSFORM-DOMAIN FILTERING

Also, numerous algorithms have been developed to separate noises in the transform domains, such as low-pass filter, Fourier Wiener filter, windowed Fourier Wiener filter, wavelet transform-based filter (WT), discrete cosine transform-based filter (DCT), etc., usually with a lower time complexity compared with spatial filtering methods.

Assume that images are piece smooth, a low-pass filter treats the high-frequency components in Fourier spectral as noise. However, as demonstrated in Figure 6.1, the sharp edges contained in the blue box represent the high-frequency component in the Fourier domain, and thus discarding high-frequency components will blur the edges. Moreover, a practical implementation of low-pass filtering methods introduces a ringing effect into the reconstructed images.

Considering that the property of image can be piecewisely described, DCT divides a noisy image X into a local $d \times d$ window x_t, transforms x_t into a linear combination of d^2 frequency squares, and cuts off those with the corresponding coefficients below a small threshold. Widely used in JPEG compression, DCT has achieved promising denoising performance. However, it is difficult to partition noisy images into local windows with suitable size, which usually causes artifacts. Similar to a low-pass filter, DCT also fails if the local window contains a high-frequency pattern.

Spatial filtering and transform-domain filtering are linked via rewriting the spatial filtering into the following convolution

$$\bar{x} = X * K, \tag{6.4}$$

where K denotes the denoised kernel, for example, $K = \mathbf{I} \in \mathbb{R}^{d \times d}$ for the mean filter. By the convolution theorem, the counterpart of Eq. (6.4) in the Fourier transform is given as follows

$$\mathcal{F}(\bar{x}) = \mathcal{F}(X) \times \mathcal{F}(K), \tag{6.5}$$

where $\mathcal{F}(X)$ and $\mathcal{F}(K)$ denote the Fourier transform of the noisy image X and kernel K, respectively. Both spatial filtering and transform-domain filtering assume images are piecewise constant and eliminate noise at the cost of blurring subtle feature structures in noisy images.

6.1.3 NONLOCAL MEANS METHODS

Instead of assuming images to be piecewise constant, nonlocal means methods observe that similar feature information is acquired repeatedly in images, for example, the sharp edge contained in the deep color box in Figure 6.2 appears elsewhere in this image, as marked in the light color box in Figure 6.2. Assume that a noisy patch x_t, as marked in the green box in Figure 6.2, is generated as follows

$$x_t = \hat{x}_t + e_t, \tag{6.6}$$

where $\hat{x}_t \in \mathbb{R}^{d \times d}$ denotes the clean patch and $e_t \in \mathbb{R}^{d \times d}$ denotes random noise with each element being independent of each other and also with $\hat{x}_t$. Given enough patches that contain the same feature information, a simple average of the searched patches will remove Gaussian noise and give a plausible estimation $\bar{x}_t$ of the clean patch $\hat{x}_t$. As shown in Figure 6.2(d), nonlocal means methods can preserve the sharp edge perfectly, compared with traditional spatial filtering and transform-domain filtering. Following this idea, several algorithms [3–7] have been proposed in the last 10 years, with promising performances.

One representative method is known as K-SVD [3,4]. By its nature, it could be implemented by iterating the step of updating the dictionary of representative patches

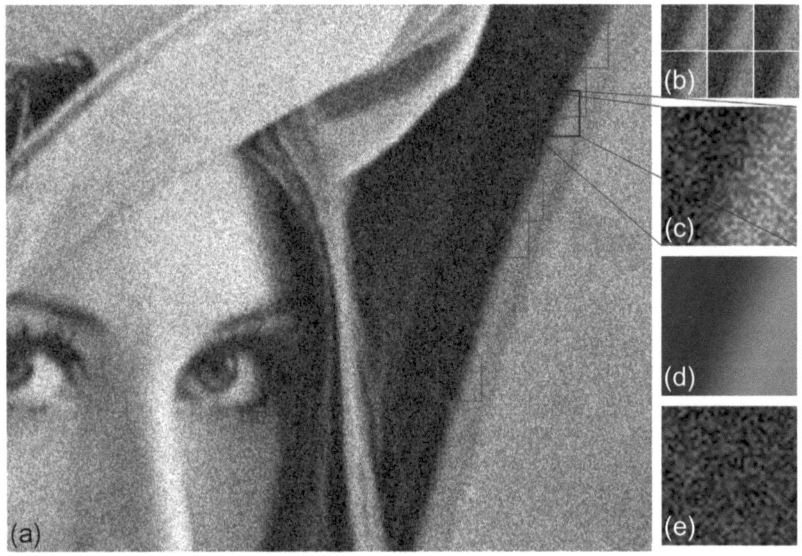

FIGURE 6.2

The motivation of nonlocal means methods. (a) A noisy image, where the predenoised patch is marked in a deep color box, (b) six representative patches similar to the deep color one, whose positions are marked in the light color boxes in (a), while (c), (d), and (e) denote the noisy patch, and the estimated clean patch, and the estimated noisy patch, respectively. As demonstrated in (d), the nonlocal means method recovers the sharp edge contained in the deep color box.

or features and the step of denoising a corrupted image based on the dictionary, such that a subset of patches is picked from the dictionary to form a sparsely weighted linear sum as a reconstruction of a corrupted patch subject to a given noise variance. Practically, to avoid high computing cost and also to improve denoising performance, K-SVD gets the dictionary obtained previously from a corpus of high-quality image database and usually with human interactive help. Then, denoising of the present corrupted image is based on this dictionary.

One other representative method is known as BM3D [5]. Those patches that match well with a reference patch under consideration are picked out and stacked to form a 3D array that is mapped into a transform domain where a collaborative filtering of the group is made by shrinkage. Then, the filtered 3D array is inversely transformed back to the estimates of those 2D patches. After processing all reference patches, the obtained estimates can overlap and then aggregate to form a final estimate of the true image.

Both K-SVD and BM3D need to know the noise variance in advance, which takes a critical role in not only getting the aggregating weights but also denoising either directly for K-SVD or via grouping and collaborative filtering for BM3D. Moreover, there are also some parameters that are heuristically chosen in advance for both K-SVD and BM3D. Furthermore, K-SVD utilizes a pretrained dictionary while BM3D uses a general DCT basis, without considering the issue of controlling the dictionary complexity. Both the methods are suitable only to a limited range of images. Tacking these limitations, we propose a new denoising method called LFA-BYY.

6.2 LFA-BYY DENOISING METHOD

Instead of considering each item in the dictionary as an image patch or feature template as K-SVD and BM3D, LFA-BYY treats it as a parametric model that represents a family of image patches or a set of features. The dictionary consists of a set of parametric models that can be provided in one of three ways:

1. Learned from the present image under processing.
2. Obtained previously by the same method and other methods as well, including human help.
3. Obtained by updating the available dictionary on the present image.

In this chapter, we focus on the first way, though the study can be further generalized to the other ways.

To facilitate mathematical formulation, we stack each $d \times d$ image patch into a d^2 dimension vector. That is, we consider Eq. (6.6) in $\mathbb{R}^{d^2}$ instead of $\mathbb{R}^{d\times d}$. Each parametric model uses a factor analysis (FA) model featured by a $d^2 \times m_i$ matrix A_i with its columns spanning a subspace that locates at $\mu_i \in \mathbb{R}^{d^2}$. Each sample vector x_t comes from a mixture of FA models at a number of locations, also called local factor analysis (LFA) [8]. Specifically, we consider that x_t comes from

$$\begin{aligned} q(x_t|\theta) &= \sum_i \alpha_i G(x_t|\mu_i, \Sigma_i),\ \theta = \{\alpha_i, \mu_i, A_i, \sigma_i^2, \Lambda_i\}_{i=1}^k, \\ G(x_t|\mu_i, \Sigma_i) &= \int G(x_t|A_i y + \mu_i, \sigma_i^2 I) G(y|0, \Lambda_i) \mathrm{d}y, \\ &\text{subject to } A_i^T A_i = I, \end{aligned} \tag{6.7}$$

where $G(u|\mu, \Sigma)$ denotes a Gaussian with the mean vector μ and the covariance matrix Σ. Equivalently, we extend Eq. (6.6) into

$$\begin{aligned} &x_t \text{ comes from } x_{i,t} \text{ with a probability } \alpha_i \text{ for } i = 1, \ldots, k, \\ &x_{i,t} = \underbrace{A_i y_{i,t} + \mu_i}_{\hat{x}_{i,t}} + e_{i,t}, \text{ subject to } A_i^T A_i = I, \\ &y_{i,t} \sim G(y_{i,t}|0, \Lambda_i),\ e_{i,t} \sim G(x_t|Ay + \mu_i, \sigma_i^2 I). \end{aligned} \tag{6.8}$$

Turning the image under processing into a set $\mathcal{X} = \{x_t\}$ of samples, we estimate the parameter set θ and determine a set of integers $\mathcal{K} = \{k, \{m_i\}_{i=1}^k\}$ by the Bayesian Ying-Yang (BYY) harmony learning, which was first proposed in 1995 [9] and systematically developed over the past two decades [10,11].

Based on the obtained θ, each x_t is transformed into its cleaned counterpart $\hat{x}_t$ as follows

$$\hat{x}_t = A_j y_{j,t} + \mu_j, \quad j = \arg\max_i p(i|x_t),$$

$$y_{j,t} = \arg\max_y [G(x_t|A_j y + \mu_j, \sigma_j^2 I) G(y|0, \Lambda_j)], \tag{6.9}$$

where $p(i|x_t)$ is already computed during learning, representing the following posteriori probability

$$p(i|x_t) = \frac{\alpha_i G(x_t|\mu_i, \Sigma_i)}{\sum_j \alpha_j G(x_t|\mu_j, \Sigma_j)}. \tag{6.10}$$

Quoting the roles taken by the LFA model and the BYY harmony learning, this denoising method is thus known by the abbreviation LFA-BYY.

Equation (6.9) implements a concept that relates to the grouping and collaborative filtering in BM3D [5]. However, the method for implementing the concept is different. First, one group of patches is actually one class of the patches represented by its corresponding FA by Eq. (6.8). All the patches are classified into k-classes based on $p(i|x_t)$ that describes the fitness of the ith FA model to the patch x_t, while BM3D searches the similar patches by Euclidean distance and groups them based on a prespecified threshold. Second, making collaborative filtering by BM3D involves heuristic settings with human help, which is no longer necessary in Eq. (6.9). The component μ_j is actually a counterpart of the result by a low-pass filtering along the stacking direction by BM3D. Since the patches in a group or class may come from different locations of the image, it is reasonable to regard them as a sequence of identically and independently distributed (i.i.d.) elements and thus only consider their mean value with other cross element relations ignored. Put together, the linear transform $x_t \to y_{j,t} \to \hat{x}_t$ is a counterpart of the collaborative filtering (i.e., 2D patches $\to$ 3D array $\to$ filtered 3D array $\to$ 2D patches). Lastly, all the patches are classified exclusively into k-classes without overlap, thus there is no need of an aggregation like that used by BM3D to smooth out inconsistency, which may cause artifacts as well.

Also, the above LFA extends the K-SVD [3,4]. In the degenerated case that there is only one FA $x_t = Ay_t + e_t$, the matrix A is the counterpart of the dictionary used in the K-SVD, while y_t is the counterpart of the sparse weights by the K-SVD. Being different, the LFA model corresponds to an organized hierarchical dictionary, with not only subsets $\{A_j\}_{j=1}^k$ in a lower layer but also $\{\mu_j\}_{j=1}^k$ in an upper layer, such that the above addressed grouping and collaborative filtering can be also performed. Moreover, the K-SVD considers weights as sparse coding that is approximately obtained by some pursuit algorithm, while the LFA considers the weights separately in each FA by $y_{j,t}$ coded by independent Gaussian distributions. Furthermore, K-SVD also uses a dictionary that is collected from a large number of clean images, while the LFA model utilizes an adaptive dictionary learned from the present noisy image.

Even critically, both K-SVD and BM3D require that the noise variance is preestimated because they have not taken the issue of controlling dictionary complexity into consideration, while estimating the noise variance is closely associated with appropriately controlling the complexity of the dictionary. A high complexity leads to an over-fitting problem and formulates noise as feature information, thus creating artifacts. In contrast, a lower complexity introduces an under-fitting problem and treats the feature information as noise, thus smoothing out the subtle feature information. The dictionary complexity is featured by $\mathcal{K}$ for the local FA model by Eqs. (6.7) and (6.8). With k and each m_i appropriately determined, we can get the noise variance σ_i^2 estimated appropriately. Therefore, we may break the limitation K-SVD and BM3D, that is, we not only need to know the noise variance but also its applicability to the heterogeneous noises, namely, different groups of image patches may be affected by noises with different variances.

Therefore, how to determine one appropriate $\mathcal{K}$ effectively is a key problem for learning LFA. In the next section, we will see that the BYY harmony learning provides a favorable solution to this problem, and also the LFA model is actually different from those existing models called mixture of factor analyzers (MFA) [12,13], because each FA model is different from the traditional FA for facilitating to determine $\mathcal{K}$ effectively.

6.3 BYY HARMONY LEARNING ALGORITHM FOR LFA

The task of determining one appropriate $\mathcal{K}$ is called model selection. A conventional model selection approach is featured by a two-stage implementation. The first stage is called parameter learning for estimating all the unknown parameters θ for every value of $\mathcal{K}$ in a prespecified set that enumerates all the candidate models under consideration. The second stage selects the best candidate by a model selection criterion, for example, Akaike's Information Criterion [14], Bayesian Information Criterion [15], Minimum Message Length [16], etc. However, a larger k and m_i imply more unknown parameters, which makes parameter estimation become less reliable such that the criterion evaluation reduces its accuracy, see Section 2.1 in [11] for a detailed discussion. Moreover, such two-stage procedure suffers a huge-computing cost.

Instead of two-stage implementation, automatic model selection is featured by determining one appropriate $\mathcal{K}$ during parameter learning on θ, with $\mathcal{K}$ starting at a value large enough and then gradually shrinking. An early effort is rival penalized competitive learning (RPCL) [17,18], featured by the cluster number automatically determined during learning. Two types of Bayesian-related approaches can perform automatic model selection. One is the BYY harmony learning and the other is variational Bayesian (VB) [19,20]. The VB introduces a function to approximate the marginal likelihood by Jensen's inequality and employs an EM-like algorithm to optimize the lower bound. The model selection of VB is realized by incorporating appropriate prior distributions of unknown parameters. As empirically demonstrated by [21,22], BYY is capable of selecting suitable model complexity automatically even without imposing any priors on the parameters, and outperforms VB with Dirichlet-Normal-Gamma prior distribution [23].

The LFA model by Eqs. (6.7) and (6.8) has already taken the issue of facilitating automatic model selection into consideration, where the FA model is actually different from the traditional FA. To avoid confusion, the traditional one is named FA-a, while each FA in Eqs. (6.7) and (6.8) is named FA-b. The orthonormal constraint of $A_i^T A_i = I$ is removed by FA-a that imposes $\Lambda_i = I$. In the studies of BYY harmony learning, FA-b was preferred as discussed in Item 9.4 in [24] and Section 3 in [25]. Two FA types are equivalent in terms of maximizing the likelihood, but behave very differently when selecting model complexity is taken into consideration, with further details referred to Section 3.2.1 in [26] for a recent overview. Extensive empirical experiments in [27] have shown that the BYY harmony learning and VB perform reliably and robustly better on FA-b than on FA-a, while BYY outperforms VB considerably, especially on FA-b. Moreover, a mixture of FA-a models is called MFA [12,13], while a mixture of FA-b models, namely the one given by Eqs. (6.7) and (6.8), is called LFA [8], also see Figures 3 and 9 in [11]. Again, empirical experiments in [28] have confirmed that learning LFA by BYY outperforms learning MFA by BYY, learning LFA by VB, and learning MFA by VB [13].

The BYY harmony learning was first proposed in [9] and systematically developed in the past two decades. BYY harmony learning on typical structures leads to new model selection criteria, new techniques for implementing learning regularization, and a class of algorithms that implement automatic model selection during parameter learning. Also, BYY harmony learning offers a theoretical explanation of RPCL. Further details and the latest systematical introduction about BYY harmony learning can be found in [10,11].

Here, we directly adopt Algorithm 5: "BYY learning for local factor analysis" given in [10], which is rewritten as Algorithm 1 in this chapter. It implements the maximization of the harmony measure in a BYY system. On the LFA model by Eqs. (6.8) and (6.7), the harmony measure is formulated as follows

$$
\begin{aligned}
&\mathrm{H}(p\|q) = \sum_{t=1}^{N}\sum_{i=1}^{k}\int p(i|x_t)p(y|i,x_t)\ln\left[\alpha_i G(x_t|A_i y + \mu_i, \sigma_i^2 I)G(y|0,\Lambda_i)\right]\mathrm{d}y, \\
&\text{s.t.:} A_i^T A_i = I, i = 1,\ldots,k, \qquad (6.11) \\
&\text{s.t.:}\mathrm{KL}\left(p(i|x_t)p(y|i,x_t)\,\middle\|\,\frac{\alpha_i G(x_t|A_i y + \mu_i, \sigma_i^2 I)G(y|0,\Lambda_i)}{q(x_t|\theta)}\right) = 0,
\end{aligned}
$$

which is maximized to determine not only $\theta, p(i|x_t)$ and $p(y|i,x_t)$, but also $\mathcal{K}$. With the help of the Lagrange method, an iterative procedure is used to implement this maximization, from which we are led to Algorithm 1.

ALGORITHM 1 BYY LEARNING FOR LOCAL FACTOR ANALYSIS

Initialize $\theta = \{\alpha_i, A_i, \Lambda_i, \sigma_i^2, \mu_i\}_{i=1}^k$, and η is controlled as described in Section 2.3 in [10].
Repeat the following two steps **until** converged.

***Yang* Step**: We get:

$p_{i,t} = p(i|x_t, \theta^{\text{old}})$ by Eq. (6.13);

$\Gamma_{i,y|x}^{\text{new}} = \frac{\eta}{1+\eta}\sigma_i^{2\text{ old}}(\mathbf{I} + \sigma_i^{2\text{ old}}\Lambda_i^{\text{old}-1})^{-1}$;

$W_i^{\text{new}} = \Gamma_{i,y|x}^{\text{old}} A_i^{\text{old}\,T} \sigma_i^{\text{old}\,-2}$.

***Ying* Step:** *get* $\alpha_i^{\text{new}}, \mu_i^{\text{new}}, A_i^{\text{new}}, \sigma_i^{2\text{ new}}, \nu_i^{\text{new}}, \Lambda_i^{\text{new}}$ by:

$\alpha_i^{\text{new}} = \frac{1}{N}\sum_{t=1}^N p_{i,t},\ \mu_i^{\text{new}} = \frac{1}{N\alpha_i^{\text{new}}}\sum_{t=1}^N p_{i,t}x_t$;

$y_{t,i} = W_i^{\text{new}}(x_t - \mu_i),\ e_{t,i} = x_t - \mu_i^{\text{new}} - A_i^{\text{old}} y_{t,i}$;

$\sigma_i^{2\text{ new}} = \frac{1}{d^2} Tr\left[A_i^{\text{old}}\Gamma_{i,y|x}^{\text{new}}A_i^{\text{old;T}} + \frac{1}{N\alpha_i^{\text{new}}}\sum_{t=1}^N p_{i,t}e_{t,i}e_{t,i}^T\right]$;

$\Lambda_i^{\text{new}} = \text{diag}\left(\Gamma_{i,y|x}^{\text{new}} + \frac{1}{N\alpha_i^{\text{new}}}\sum_{t=1}^N p_{i,t}y_{t,i}y_{t,i}^T\right)$;

$A_i^{\text{new}} = G_S\left[\sum_{t=1}^N p_{i,t}(x_t - \mu_i^{\text{new}} y_{t,i}^T)\right]\left[\sum_{t=1}^N p_{i,t}(y_{t,i}y_{t,i}^T + \Gamma_{i,y|x}^{\text{new}})\right]^{-1}$.

where $G_S[\phi]$ denotes a Gram-Schmidt operator that orthogonalizes ϕ, i.e., $\phi\phi^T = I$.
TRIMMING:

if one $\lambda_{i,i}^{\text{new}}$ of $\Lambda_i = \text{diag}[\lambda_{i,1}, \ldots, \lambda_{i,m_i}]$ tends to 0, discard the ith column of A_i, let $m_i = m_i - 1$; if $\alpha_i^{\text{new}} \to 0$ or $\alpha_i^{\text{new}}\sigma_i^{2\text{ new}} \to 0$, discard $G(x|A_i y + \mu_i, \sigma_i^2 I)$ and $G(y|0, \Lambda_i)$, let $k = k - 1$.

The above $H(p\|q)$ is a specific derivation of the following general form

$$H(p||q) = \int p(R|X)p(X)\ln[q(X|R)q(R)]\mathrm{d}X\mathrm{d}R, \tag{6.12}$$

where R consists of not only θ, y, i explicitly but also $\mathcal{K}$ implicitly. Maximizing $H(p||q)$ forces $q(X|R)q(R)$ (called the Ying structure) to match $p(R|X)p(X)$ (called the Yang structure). There are always certain structural constraints imposed on the Ying-Yang structures with an additional one coming from $p(X)$ to accommodate a finite size of samples, because a perfect equality $q(X|R)q(R) = p(R|X)p(X)$ may not be really reached but still be approached as close as possible. At this equality, $H(p||q)$ becomes the negative entropy that describes the complexity of the BYY system. Further maximizing it will decrease the system complexity and thus provides an ability for determining an appropriate $\mathcal{K}$.

Observing Algorithm 1, such a model selection ability is reflected mainly in its Ying step, namely

$$p(i|x_t, \theta) = \frac{[\alpha_i G(x_t|\mu_i, \Sigma_i)]^{\frac{1+\eta}{\eta}}}{\sum_{i=1}^k [\alpha_i G(x_t|\mu_i, \Sigma_i)]^{\frac{1+\eta}{\eta}}},$$

$$p(y|i, x_t) = G\left(y|y^*, \frac{\eta}{1+\eta}\Sigma_{p(y|i,x_t)}\right),$$

$$y^* = (\mathbf{I} + \sigma_i^2 \Lambda_i^{-1})^{-1} A_i^T (x_t - \mu_i),$$

$$\Sigma_{p(y|i,x_t)} = \sigma_i^2 (\mathbf{I} + \sigma_i^2 \Lambda_i^{-1})^{-1}, \tag{6.13}$$

where μ_i, Σ_i are the same as in Eq. (6.7). The Lagrange parameter η reflects an agreement of balance between the Ying and the Yang. Not only the difference between the above $p(i|x_t, \theta)$ and its posteriori probability counterpart by Eq. (6.10) but also the difference between the above $p(y|i, x_t)$ and its posteriori probability counterpart $G(y|y^*, \Sigma_{p(y|i,x_t)})$ are featured by η, which makes $p(i|x_t, \theta)$ and $p(y|i, x_t)$ become more selective for automatic model selection on $\mathcal{K}$. As a result, the sparsity for each observable sample x_t is realized by $\hat{x}_t$ via the linear transform $x_t \to y_{j,t} \to \hat{x}_t$ by Eq. (6.9).

When $\eta = \infty$, the Yang step will degenerate into the E-step of the classical expectation-maximization (EM) algorithm. With $p(i|x_t), p(y|i, x_t)$ replaced by their posteriori probability counterparts, or letting $\eta = \infty$, the Ying step of Algorithm 1 also becomes the same as the same M-step with the classical EM algorithm. Further discussions can be found in Section 2.3 in [10].

6.4 EXPERIMENTS AND DISCUSSION

6.4.1 COMPARATIVE RESULTS WITH COMPETING ALGORITHMS

To make a systematical assessment, we compare LFA-BYY with four state-of-the-art image denoising algorithms: BM3D [5], K-SVD for color image denoising [29], expected patch log likelihood (EPLL) [6], and multispectral image denoising (Msi) [30], on the benchmark Kodak image processing dataset that contains 24 natural images. Each image is polluted by Gaussian noise with 10 different intensities from $\sigma = 10$ to $\sigma = 100$.

We perform each of the competitive algorithms with the code provided with its published paper and websites as follows:

- BM3D (http://www.cs.tut.fi/~foi/GCF-BM3D/);
- K-SVD (http://www.ipol.im/pub/art/2012/llm-ksvd/);
- EPLL (http://people.csail.mit.edu/danielzoran/); and
- Msi (www.cs.cmu.edu/yiyang/Publications.html).

All the free parameters of these approaches are set as suggested in the original papers. We also provide the exact noise intensity of each polluted image as input to all the competing algorithms but not to LFA-BYY as it is able to estimate the noise intensity of a polluted image during the denoising procedure. The patch size of our algorithm is set to be 8×8.

The performance of each algorithm is evaluated by comparing the similarity between the original clean images and denoised ones with the measurements of peak

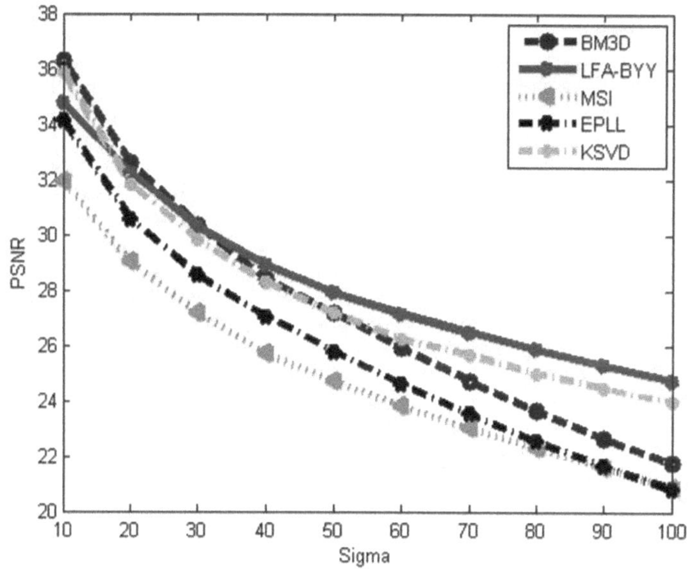

FIGURE 6.3

Average PSNR value of 24 natural images over different noise intensities.

signal-to-noise ratio (PSNR), structure similarity (SSIM) [31], and feature similarity (FSIM) [32]. The larger these measurements are, the more similar a denoised image is with the original.

Figures 6.3, 6.4, and 6.5 illustrate the experimental results. It can be observed that LFA-BYY consistently produces promising experimental results. Especially, on images with large noise intensity ($\sigma > 20$), LFA-BYY consistently produces the most robust and superior performances over the competing methods. Figure 6.6 demonstrates results on an image with noise intensity $\sigma = 100$, from which we notice that LFA-BYY can better preserve the detailed features of images polluted with large noise. The experiments echo that LFA-BYY can appropriately control the complexity of the LFA model (i.e., the dictionary) and learn the noise intensity per image under processing. In comparison, other methods either utilize a pre-trained dictionary (K-SVD, EPLL, Msi) or actually a general DCT basis (BM3D), without considering the issue of controlling the dictionary complexity. This is the reason why they all not only need to know the noise intensities in advance but also under-perform LFA-BYY, especially when the patches are polluted by large noises. In such cases, LFA-BYY will accordingly learn an LFA model with a reduced complexity to ignore unreliable details, while the competing methods still use the same dictionaries.

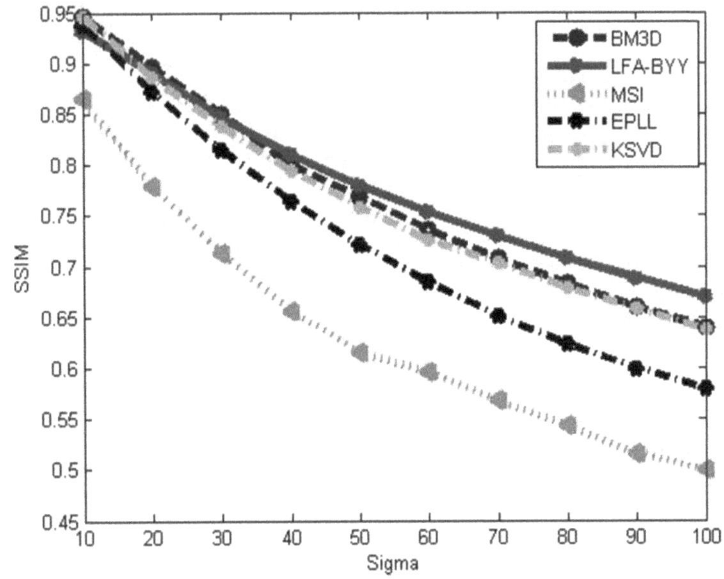

FIGURE 6.4

Average SSIM value of 24 natural images over different noise intensities.

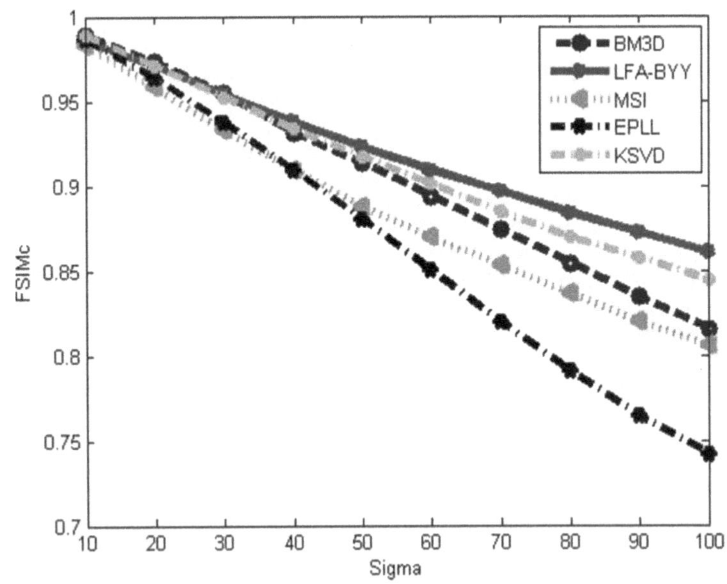

FIGURE 6.5

Average FSIM value of 24 natural images over different noise intensities.

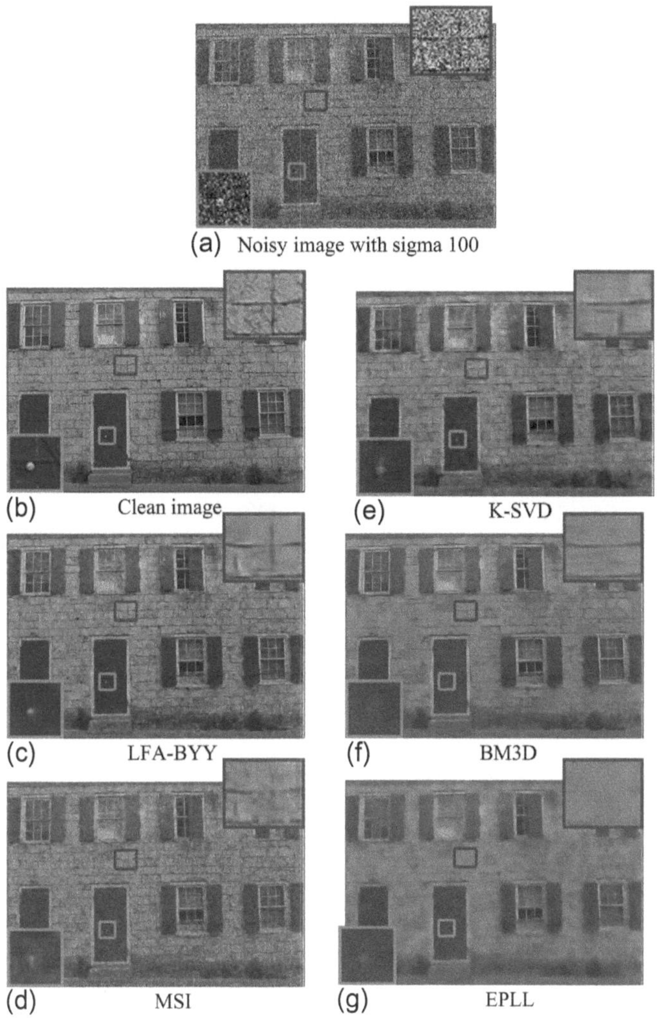

FIGURE 6.6

Results on an image with a real noise intensity $\sigma = 100$.

6.4.2 FAVORABLE FEATURES OF LFA-BYY

6.4.2.1 Robustness to different image datasets

LFA-BYY learns the dictionary of features from a noisy image adaptively per image under processing and thus the performances remain robust to different image datasets. Also, other algorithms use either a pretrained dictionary or a general basis, and thus are only suitable to a limited range of images. For example, as BM3D utilizes a DCT basis to describe features, it is more suitable for processing

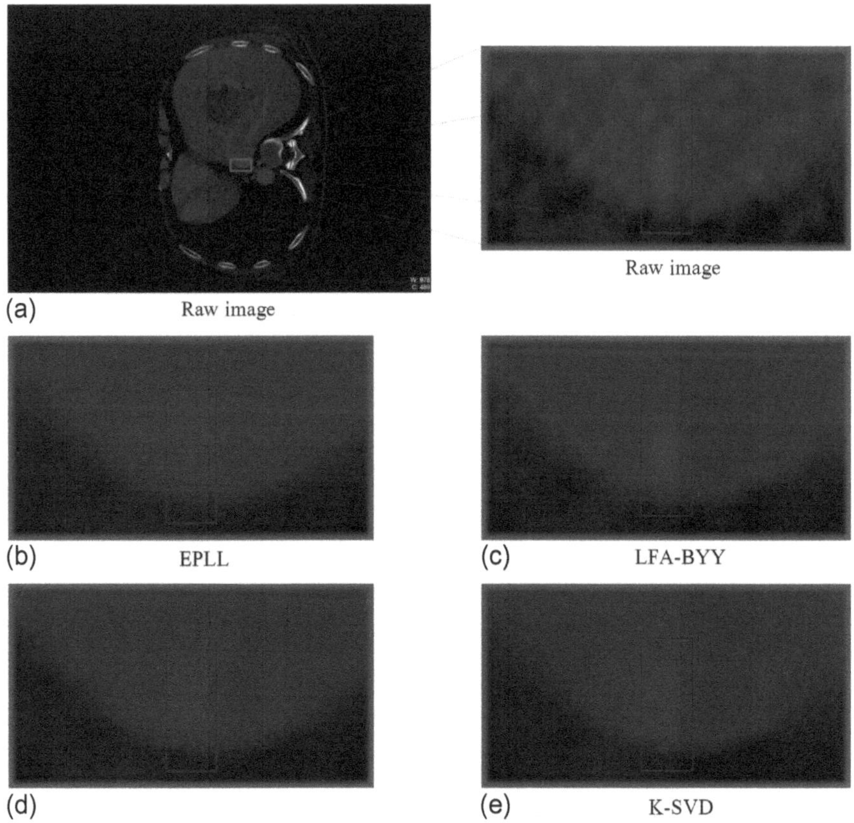

FIGURE 6.7

Comparative result on practical medical images. The structure shown in the box has some clinical meaning, but the noise around this structure may disturb the judgment of doctors. LFA-BYY enhances the clinical structure and removes the surrounding noise concurrently, while other algorithms remove the noise at the cost of smoothing detailed structures.

natural images. Figure 6.7 illustrates comparative results of denoising medical images. Significant detailed feature is smoothed by other algorithms while preserved by LFA-BYY.

6.4.2.2 Robustness to unknown noise intensity

LFA-BYY is not only able to learn the noise intensity per image under processing applicable to the heterogeneous noises on one image. In addition, another competing method needs a two-stage procedure, namely noise intensity estimation and image denoising with the estimated noise intensity. Not only can this procedure be highly

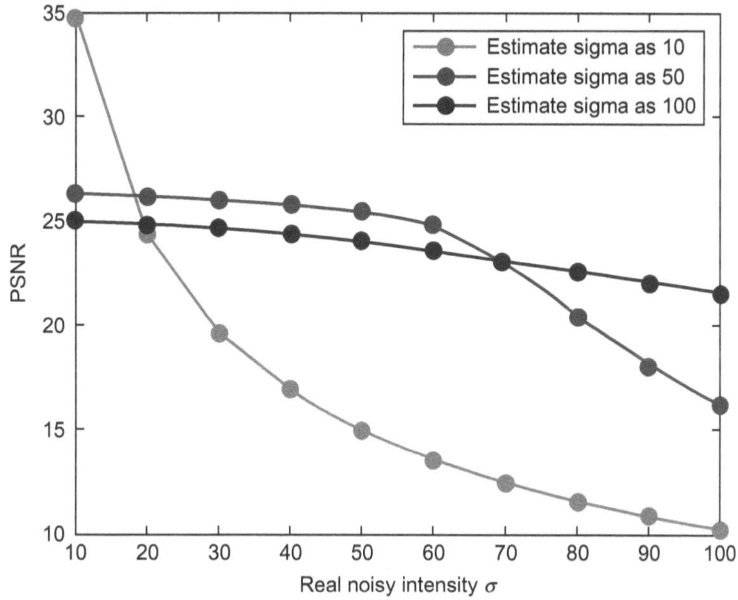

FIGURE 6.8

The performance of BM3D under different noise intensities where the estimated noise intensity is supposed to be 10, 20, and 100, respectively.

time-consuming but also noise intensity estimation is a difficult task while a poor estimation can seriously affect their performances.

Figure 6.8 demonstrates the sensitivity of BM3D to a noise intensity estimation. With a poor estimated noise intensity, the denoising result can be highly affected. When the estimated noise intensity is higher than the real one, the detailed features of an image will be considered as noise, resulting in an over-smoothed image as shown in Figures 6.9(c) and 6.10(c). On the other hand, when the estimated noise intensity is lower than the real one, the noise cannot be eliminated efficiently as shown in Figure 6.11. Moreover, the results shown in Figure 6.10 coincide with the results given in Figures 6.3, 6.4, and 6.5, which state that the LFA-BYY algorithm can preserve much more detail than BM3D when real σ is large.

6.4.2.3 No free parameters

There are usually many heuristically picked parameters in the existing denoising algorithms, and the performance of these algorithms can be highly affected by the inappropriate setting of these parameters. In contrast, LFA-BYY does not contain such parameters. All the knowledge comes from the image under processing, producing consistent performances on different images.

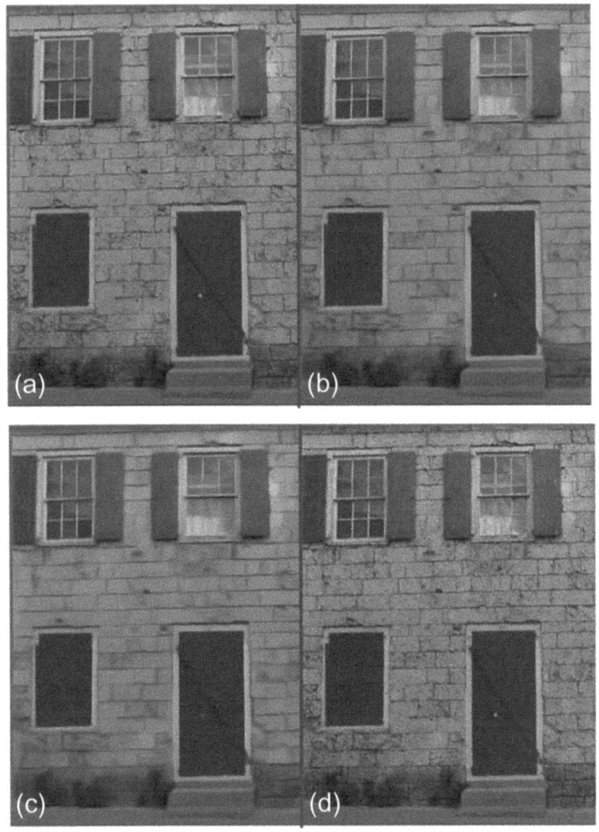

FIGURE 6.9

The denoising performances on one image with the true noise intensity $\sigma = 10$. (a) The denoised image by BM3D provided with the estimated intensity $\sigma = 10$; (b) the denoised image by BM3D provided with the estimated intensity $\sigma = 50$; (c) the denoised image by BM3D provided with the estimated intensity $\sigma = 100$; and (d) the denoised image by LFA-BYY with noise intensity determined automatically.

6.5 CONCLUDING REMARKS

The proposed novel image denoising method LFA-BYY learns the LFA model (i.e., the dictionary) per image whilst processing. With the help of the BYY harmony learning, LFA-BYY can appropriately control the dictionary complexity and learn the noise intensity from the present image under processing, while existing state-of-the-art methods have not considered the issues. In comparison with

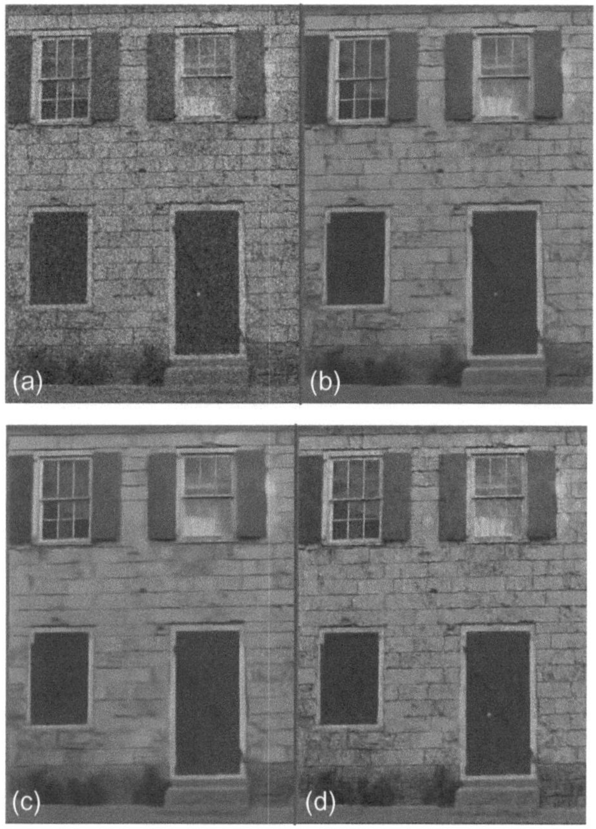

FIGURE 6.10

The denoising performances on one image with the true noise intensity $\sigma = 50$. (a) The denoised image by BM3D provided with the estimated intensity $\sigma = 10$; (b) the denoised image by BM3D provided with the estimated intensity $\sigma = 50$; (c) the denoised image by BM3D provided with the estimated intensity $\sigma = 100$; and (d) the denoised image by LFA-BYY with noise intensity determined automatically.

four methods, BM3D, K-SVD, EPLL, and Msi, on the benchmark Kodak image processing dataset that contains 24 natural images and also additional medical data, experiments have shown that LFA-BYY has not only obtained competitive results on images polluted by a small noise but also outperformed these competing methods when the noise intensity increases beyond a point, especially with significant improvements as the noise intensity becomes large.

The patch size $d \times d$ in this chapter is chosen following the previous nonlocal means methods. This size will influence the performance of image denoising

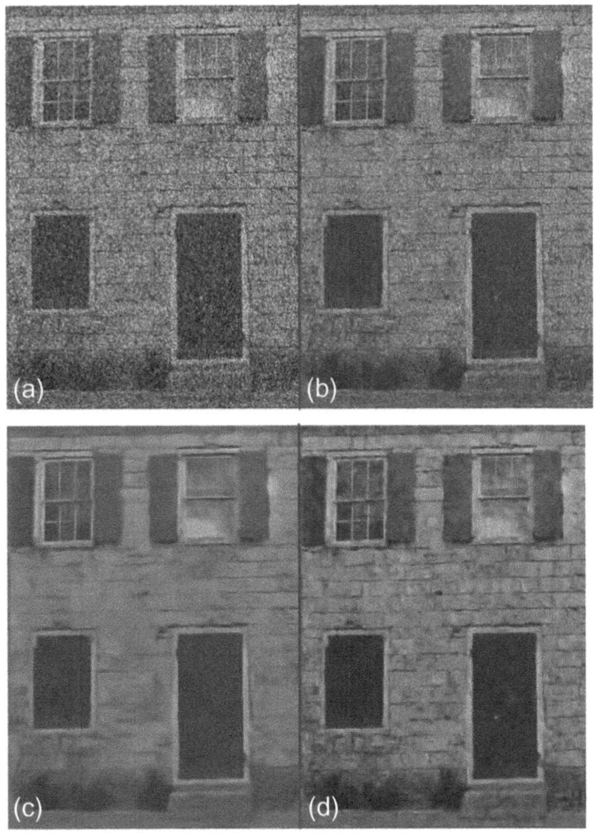

FIGURE 6.11

The denoising performances on one image with the true noise intensity $\sigma = 100$. (a) The denoised image by BM3D provided with the estimated intensity $\sigma = 10$; (b) the denoised image by BM3D provided with the estimated intensity $\sigma = 50$; (c) the denoised image by BM3D provided with the estimated intensity $\sigma = 100$; and (d) the denoised image by LFA-BYY with noise intensity determined automatically.

especially when images are polluted with strong noise. Further investigation for appropriate patch sizes and other possible improvements are left for further work.

REFERENCES

[1] A. Buades, B. Coll, J.-M. Morel, A review of image denoising algorithms, with a new one, Multiscale Model. Simul. 4 (2) (2005) 490-530.

[2] R.C. Gonzalez, R.E. Woods, Digital image processing (2002). Prentice Hall, Upper Saddle River, NJ.

[3] M. Aharon, M. Elad, A. Bruckstein, K-SVD: an algorithm for designing overcomplete dictionaries for sparse representation, IEEE Trans. Signal Process. 54 (11) (2006) 4311-4322.
[4] M. Elad, M. Aharon, Image denoising via sparse and redundant representations over learned dictionaries, IEEE Trans. Image Process. 15 (12) (2006) 3736-3745.
[5] K. Dabov, A. Foi, V. Katkovnik, K. Egiazarian, Image denoising by sparse 3-D transform-domain collaborative filtering, IEEE Trans. Image Process. 16 (8) (2007) 2080-2095.
[6] D. Zoran, Y. Weiss, From learning models of natural image patches to whole image restoration, in: IEEE International Conference on Computer Vision (ICCV), 2011, pp. 479-486.
[7] S. Gu, L. Zhang, W. Zuo, X. Feng, Weighted nuclear norm minimization with application to image denoising, in: IEEE Conference on Computer Vision and Pattern Recognition, 2014.
[8] L. Xu, BYY harmony neural networks, structural RPCL, and topological self-organizing on mixture models, Neural Netw. 15 (2002) 1125-1151.
[9] L. Xu, Bayesian-Kullback coupled Ying-Yang machines: unified learnings and new results on vector quantization, in: Proceedings of International Conference on Neural Information Processing, 1995, pp. 977-988.
[10] L. Xu, Further advances on Bayesian Ying-Yang harmony learning, Appl. Inform. (in press).
[11] L. Xu, Bayesian Ying-Yang system, best harmony learning, and five action circling, Front. Electr. Electron. Eng. China 5 (3) (2010) 281-328.
[12] G.E. Hinton, P. Dayan, M. Revow, Modeling the manifolds of images of handwritten digits, IEEE Trans. Neural Netw. 8 (1) (1997) 65-74.
[13] Z. Ghahramani, M.J. Beal, Variational inference for Bayesian mixtures of factor analysers, in: NIPS, 1999, pp. 449-455.
[14] H. Akaike, A new look at the statistical model identification, IEEE Trans. Autom. Control 19 (6) (1974) 716-723.
[15] G. Schwarz, Estimating the dimension of a model, Ann. Stat. 6 (2) (1978) 461-464.
[16] C.S. Wallace, D.M. Boulton, An information measure for classification, Comput. J. 11 (2) (1968) 185-194.
[17] L. Xu, A. Krzyzak, E. Oja, Unsupervised and supervised classifications by rival penalized competitive learning, in: Proceedings. 11th IAPR International Conference on Pattern Recognition. Vol. II. Conference B: Pattern Recognition Methodology and Systems, IEEE, 1992, pp. 496-499.
[18] L. Xu, A. Krzyzak, E. Oja, Rival penalized competitive learning for clustering analysis, RBF net, and curve detection, IEEE Trans. Neural Netw. 4 (4) (1993) 636-649.
[19] R.M. Neal, G.E. Hinton, A view of the EM algorithm that justifies incremental, sparse, and other variants, in: Learning in Graphical Models, Springer, 1998, pp. 355-368.
[20] H. Attias, A variational Bayesian framework for graphical models, Adv. Neural Inform. Process. Syst. 12 (1-2) (2000) 209-215.
[21] L. Shi, S. Tu, L. Xu, Learning Gaussian mixture with automatic model selection: a comparative study on three Bayesian related approaches, Front. Electr. Electron. Eng. China 6 (2) (2011) 215-244.
[22] G. Chen, H. Pheng, H. Ann, L. Xu, Projection embedded BYY learning algorithm for Gaussian mixture based clustering, Appl. Inform. 1 (2014), 2.

[23] A. Corduneanu, C.M. Bishop, Variational Bayesian model selection for mixture distributions, in: Artificial Intelligence and Statistics, vol. 2001, Morgan Kaufmann, Waltham, MA, 2001, pp. 27-34.
[24] L. Xu, Bayesian Ying-Yang system and theory as a unified statistical learning approach: (i) unsupervised and semi-unsupervised learning, in: Brain-Like Computing and Intelligent Information Systems, Springer-Verlag, Heidelberg, 1997, 241-274.
[25] L. Xu, Bayesian Ying-Yang system and theory as a unified statistical learning approach: (iii) models and algorithms for dependence reduction, data dimension reduction, ICA and supervised learning, in: Lecture Notes in Computer Science: Proc. of International Workshop on Theoretical Aspects of Neural Computation, 1997, pp. 43-60.
[26] L. Xu, On essential topics of BYY harmony learning: current status, challenging issues, and gene analysis applications, Front. Electr. Electron. Eng. China 7 (2012) 147-196.
[27] S.K. Tu, L. Xu, Parameterizations make different model selections: empirical findings from factor analysis, Front. Electr. Electron. Eng. China 6 (2011) 256-274 (a special issue on Machine Learning and Intelligence Science: IScIDE2010 (B)).
[28] L. Shi, Z.-Y. Liu, S. Tu, L. Xu, Learning local factor analysis versus mixture of factor analyzers with automatic model selection, Neurocomputing 139 (2014) 3-14.
[29] J. Mairal, M. Elad, G. Sapiro, Sparse representation for color image restoration, IEEE Trans. Image Process. 17 (1) (2008) 53-69.
[30] Y. Peng, D. Meng, Z. Xu, C. Gao, Y. Yang, B. Zhang, Decomposable nonlocal tensor dictionary learning for multispectral image denoising, in: Proceedings of the IEEE Conference on Computer Vision and Pattern Recognition, 2013, pp. 2949-2956.
[31] Z. Wang, A.C. Bovik, H.R. Sheikh, E.P. Simoncelli, Image quality assessment: from error visibility to structural similarity, IEEE Trans. Image Process. 13 (4) (2004) 600-612.
[32] L. Zhang, D. Zhang, X. Mou, FSIM: a feature similarity index for image quality assessment, IEEE Trans. Image Process. 20 (8) (2011) 2378-2386.

CHAPTER

Unsupervised deep learning: A short review

7

Juha Karhunen[1], Tapani Raiko[1] and KyungHyun Cho[2]
[1]Department of Computer Science, Aalto University, Espoo, Finland, [2]University of Montreal, Montreal, Canada

7.1 INTRODUCTION

In the late 1980s, neural networks became a hot topic in machine learning due to the invention of several efficient learning methods and network structures. These new methods included multilayer perceptron (MLP) networks trained by backpropagation-type algorithms, self-organizing maps, and radial basis function networks [1,2]. While neural networks are successfully used in many applications, interest in their research decreased later on. The emphasis in machine learning research moved to other areas, such as kernel methods and Bayesian graphical models.

Deep learning was introduced by Hinton and Salakhutdinov in 2006 [3]. Deep learning has since become a hot topic in machine learning, leading to a renaissance of neural networks research. This is because when trained properly, deep networks have achieved world-record results in many classification and regression problems.

Deep learning is quite an advanced topic. In this short review, we do not discuss most of their learning algorithms in detail. However, the NADE-k method introduced recently by two of the authors in [4] is discussed in more detail. There exist different types of reviews on deep learning containing more information. An older review is [5], and the doctoral theses [6,7] are good introductions to deep learning. Schmidhuber [8] lists in his recent review over 700 references on deep learning, but the review itself is very short with no formulas. The book [9], in preparation, will probably become a quite popular reference on deep learning, but it is still a draft, with some chapters lacking. In general, research on deep learning is advancing very rapidly, with new ideas and methods introduced all the time.

In the following, we first briefly discuss MLP networks and restricted Boltzmann machines (RBMs) as starting points to deep learning, and move then to various deep networks.

E. Bingham, S. Kaski, J. Laaksonen and J. Lampinen (Eds) Advances in Independent Component Analysis and Learning Machines.
http://dx.doi.org/10.1016/B978-0-12-802806-3.00007-5

7.2 MULTILAYER PERCEPTRON NETWORKS

Figure 7.1 shows an MLP network having an input layer, $L \geq 1$ hidden layers, and the output layer. In general, the numbers of neurons (nodes) in each layer can vary. Usually the processing in the hidden layers is nonlinear, while the output layer can be linear or nonlinear. In the input layer, no computations take place, only the components of the input vector are inputted there, one component in each neuron.

The operation of neuron k in the lth hidden layer is described by the equation

$$h_k^{[l]} = \phi\left(\sum_{j=1}^{m^{[l-1]}} w_{kj}^{[l]} h_j^{[l-1]} + b_k^{[l]}\right), \tag{7.1}$$

where $h_j^{[l-1]}, j = 1, \ldots, m^{[l-1]}$ are the $m^{[l-1]}$ input signals coming to the neuron k, and $w_{kj}^{[l]}, j = 1, \ldots, m^{[l-1]}$ are the respective weights multiplying the input signals. The number of neurons in the lth layer is $m^{[l]}$. The input signals to the MLP network and to its first hidden layer are $x_1, \ldots, x_p$. The constant bias term b_k is added to the weighted sum. The components of the output vector $\mathbf{y}$ are computed similarly as the outputs of the lth hidden layer in Eq. (7.1). The function $\phi(t)$ is the nonlinearity applied to the weighted sum. It is typically chosen to be the hyperbolic tangent $\phi(t) = \tanh(at)$ where a is constant or the logistic sigmoidal function $\phi(t) = 1/(1 + \mathrm{e}^{-at})$. If the operation of a neuron is linear, $\phi(t) = at$ [1,2].

Even though processing in a single neuron is simple, it is nonlinear. These distributed nonlinearities in each neuron of the hidden layers and possibly also in the output layer of the MLP network give to it a high representation power, but on the other hand make its exact mathematical analysis impossible and cause other problems such as local minima in the cost function. However, an MLP network having enough neurons in a single hidden layer can approximate any smooth enough nonlinear input-output mapping [1,2].

With detailed notations the learning algorithms of MLP networks become quite complicated. We do not here go into details but just present an overall view; for details see the books [1,2]. In general, MLP networks are trained in a supervised manner using N known training pairs $\{\mathbf{x}_i, \mathbf{d}_i\}$ where $\mathbf{x}_i$ is the ith input vector and

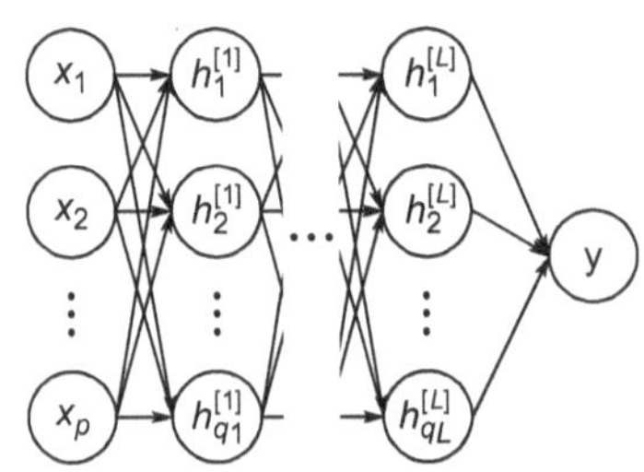

FIGURE 7.1

The architecture of an MLP network with L hidden layers.

$\mathbf{d}_i$ is the corresponding desired response (output). The vector $\mathbf{x}_i$ is inputted to the MLP network and the corresponding output $\mathbf{y}_i$ is vector computed. The criterion used to learn the weights of the MLP network is typically the mean-square error $E = E\{\| \mathbf{d}_i - \mathbf{y}_i \|^2\}$, which is minimized.

The steepest descent learning rule for a weight w_{ji} in any layer is given by

$$\Delta w_{ji} = -\mu \frac{\partial E}{\partial w_{ji}}. \tag{7.2}$$

In practice, the steepest descent is replaced by an instantaneous gradient or a mini batch over 100-1000 training pairs. The required gradients are computed first for the neurons in the output layer using their local errors. These local errors are then propagated backwards to the previous layer, and the weights of its neurons can be updated, and so on. The name backpropagation of the basic learning algorithm for MLP networks comes from this. Usually numerous iterations and sweeps over the training data are required for convergence, especially if an instantaneous stochastic gradient is used. Many variants of backpropagation learning and faster converging alternatives have been introduced [1,2].

Usually MLP networks are designed to have only one or two hidden layers, because training more hidden layers using backpropagation-type algorithms using steepest descent directions has proven to be unsuccessful. The additional hidden layers do not learn useful features easily because the gradients with respect to them decay exponentially [10,11]. A learning algorithm using only the steepest descent update directions often leads to poor local optima or saddle points [12], potentially due to its inability to break symmetries among multiple neurons in each hidden layer [13].

7.3 DEEP LEARNING

However, it would be desirable to have deep neural networks having several hidden layers. The idea is that the layer closest to the data vectors learns simple features, while the higher layers should learn higher-level features. For example, in digital images the first low-level features such as edges and lines in different directions are learned in the first hidden layer. They are followed by shapes, objects, etc., in higher-level layers. An example is shown in Figure 7.2. Human brains, especially cortex, contain deep biological neural networks working in this way. They are very efficient in tasks that are difficult for computers such as various applications of pattern recognition.

Deep learning addresses problems encountered when applying backpropagation-type algorithms to deep networks with many layers. A key idea is to learn not only the nonlinear mapping between input and output vectors but also the underlying structure of data (input) vectors. To achieve this goal, unsupervised pretraining is used. This is achieved in practice by using RBMs or autoencoders in each hidden layer as building blocks in forming deep neural networks.

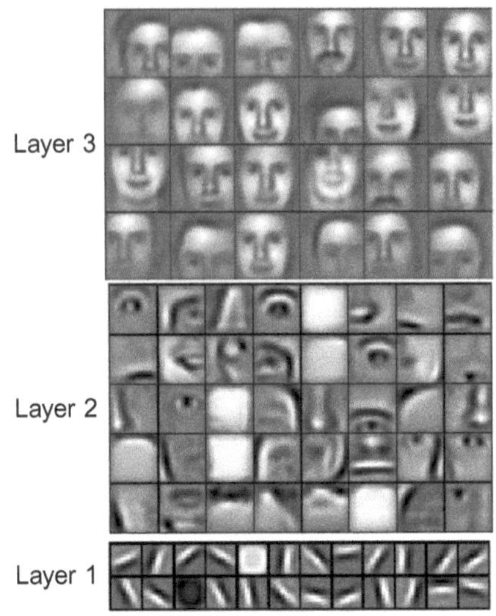

FIGURE 7.2

Higher layers extract more general features of face images. The input to the network consists of pixels. Reprinted from [14].

7.4 RESTRICTED BOLTZMANN MACHINES

Boltzmann machines are a class of neural networks introduced already in the late 1980s. They are based on statistical physics, and they use stochastic neurons contrary to most other neural network methods. RBMs are simplified versions of Boltzmann machines; see Figure 7.3. In RBMs, the connections between the hidden neurons (top) and between the visible neurons (bottom) in the original Boltzmann machines are removed. Only the connections between the neurons in the visible layer and the hidden layer remain. Their weights are collected to the matrix **W**. This simplification makes learning in RBMs tractable compared with Boltzmann machines, where it soon becomes intractable due to the many connections except for small-scale toy problems.

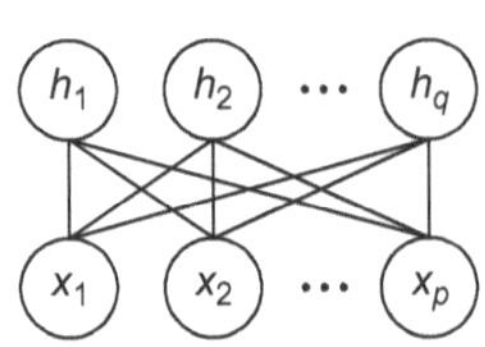

FIGURE 7.3

Restricted Boltzmann machine.

7.4.1 MODELING BINARY DATA

In an RBM, the top layer represents a vector of stochastic binary features $\mathbf{h}$. That is, the value of the state of each neuron can be either 0 or 1 with a certain probability. The bottom layer contains stochastic binary "visible" variables $\mathbf{x}$. Their joint Boltzmann distribution is [6]

$$p(\mathbf{x},\mathbf{h}) = \frac{1}{Z}\exp(-E(\mathbf{x},\mathbf{h})),$$

where $E(\mathbf{x},\mathbf{h})$ is an energy term given by

$$E(\mathbf{x},\mathbf{h}) = -\sum_i b_i x_i - \sum_j b_j h_j - \sum_{i,j} x_i h_j W_{ij},$$

and the normalization constant is

$$Z = \sum_x \sum_h \exp(-E(\mathbf{x},\mathbf{h})).$$

From these equations, one can derive the conditional Bernoulli distributions

$$p(h_j = 1 \mid \mathbf{x}) = \sigma\left(b_j + \sum_i W_{ij} x_i\right)$$

$$p(x_i = 1 \mid \mathbf{h}) = \sigma\left(b_i + \sum_i W_{ij} h_j\right).$$

There $\sigma(z)$ is the logistic sigmoidal function

$$\sigma(z) = \frac{1}{1+\mathrm{e}^{-z}}$$

W_{ij} is a symmetric interaction term between input i and feature j, and b_i, b_j are bias terms. The marginal distribution over visible vector $\mathbf{x}$ is

$$p(\mathbf{x}) = \sum_{\mathbf{h}} \frac{\exp(-E(\mathbf{x},\mathbf{h}))}{\sum_{\mathbf{u},\mathbf{g}} \exp(-E(\mathbf{u},\mathbf{g}))}. \tag{7.3}$$

The parameter update required to perform gradient ascent in the log-likelihood becomes ($\langle\cdot\rangle$ denotes expectation)

$$\Delta W_{ij} = \epsilon\left(\langle x_i h_j\rangle_{\text{data}} - \langle x_i h_j\rangle_{\text{model}}\right). \tag{7.4}$$

In the data distribution, $\mathbf{x}$ is taken from the data set and $\mathbf{h}$ from the conditional distribution $p(\mathbf{h} \mid \mathbf{x},\boldsymbol{\theta})$ given by the model. In the model distribution, both are taken from the joint distribution $p(\mathbf{x},\mathbf{h})$ of the model. For the bias terms, one gets a similar but simpler equation. The expectations can be estimated using Gibbs sampling in which samples are generated from the respective probability distributions.

7.4.2 MODELING REAL-VALUED DATA

RBMs can be generalized to exponential family distributions [6,15]. For example, digital images with real-valued pixels can be modeled by visible units that have a Gaussian distribution. The mean of this distribution is determined by the hidden units:

$$p(x_i \mid \mathbf{h}) = \frac{1}{\sqrt{2\pi}\sigma_i} \exp\left(-\frac{(x - b_i - \sum_j h_j w_{ij})^2}{2\sigma_i^2}\right)$$

$$p(h_j = 1 \mid \mathbf{x}) = \sigma\left(b_j + \sum_i W_{ij}\frac{x_i}{\sigma_i^2}\right).$$

The marginal distribution over visible units $\mathbf{x}$ is given by Eq. (7.3) with an energy term

$$E(\mathbf{x}, \mathbf{h}) = \sum_i \frac{(x_i - b_i)^2}{2\sigma_i^2} - \sum_j b_j h_j - \sum_{i,j} h_j w_{ij}\frac{x_i}{\sigma_i^2}.$$

If the variances are set to $\sigma_i^2 = 1$ for all visible units i, the parameter updates are the same as defined in Eq. (7.4).

7.5 DEEP BELIEF NETWORKS

Deep belief networks (DBNs) are generative models with many layers of hidden causal variables. Each layer of a DBN consists of an RBM. Hinton et al. derived a way to perform fast, greedy learning of DBNs one layer at a time [3]. When an RBM has learned, its feature activations are used as the "data" for training the next RBM in the DBNs, see Figure 7.4.

An important aspect of this layer-wise learning procedure is that each extra layer increases a lower bound on the log probability of the data, provided that the number of features per layer does not decrease. This layer-by-layer training can be repeated several times for learning a deep, hierarchical model of the data. Each layer of features captures strong high-order correlations between the activities of the features in the layer below.

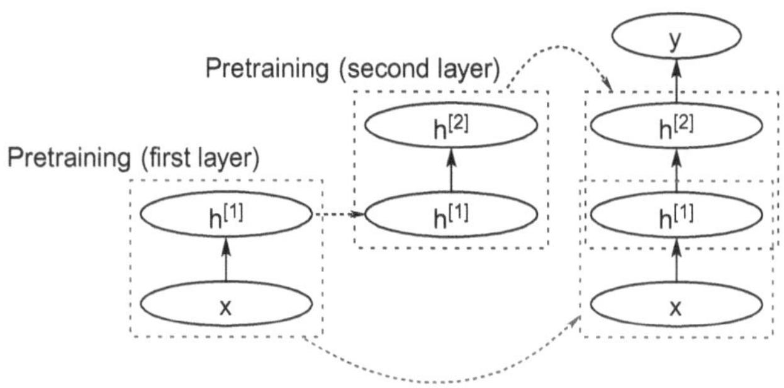

FIGURE 7.4

Pretraining of a deep belief network.

This representation is more efficient than using a single hidden layer with many units. Using the greedy algorithm, one can learn a relatively good hierarchical representation of the data. But it is not yet an optimal representation, because the weights of each layer are learned independently of the weights of the next layers. Therefore, the representation found by the greedy learning algorithm can be improved using a fine tuning algorithm for the weights. To this end, one can use a variant of the standard backpropagation algorithm which is good at fine tuning.

Recursive learning of a deep generative model in this manner can be summarized as follows:

1. Learn the parameter vector W^1 of the first layer of a Bernoulli or Gaussian model.
2. Freeze the parameters of the lower-level model. Use as the data for training the next layer of binary features and the activation probabilities of the binary features, when they are driven by the training data.
3. Freeze the parameters W^2 that define the second layer of features, and use the activation probabilities of those features as data for training the third layer of features.
4. Proceed recursively for as many layers as desired.

7.6 DEEP BOLTZMANN MACHINES

In the DBN discussed above, the connections between the layers are directed except for the two upmost layers. However, for better flow of information, it would be desirable to have undirected connections everywhere. Salakhutdinov and Hinton introduced such deep Boltzmann machines (DBM) in 2009 [6,16]. In Figure 7.5, the vectors of the activities of the neurons in the visible and hidden layers are denoted, respectively, by $\mathbf{v}, \mathbf{h}^1, \ldots, \mathbf{h}^L$. The state vector of the DBM network having L hidden

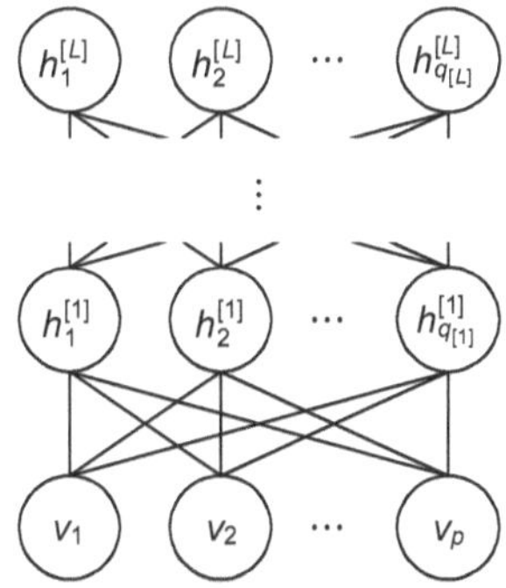

FIGURE 7.5

Deep Boltzmann machine has undirected connections.

layers is denoted by $\mathbf{x} = [\mathbf{v}, \mathbf{h}^1, \ldots, \mathbf{h}^L] = [\mathbf{v}, \mathbf{h}]$. Note here the difference in notation: in RBMs, $\mathbf{x}$ denoted the activities of visible units.

The Boltzmann distribution

$$p(\mathbf{x} \mid \boldsymbol{\theta}) = \frac{1}{Z(\boldsymbol{\theta})} \exp(-E(\mathbf{x} \mid \boldsymbol{\theta}))$$

is still used, where $\boldsymbol{\theta}$ denotes all the parameters. The energy function $-E(\mathbf{x} \mid \boldsymbol{\theta})$ is pretty complicated, consisting of products of activities multiplied by the respective weights. They are summed up together with activities multiplied by biases. Each parameter θ is updated according to the rule

$$\Delta\theta = \epsilon(\langle T_1 \rangle_{\text{data}} - \langle T_2 \rangle_{\text{model}})$$

The expectation $\langle T_1 \rangle_{\text{data}}$ of the term

$$T_1 = -\frac{\partial E(\mathbf{v}^{(n)}, \mathbf{h} \mid \boldsymbol{\theta})}{\partial \theta}$$

is computed over the data distribution $P(\mathbf{h} \mid \{\mathbf{v}^{(n)}\}, \boldsymbol{\theta})$. Here, $\mathbf{v}^{(n)}$ denotes the nth visible data vector. The expectation $\langle T_2 \rangle_{\text{model}}$ of the term

$$T_2 = -\frac{\partial E(\mathbf{v}, \mathbf{h} \mid \boldsymbol{\theta})}{\partial \theta}$$

is computed over the model distribution $P(\mathbf{v}, \mathbf{h} \mid \boldsymbol{\theta})$. The expectation over the data distribution can be estimated using a variational approximation, while the expectation over the model distribution can be computed using Gibbs sampling [6,16].

7.6.1 DIFFERENCES BETWEEN DBNs AND DBMs

Thus far we have discussed only learning, but not using these networks for inference and generation of new samples. Considering first DBN, it actually provides two networks that have common weights. In the recognition network, data vector is inputted to the visible layer, and information then proceeds upwards. One feedforward pass is required for inference. New data vectors resembling training vectors can be generated by sampling the uppermost hidden layer with Markov chain Monte Carlo (MCMC) methods. Then the information proceeds downwards to the visible layer, and every layer tries to represent all the dependencies in the layer below it.

On the contrary, DBMs have only one undirected network. Both recognition and generation of new samples require inference over the entire network. For recognition mean field methods are typically used, and MCMC for generation. The upper layers represent only such information that lower layers have not managed to represent.

7.7 NONLINEAR AUTOENCODERS

DBNs can be used for training nonlinear autoencoders [7]. Autoencoder is a neural network (or mapping method) where the desired output is the input (data) vector itself.

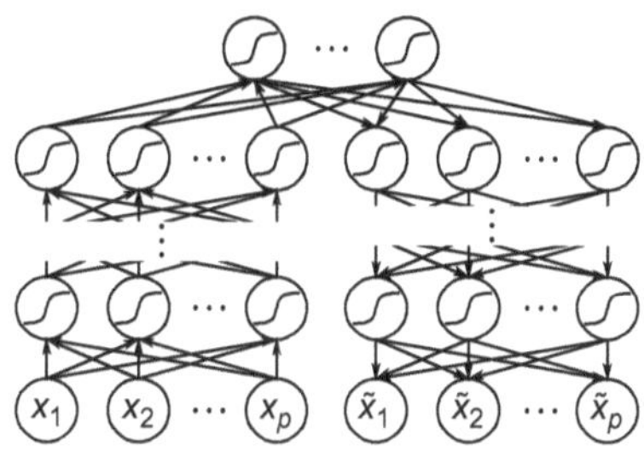

FIGURE 7.6

An MLP network acting as an autoencoder.

This is meaningful because in the middle of an autoencoder, there is a data compressing bottleneck layer having fewer neurons than in the input and output layers. Therefore, the output vector of an autoencoder network is usually an approximation of the input vector only. Comparing it with the input vector provides the error vector needed in training the autoencoder network. Figure 7.6 shows a simple example of an autoencoder.

Autoencoders were first studied in the 1990s for nonlinear data compression [17,18] as a nonlinear extension of standard linear principal component analysis (PCA). Traditional autoencoders have five layers: a hidden layer between the input layer and the data compressing middle bottleneck layer, as well as a similar hidden layer with many neurons between the middle bottleneck layer and output layer [2]. Output vector of the middle bottleneck layer in autoencoders can be used for nonlinear data compression. They were trained using the backpropagation algorithm by minimizing the mean-square error, but this is difficult for multiple hidden layers with millions of parameters.

The greedy learning algorithm for RBMs can be used to pretrain autoencoders also for large problems. It performs a global search for a good, sensible region in the parameter space. The fine-tuning of model parameters is carried out using a variant of standard backpropagation. Generally speaking backpropagation is better at local fine-tuning of the model parameters than global search. So further training of the entire autoencoder using backpropagation will result in a good local optimum. Nonlinear autoencoders trained in this way perform considerably better than linear data compression methods such as PCA.

Autoencoders must be regularized for preventing them to learn identity mapping. Instead of a middle bottleneck layer, one can add noise to input vectors or put some of their components zero [19]. Or one can impose sparsity by penalizing hidden unit activations near zero. Still another possibility is to force the encoder to have small derivatives with respect to the inputs $\mathbf{x}$ (contractive constraint) [20,21]. Discrete inputs can be handled by using a cross-entropy or log-likelihood reconstruction criterion.

Instead of stacking RBMs, one can use a stack of shallow autoencoders to train DBNs, DBMs, or deep autoencoders [22].

7.8 NEURAL AUTOREGRESSIVE DENSITY ESTIMATOR (NADE)

7.8.1 BACKGROUND

Traditional building blocks for deep learning have, however, some unsatisfactory properties. For example, Boltzmann machines are difficult to train due to the intractability of computing the statistics of the model distribution. This may lead to potentially high-variance MCMC estimators during training [23] and the computationally intractable objective function. Autoencoders have a simpler objective function such as denoising reconstruction error [19], which can be used for model selection but not for the important choice of the corruption function.

Larochelle and Murray [24] introduced the so-called neural autoregressive distribution estimator (NADE), which specializes in previous neural auto-regressive density estimators [25] and was recently extended [26] to deeper architectures. It is appealing because both the training criterion (just log-likelihood) and its gradient can be computed tractably and used for model selection, and the model can be trained by stochastic gradient descent with backpropagation. However, it has been observed that the performance of NADE still has room for improvement.

Training of the neural autoregressive density estimator (NADE) can be viewed as performing one step of probabilistic inference on missing values in data for reconstructing missing values in data. The idea of using missing value imputation as a training criterion has appeared in three recent papers. This approach can be seen either as training an energy-based model to impute missing values well [27], as training a generative probabilistic model to maximize a generalized pseudo-log-likelihood [28], or as training a denoising autoencoder with a masking corruption function [26].

The NADE model involves an ordering over the components of the data vector. The core of the model is the reconstruction of the next component given all the previous ones. In the following, we describe a new model and method called NADE-k, proposed recently by two of the authors in [4]. It is an extension of the NADE model based on the reinterpretation of the reconstruction procedure as a single iteration in a variational inference algorithm.

7.8.2 ITERATIVE NADE-k METHOD

The NADE-k method extends the inference scheme of the original NADE method in [24] to multiple steps [4]. We argue that it is easier to learn to improve a reconstruction iteratively in k steps rather than to learn to reconstruct in a single inference step. The proposed model is an unsupervised building block for deep learning that combines the desirable properties of NADE and multi-prediction training: (1) its test likelihood can be computed analytically, (2) it is easy to generate independent samples from it, and (3) it uses an inference engine that is a superset of variational inference for Boltzmann machines.

The NADE-k method is introduced and discussed in more detail in [4], with experiments on two datasets, MNIST handwritten numbers and Caltech-101 Silhouettes. In these experiments, the NADE-k method outperformed the original NADE [24] as well as NADE trained with the order-agnostic training algorithm [26].

In the probabilistic NADE-k method D-dimensional binary data vectors $\mathbf{x}$ are considered. We start by defining conditional distribution p_θ for imputing missing values using a fully factorial conditional distribution:

$$p_\theta(\mathbf{x}_{\text{mis}} \mid \mathbf{x}_{\text{obs}}) = \prod_{i \in \text{mis}} p_\theta(x_i \mid \mathbf{x}_{\text{obs}}), \tag{7.5}$$

where the subscripts “mis” and “obs” denote missing and observed components of $\mathbf{x}$. From the conditional distribution p_θ we compute the joint probability distribution over $\mathbf{x}$ given an ordering o (a permutation of the integers from 1 to D) by

$$p_\theta(\mathbf{x} \mid o) = \prod_{d=1}^{D} p_\theta(x_{o_d} \mid \mathbf{x}_{o_{<d}}), \tag{7.6}$$

where $o_{<d}$ stands for indices $o_1 \ldots o_{d-1}$.

The model is trained to minimize the negative log-likelihood averaged over all possible orderings o

$$\mathcal{L}(\boldsymbol{\theta}) = \mathrm{E}_{o \in D!}\left[\mathrm{E}_{\mathbf{x} \in \text{data}}\left[-\log p_\theta(\mathbf{x} \mid o)\right]\right]. \tag{7.7}$$

using an unbiased, stochastic estimator of $\mathcal{L}(\boldsymbol{\theta})$

$$\hat{\mathcal{L}}(\boldsymbol{\theta}) = -\frac{D}{D-d+1} \log p_\theta(\mathbf{x}_{o_{\geq d}} \mid \mathbf{x}_{o_{<d}}) \tag{7.8}$$

by drawing o uniformly from all $D!$ possible orderings and d uniformly from $1 \ldots D$ [26]. Note that while the model definition in Eq. (7.6) is sequential in nature, the training criterion (Eq. 7.8) involves reconstruction of all the missing values in parallel. In this way, training does not involve picking or following specific orders of indices.

We define the conditional model $p_\theta(\mathbf{x}_{\text{mis}} \mid \mathbf{x}_{\text{obs}})$ using a deep feedforward neural network with nk layers, where we use n weight matrices k times. This can also be interpreted as running k successive inference steps with an n-layer neural network.

The input to the network is

$$\mathbf{v}^{\langle 0 \rangle} = \mathbf{m} \odot \mathrm{E}_{\mathbf{x} \in \text{data}}[\mathbf{x}] + (\mathbf{1} - \mathbf{m}) \odot \mathbf{x}, \tag{7.9}$$

where $\mathbf{m}$ is a binary mask vector indicating missing components with 1, and $\odot$ is an element-wise multiplication. $\mathrm{E}_{\mathbf{x} \in \text{data}}[\mathbf{x}]$ is an empirical mean of the observations. For simplicity, we give equations for a simple structure with $n = 2$, shown in Figure 7.7 (left).

In this case, the activations of the layers at the tth step are

$$\mathbf{h}^{\langle t \rangle} = \phi\left(\mathbf{W}\mathbf{v}^{\langle t-1 \rangle} + \mathbf{c}\right) \tag{7.10}$$

$$\mathbf{v}^{\langle t \rangle} = \mathbf{m} \odot \sigma\left(\mathbf{V}\mathbf{h}^{\langle t \rangle} + \mathbf{b}\right) + (\mathbf{1} - \mathbf{m}) \odot \mathbf{x}, \tag{7.11}$$

where ϕ is an element-wise nonlinearity, σ is a logistic sigmoid function, and the iteration index t runs from 1 to k. The conditional probabilities of the variables (see Eq. 7.5) are read from the output $\mathbf{v}^{\langle k \rangle}$ as

$$p_\theta(x_i = 1 \mid \mathbf{x}_{\text{obs}}) = v_i^{\langle k \rangle}. \tag{7.12}$$

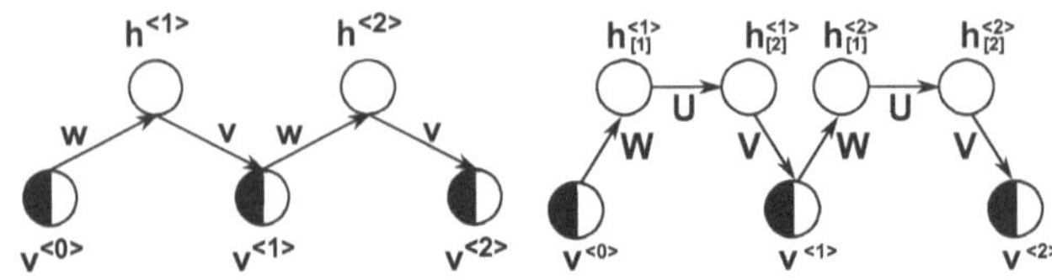

FIGURE 7.7

The choice of a structure for NADE-k is very flexible. *Left*: Basic structure corresponding to Eqs. (7.10) and (7.11) with $n = 2$ and $k = 2$. *Right*: Depth added as in NADE by [26] with $n = 3$ and $k = 2$. These two structures are used in the experiments.

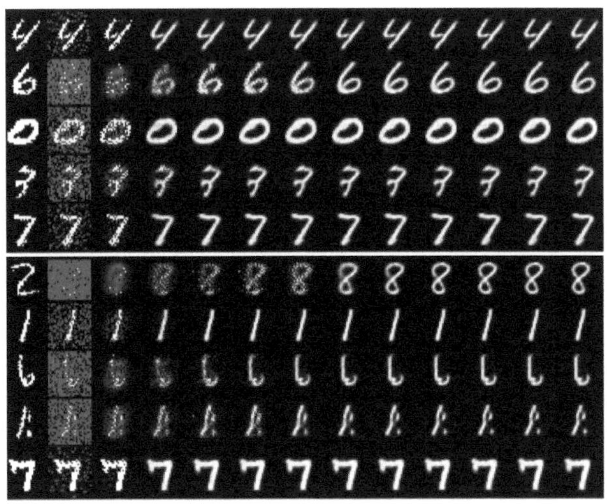

FIGURE 7.8

The inner working mechanism of NADE-k. The left most column shows the data vectors $\mathbf{x}$, the second column shows their masked version, and the subsequent columns show the reconstructions $\mathbf{v}^{\langle 0 \rangle} \ldots \mathbf{v}^{\langle 10 \rangle}$ (see Eq. 7.11).

Figure 7.8 shows examples of how $\mathbf{v}^{\langle t \rangle}$ evolves over iterations, with the trained model.

The parameters $\boldsymbol{\theta} = \{\mathbf{W}, \mathbf{V}, \mathbf{c}, \mathbf{b}\}$ can be learned by stochastic gradient descent to minimize $-\mathcal{L}(\boldsymbol{\theta})$ in Eq. (7.7), or its stochastic approximation $-\hat{\mathcal{L}}(\boldsymbol{\theta})$ in Eq. (7.8), with the stochastic gradient computed by back-propagation.

Once the parameters $\boldsymbol{\theta}$ are learned, one can define a mixture model by using a uniform probability over a set of orderings O. The probability of a given vector $\mathbf{x}$ as a mixture model can be computed

$$p_{\text{mixt}}(\mathbf{x} \mid \boldsymbol{\theta}, O) = \frac{1}{|O|} \sum_{o \in O} p_{\boldsymbol{\theta}}(\mathbf{x} \mid o) \tag{7.13}$$

with Eq. (7.6). One can draw independent samples from the mixture by first drawing an ordering o and then sequentially drawing each variable using $x_{o_d} \sim p_{\boldsymbol{\theta}}(x_{o_d} \mid \mathbf{x}_{o_{<d}})$. Furthermore, samples can be drawn from the conditional $p(\mathbf{x}_{\text{mis}} \mid \mathbf{x}_{\text{obs}})$ easily by considering only orderings where the observed indices appear before the missing ones.

It is well known that training deep networks is difficult without pretraining, and in the experiments described in more detail in [4], the networks are trained up to $kn = 7 \times 3 = 21$ layers. When pretraining, the model is trained to produce good reconstructions $\mathbf{v}^{\langle t \rangle}$ at each step $t = 1 \ldots k$. More formally, in the pretraining phase, Eqs. (7.8) and (7.12) are replaced by

$$\hat{\mathcal{L}}_{\text{pre}}(\boldsymbol{\theta}) = -\frac{D}{D-d+1}\frac{1}{k}\sum_{t=1}^{k} \log \prod_{i \in o_{\geq d}} p_{\boldsymbol{\theta}}^{\langle t \rangle}(x_i \mid \mathbf{x}_{o_{<d}}) \tag{7.14}$$

$$p_{\boldsymbol{\theta}}^{\langle t \rangle}(x_i = 1 \mid \mathbf{x}_{\text{obs}}) = v_i^{\langle t \rangle}. \tag{7.15}$$

7.8.3 RELATED METHODS AND APPROACHES

Order-agnostic NADE. The proposed method follows closely the order-agnostic version of NADE [26], which may be considered as the special case of NADE-k with $k = 1$. On the other hand, NADE-k can be seen as a deep NADE with some specific weight sharing (matrices $\mathbf{W}$ and $\mathbf{V}$ are reused for different depths) and gating in the activations of some layers (see Eq. 7.11).

Additionally, in [26], it was found crucial to give the mask $\mathbf{m}$ as an auxiliary input to the network, and initialized missing values to zero instead of the empirical mean (see Eq. 7.9). Due to these differences, the approach in [26] is called here NADE-mask. One should note that NADE-mask has more parameters due to using the mask as a separate input to the network, whereas NADE-k is roughly k times more expensive to compute.

Probabilistic inference. Consider the task of missing value imputation in a probabilistic latent variable model. The conditional probability of interest is obtained by marginalizing out the latent variables from the posterior distribution:

$$p(\mathbf{x}_{\text{mis}} \mid \mathbf{x}_{\text{obs}}) = \int_{\mathbf{h}} p(\mathbf{h}, \mathbf{x}_{\text{mis}} \mid \mathbf{x}_{\text{obs}}) d\mathbf{h}.$$

Accessing the joint distribution $p(\mathbf{h}, \mathbf{x}_{\text{mis}} \mid \mathbf{x}_{\text{obs}})$ directly is often harder than alternatively updating $\mathbf{h}$ and $\mathbf{x}_{\text{mis}}$ based on the conditional distributions $p(\mathbf{h} \mid \mathbf{x}_{\text{mis}}, \mathbf{x}_{\text{obs}})$ and $p(\mathbf{x}_{\text{mis}} \mid \mathbf{h})$. Variational inference is one of the representative examples that exploit this.

In variational inference, a factorial distribution $q(\mathbf{h}, \mathbf{x}_{\text{mis}}) = q(\mathbf{h})q(\mathbf{x}_{\text{mis}})$ is iteratively fitted to $p(\mathbf{h}, \mathbf{x}_{\text{mis}} \mid \mathbf{x}_{\text{obs}})$ such that the KL-divergence between q and p

$$\text{KL}[q(\mathbf{h}, \mathbf{x}_{\text{mis}}) || p(\mathbf{h}, \mathbf{x}_{\text{mis}} \mid \mathbf{x}_{\text{obs}})] = -\int_{\mathbf{h}, \mathbf{x}_{\text{mis}}} q(\mathbf{h}, \mathbf{x}_{\text{mis}}) \log \left[\frac{p(\mathbf{h}, \mathbf{x}_{\text{mis}} \mid \mathbf{x}_{\text{obs}})}{q(\mathbf{h}, \mathbf{x}_{\text{mis}})} \right] d\mathbf{h}\, d\mathbf{x}_{\text{mis}}$$

is minimized. The algorithm alternates between updating $q(\mathbf{h})$ and $q(\mathbf{x}_{\text{mis}})$, while considering the other one fixed.

As an example, consider an RBM defined by

$$p(\mathbf{v}, \mathbf{h}) \propto \exp(\mathbf{b}^\top \mathbf{v} + \mathbf{c}^\top \mathbf{h} + \mathbf{h}^\top \mathbf{W} \mathbf{v}).$$

One can fit an approximate posterior distribution parameterized as $q(v_i = 1) = \bar{v}_i$ and $q(h_j = 1) = \bar{h}_j$ to the true posterior distribution by iteratively computing

$$\begin{aligned} \bar{\mathbf{h}} &\leftarrow \sigma(\mathbf{W}\bar{\mathbf{v}} + \mathbf{c}) \\ \bar{\mathbf{v}} &\leftarrow \mathbf{m} \odot \sigma(\mathbf{W}^\top \mathbf{h} + \mathbf{b}) + (\mathbf{1} - \mathbf{m}) \odot \mathbf{v}. \end{aligned}$$

Notice the similarity to Eqs. (7.10) and (7.11): if one assumes $\phi = \sigma$ and $\mathbf{V} = \mathbf{W}^\top$, the inference in the NADE-k is equivalent to performing k iterations of variational inference on an RBM for the missing values [29].

Multi-predictive DBM. Goodfellow et al. [28] and Brakel et al. [27] use back-propagation through variational inference steps to train a DBM. This is very similar to NADE-k, except that they approach the problem from the view of maximizing the generalized pseudo-likelihood [30]. The DBM also lacks the tractable probabilistic interpretation similar to NADE-k, see Eq. (7.6), that would allow to compute a probability or to generate independent samples without resorting to a Markov chain. NADE-k is also somewhat more flexible in the choice of model structures, as can be seen in Figure 7.7. For instance, encoding and decoding weights do not have to be shared and any type of nonlinear activations, other than a logistic sigmoid function, can be used.

Product and mixture of experts. One can ask what would happen if we would define an ensemble likelihood along the line of the training criterion in Eq. (7.7). That is,

$$-\log p_{\text{prod}}(\mathbf{x} \mid \boldsymbol{\theta}) \propto \mathrm{E}_{o \in D!} \left[-\log p(\mathbf{x} \mid \boldsymbol{\theta}, o) \right].$$

Maximizing this likelihood directly will correspond to training a product-of-experts model [31]. However, this requires evaluation of the intractable normalization constant during training as well as in the inference, making the model not tractable any more.

On the other hand, one can consider using the log-probability of a sample under the mixture-of-experts model as the training criterion

$$-\log p_{\text{mixt}}(\mathbf{x} \mid \boldsymbol{\theta}) = -\log \mathrm{E}_{o \in D!} \left[p(\mathbf{x} \mid \boldsymbol{\theta}, o) \right].$$

This criterion resembles clustering, where individual models may specialize in only a fraction of the data. In this case, however, the simple estimator such as in Eq. (7.8) would not be available.

7.8.4 EXPERIMENTAL RESULTS

The NADE-k model has been studied with two datasets: binarized MNIST handwritten digits and Caltech 101 silhouettes. NADE-k was trained with one or two hidden layers (see Figure 7.7 left and right) with a hyperbolic tangent as the activation function $\phi(\cdot)$. Stochastic gradient descent was used on the training set with a mini batch size fixed to 100. The AdaDelta method [32] was used to adaptively choose a learning rate for each parameter update on-the-fly. We used a validation set for early stopping and to select the hyperparameters. With the best model on the validation set, we report the log-probability computed on the test set.

In [4], the NADE-k method is tested extensively with the MNIST data and compared favorably with the NADE-mask method introduced in [26]. We mostly skip here these experiments, but in Figure 7.8 we present how each iteration $t = 1 \dots k$ improves the corrupted input $\mathbf{v}^{\langle t \rangle}$ from Eq. (7.9). We also investigated what happens with test time k being larger than the training $k = 5$. We can see that in all cases, the iteration – which is a fixed point update – seems to converge to a point that is in most cases close to the ground-truth sample. However, the experiments in [4] show that the generalization performance drops after $k = 5$ when training with $k = 5$. From Figure 7.8, we can see that the reconstruction continues to be sharper even after $k = 5$, which seems to be the underlying reason for this phenomenon.

We also evaluate the proposed NADE-k method on Caltech-101 Silhouettes [33], using the standard split of 4100 training samples, 2264 validation samples, and 2307 test samples. We demonstrate the advantage of NADE-k compared with NADE-mask under the constraint that they have a matching number of parameters. In particular, we compare NADE-k with 1000 hidden units with NADE-mask with 670 hidden units. We also compare NADE-k with 4000 hidden units with NADE-mask with 2670 hidden units.

We optimized the hyper-parameter $k \in \{1, 2, \dots, 10\}$ in the case of NADE-k. In both NADE-k and NADE-mask, we experimented without regularizations, with weight decays, or with dropout. We did not use the pretraining scheme described in Eq. (7.14).

As one can see from Table 7.1, NADE-k outperforms the NADE-mask regardless of the number of parameters. In addition, NADE-2 with 1000 hidden units matches

Table 7.1 Average Log-Probabilities of Test Samples of Caltech-101 Silhouettes

Model	Test LL	Model	Test LL
RBM* (2000h, 1.57M)	−108.98	RBM * (4000h, 3.14M)	**−107.78**
NADE-mask (670h, 1.58M)	−112.51	NADE-mask (2670h, 6.28M, L2=0.00106)	−110.95
NADE-2 (1000h, 1.57M, L2=0.0054)	−108.81	NADE-5 (4000h, 6.28M, L2=0.0068)	**−107.28**

Notes: The results marked with ⋆ are from [34]. The terms in parentheses indicate the number of hidden units, the total number of parameters (M for million), and the L2 regularization coefficient. NADE-mask 670h achieves the best performance without any regularizations.

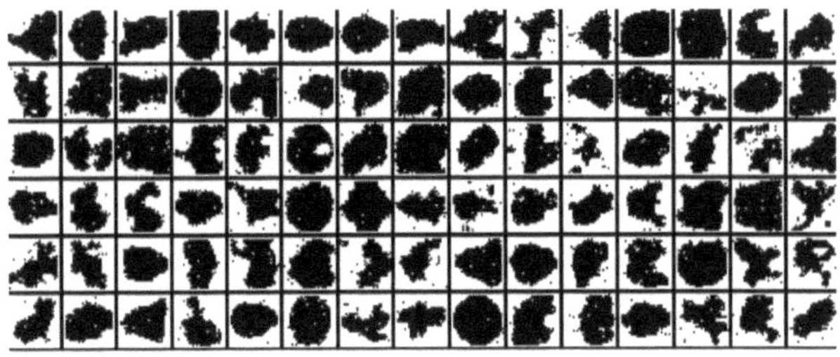

FIGURE 7.9

Samples generated from NADE-k trained on Caltech-101 Silhouettes.

the performance of an RBM with the same number of parameters. Furthermore, NADE-5 has outperformed the previous best result obtained with the RBMs in [34], achieving the state-of-the-art result on this dataset. We can see from the samples generated by the NADE-k shown in Figure 7.9 that the model has learned the data well.

7.9 CONCLUSIONS

Deep learning has become a hot topic in machine learning, because it can provide world-record results in different classification and regression problems and datasets. Many corporations including Google, Microsoft, Nokia, etc., study it actively. Understanding deep learning well requires mathematical maturity and good knowledge of probabilistic modeling. Learning algorithms are complicated, and good initialization is important. The field is developing quite rapidly, with new structures and learning methods introduced all the time.

In this chapter, we have reviewed some of the most widely studied and used deep learning models for unsupervised learning tasks. Also, we have discussed in more detail a new model called iterative neural autoregressive distribution estimator NADE-k [4], which extends the conventional NADE [26] and its training procedure. The proposed NADE-k method maintains the tractability of the original NADE, while we showed that it outperforms the original NADE as well as similar, but intractable, generative models such as RBMs and DBNs.

The list of unsupervised models we have reviewed in this chapter is not exhaustive. During the last few years, a number of new deep learning models for unsupervised learning have been proposed. For instance, Kingma and Welling [35] proposed a so-called variational autoencoder, where they proposed to train an autoencoder to maximize a variational lower bound of a directed belief network. Bengio et al. [36] recently proposed a rather distinct deep learning-based framework for unsupervised learning, called generative stochastic network, which aims to learn an MCMC transition operator instead of a full probability distribution.

REFERENCES

[1] S. Haykin, Neural Networks and Learning Machines, third ed., Pearson, Upper Saddle River, NJ, 2009.

[2] K.-L. Du, M. Swamy, Neural Networks and Statistical Learning, Springer-Verlag, Dordrecht, 2014.

[3] G. Hinton, S. Osindero, Y.-W. Teh, A fast learning algorithm for deep belief nets, Neural Comput. 18 (2006) 1527-1554.

[4] T. Raiko, L. Yao, K. Cho, Y. Bengio, Iterative neural autoregressive distribution estimator (NADE-k), in: Neural Information Processing Systems 2014 Conference (NIPS 2014), Montreal, Canada, December, 2014.

[5] Y. Bengio, Learning deep architectures for AI, Found. Trends Mach. Learn. 2 (1) (2009) 1-127.

[6] R. Salakhutdinov, Learning deep generative models (Doctoral thesis), MIT, 2009. Available from: http://www.mit.edu/$\sim$rsalakhu/papers/Russ_thesis.pdf.

[7] K. Cho, Foundations and advances in deep learning (Doctoral thesis), Aalto University School of Science, Espoo, Finland, 2014.

[8] J. Schmidhuber, Deep learning in neural networks: an overview, Technical report IDSIA-03-14, Switzerland, arXiv:1404.7828 [cs.NE].

[9] Y. Bengio, I. Goodfellow, A. Courville, Deep Learning. An MIT Press book in preparation. Draft chapters available from: http://www.iro.umontreal.ca/$\sim$bengioy/dlbook/.

[10] Y. Bengio, P. Simard, P. Frasconi, Learning long-term dependencies with gradient descent is difficult, IEEE Trans. Neural Netw. 5 (2) (1994) 157-166.

[11] X. Glorot, Y. Bengio, Understanding the difficulty of training deep feedforward neural networks, in: Proc. of the Int. Conf. on Artificial Intelligence and Statistics (AISTATS 2010), 2010, pp. 249-256.

[12] Y. Dauphin, R. Pascanu, C. Gulcehre, K. Cho, S. Ganguli, Y. Bengio, Identifying and attacking the saddle point problem in high-dimensional non-convex optimization, in: Advances in Neural Information Processing Systems (NIPS), 27, 2014.

[13] T. Raiko, H. Valpola, Y. LeCun, Deep learning made easier by linear transformations in perceptrons, in: Proc. of the 15th Int. Conf. on Artificial Intelligence and Statistics (AISTATS 2012), 2012, pp. 924-932.

[14] H. Lee, R. Grosse, R. Ranganath, A. Ng, Convolutional deep belief networks for scalable unsupervised learning of hierarchical representations, in: Proc. of the 26th Int. Conf. on Machine Learning (ICML 2009), 2009, pp. 609-616.

[15] K. Cho, A. Ilin, T. Raiko, Improved learning of Gaussian-Bernoulli restricted Boltzmann machines, in: Lecture Notes in Computer Science, vol. 6791, Artificial Neural Networks and Machine Learning (ICANN 2011), 2011, pp. 10-17.

[16] R. Salakhutdinov, G. Hinton, Deep Boltzmann machines, in: Proc. of 12th Int. Conf. on Artificial Intelligence and Statistics (AISTATS 2009), Clearwater Beach, Florida, USA, 2009, pp. 448-455.

[17] M. Kramer, Nonlinear principal component analysis using autoassociative neural networks, AIChE J. 37 (2) (1991) 233-243.

[18] E. Oja, Data compression, feature extraction, and autoassociation in feedforward neural networks, in: Proc. of the Int. Conf. on Artificial Neural Networks (ICANN-91), Helsinki, Finland, Elsevier, June 1991, pp. 737-745.

[19] P. Vincent, H. Larochelle, I. Lajoie, Y. Bengio, P.-A. Manzagol, Stacked denoising autoencoders: learning useful representations in a deep network with a local denoising criterion, J. Mach. Learn. Res. 11 (2010) 3371-3408.

[20] S. Rifai, P. Vincent, X. Muller, X. Glorot, Y. Bengio, Contractive auto-encoders: explicit invariance during feature extraction, in: Proc. of the 28th Int. Conf. on Machine Learning (ICML 2011), 2011.

[21] H. Schulz, K. Cho, T. Raiko, S. Behnke, Two-layer contractive encodings for learning stable nonlinear features, Neural Netw. (2014) (in press).

[22] Y. Bengio, P. Lamblin, D. Popovici, H. Larochelle, Greedy layer-wise training of deep networks, in: Advances in Neural Information Processing Systems (NIPS), 2007.

[23] Y. Bengio, G. Mesnil, Y. Dauphin, S. Rifai, Better mixing via deep representations, in: Proc. of the 30th Int. Conf. on Machine Learning (ICML 2013), arXiv preprint arXiv:1207.4404, 2013.

[24] H. Larochelle, I. Murray, The neural autoregressive distribution estimator, J. Mach. Learn. Res. 15 (2011) 29-37.

[25] Y. Bengio, S. Bengio, Modeling high-dimensional discrete data with multi-layer neural networks, in: Advances in Neural Information Processing Systems, MIT Press, 2000, pp. 400-406.

[26] B. Uria, I. Murray, H. Larochelle, A deep and tractable density estimator, in: Proc. of the 31st Int. Conf. on Machine Learning (ICML 2014), arXiv preprint arXiv:1310.1757, 2014.

[27] P. Brakel, D. Stroobandt, B. Schrauwen, Training energy-based models for time-series imputation, J. Mach. Learn. Res. 14 (2013) 2771-2797.

[28] I. Goodfellow, M. Mirza, A. Courville, Y. Bengio, Multi-prediction deep Boltzmann machines, in: Advances in Neural Information Processing Systems, 2013, pp. 548-556.

[29] C. Peterson, J. Anderson, A mean field theory learning algorithm for neural networks, Complex Syst. 1 (5) (1987) 995-1019.

[30] F. Huang, Y. Ogata, Generalized pseudo-likelihood estimates for Markov random fields on lattice, Ann. Instit. Stat. Math. 54 (1) (2002) 1-18.

[31] G. Hinton, Training products of experts by minimizing contrastive divergence, Technical report GCNU TR 2000-004, Gatsby Unit, University College, London, 2000.

[32] M. Zeiler, ADADELTA: an adaptive learning rate method, Technical report, arXiv 1212.5701, 2012.

[33] B. Marlin, K. Swersky, B. Chen, N. de Freitas, Inductive principles for restricted Boltzmann machine learning, in: Proc. of the 13th Int. Conf. on Artificial Intelligence and Statistics (AISTATS 2010), 2010, pp. 509-516.

[34] K. Cho, T. Raiko, A. Ilin, Enhanced gradient for training restricted Boltzmann machines, Neural Comput. 25 (3) (2013) 805-831.

[35] D.P. Kingma, M. Welling, Auto-encoding variational Bayes, in: Proc. of the Int. Conf. on Learning Representations (ICLR 2014), 2014.

[36] Y. Bengio, E. Thibodeau-Laufer, G. Alain, J. Yosinski, Deep generative stochastic networks trainable by Backprop, in: Proc. of the 31st Int. Conf. on Machine Learning (ICML 2014), 2014.

CHAPTER

From neural PCA to deep unsupervised learning

8

Harri Valpola
ZenRobotics Ltd., Helsinki, Finland

8.1 INTRODUCTION

Ever since Hubel and Wiesel [1] published their findings about a hierarchy of increasingly abstract invariant visual features in the cat neocortex, researchers have been trying to mimic the same hierarchy of feature extraction stages with artificial neural networks. An early example is the neocognitron by Fukushima [2].

Nowadays the art of estimating deep feature extraction hierarchies is called deep learning (for a thorough overview of the history of the field, including recent developments, see the review by Schmidhuber [3]). The topic has received a lot of attention after Hinton et al. [4] and Hinton and Salakhutdinov [5] proposed an unsupervised[a] pretraining scheme which made subsequent supervised learning efficient for a deeper network than before.

What is somewhat embarrassing for the field, though, is that recently purely supervised learning has achieved as good or better results as unsupervised pretraining schemes (e.g., [6,7]). In most classification problems, finding and producing labels for the samples are hard. In many cases, plenty of unlabeled data exist and it seems obvious that using them should improve the results. For instance, there are plenty of unlabeled images available and in most image classification tasks there are vastly more bits of information in the statistical structure of input images than in their labels.[b]

It is argued here that the reason why unsupervised learning has not been able to improve results is that most current versions are incompatible with supervised learning. The problem is that many unsupervised learning methods try to represent as much information about the original data as possible whereas supervised learning tries to filter out all the information which is irrelevant for the task at hand.

This chapter presents an unsupervised learning network whose properties make it a good fit with supervised learning. First, learning is based on minimizing a cost function much the same way as in stochastic gradient descent of supervised feedforward networks. Learning can therefore continue alongside supervised learning rather than be restricted to a pretraining phase. Second, the network can discard

E. Bingham, S. Kaski, J. Laaksonen and J. Lampinen (Eds) Advances in Independent Component Analysis and Learning Machines.
http://dx.doi.org/10.1016/B978-0-12-802806-3.00008-7

information from the higher layers and leave the details for lower layers to represent. This means that the approach allows supervised learning to select the relevant features. The proposed unsupervised learning can filter out noise from the selected features and come up with new features that are related to the selected features.

The two main new ideas presented in this chapter are as follows:

1. Section 8.2 explains how adding lateral shortcut connections to an autoencoder gives it the same representational capacity as hierarchical latent variable models. This means that higher layers no longer need to represent all the details but can concentrate on abstract invariant representations. The model structure is called a ladder network because two vertical paths are connected by horizontal lateral connections at regular intervals.
2. Learning of deep autoencoders can be slow since training signals need to travel a long distance from the decoder output through both the decoder and the encoder. Lateral shortcuts tend to slow it down even further since the shortcuts learn first, shunting the training signals along the longer paths. Section 8.3 explains how this can be remedied by adding training targets to each level of the hierarchy. The novel idea is to combine denoising source separation (DSS) framework [8] with training denoising functions to remove injected noise [9].

The experiments presented in Section 8.4 demonstrate that the higher levels of a ladder network can discard information and focus on invariant representations and that the training targets on higher layers speed up learning. The results presented here are promising but preliminary. Section 8.5 discusses potential extensions and related work.

8.2 LADDER NETWORK: AN AUTOENCODER WHICH CAN DISCARD INFORMATION

As argued earlier, unsupervised learning needs to tolerate discarding information in order to work well with supervised learning. Many unsupervised learning methods are not good at this but one class of models stands out as an exception: hierarchical latent variable models. Unfortunately their derivation can be quite complicated and often involves approximations which compromise their performance. A simpler alternative is offered by autoencoders which also have the benefit of being compatible with standard supervised feedforward networks. They would be a promising candidate for combining supervised and unsupervised learning but unfortunately autoencoders normally correspond to latent variable models with a single layer of stochastic variables, that is, they do not tolerate discarding information.

This section summarizes the complementary roles of supervised and unsupervised learning, reviews latent variable models and their relation to standard autoencoder networks, and proposes a new network structure, the ladder network, whose lateral shortcut connections give it the same representational capacity as hierarchical latent variable models.

8.2.1 COMPLEMENTARY ROLES OF SUPERVISED AND UNSUPERVISED LEARNING

Consider the roles of supervised and unsupervised learning in a particular task, such as classification, prediction, or regression. Further assume that (1) there are input-output pairs which can be used for supervised learning but far more unlabeled samples that are lacking the output, with only the inputs available and (2) inputs have far more information than the outputs.

In general, this typical setup means that unsupervised learning should be used for anything it works for because the precious bits of information in the available output samples should be reserved for those tasks that unsupervised learning cannot handle.

The main role for supervised learning is clear enough: figure out which type of representations are relevant for the task at hand. Only supervised learning can do this because, by definition, unsupervised learning does not have detailed information about the task.

One obvious role, the traditional one, for unsupervised learning is to act as a preprocessing or pretraining step for supervised learning. However, the key question is: what can unsupervised learning do after supervised learning has kicked in. This is important because in many problems there is so much information in the inputs that it cannot possibly be fully summarized by unsupervised learning first. Rather, it would be useful for unsupervised learning to continue tuning the representations even after supervised learning has started to tune the relevant features and filter out the irrelevant ones.

The combination of supervised and unsupervised learning is known as semi-supervised learning. As argued earlier, it can be a happy marriage only if unsupervised learning is content with discarding information and concentrating on the features which supervised learning deems relevant.

What unsupervised learning should be able to do efficiently is to find new features which correlate with and predict the features selected by supervised learning. This improves generalization to new samples. As an example, consider learning to recognize a face. Suppose supervised learning has figured out that an eye is an important feature for classifying faces versus nonfaces from a few samples. What unsupervised learning can do with all the available unlabeled samples is find other features which correlate with the selected one. Such new features could be, for instance, a detector for nose, eye brow, ear, mouth, and so on. These features improve the generalization of a face detector in cases where the eye feature is missing, for instance due to eyes being closed or occluded by sunglasses.

Specifically, what unsupervised learning must *not* do is keep pushing new features to the representation intended for supervised learning simply because these features carry information about the inputs. While such behavior may be reasonable as a pretraining or preprocessing step, it is not compatible with semi-supervised learning. In other words, before unsupervised learning knows which features are relevant, it is reasonable to select features which carry as much information as possible about the inputs. However, after supervised learning starts showing a preference of some features over some others, unsupervised learning should follow suite and present more of the kind that supervised learning seems to be interested in.

8.2.2 LATENT VARIABLE MODELS

Many unsupervised learning methods can be framed as latent variable models (e.g., [10]) where unknown latent variables $s(t)$ are assumed to generate the observed data $\boldsymbol{x}(t)$. A common special case with continuous variables is that the latent variables predict the mean of the observations:

$$\boldsymbol{x}(t) = g(s(t); \boldsymbol{\xi}) + \boldsymbol{n}(t), \tag{8.1}$$

where $\boldsymbol{n}(t)$ denotes the noise or modeling error and $\boldsymbol{\xi}$ the parameters of mapping g. Alternatively, the same can be expressed through a probability model

$$p_{\mathrm{x}}(\boldsymbol{x}(t)|s(t), \boldsymbol{\xi}) = p_{\mathrm{n}}(\boldsymbol{x}(t) - g(s(t); \boldsymbol{\xi})), \tag{8.2}$$

where p_{n} denotes the probability density function (p.d.f.) of the noise term $\boldsymbol{n}(t)$. Inference of the unknown latent variables $\boldsymbol{s}(t)$ and parameters $\boldsymbol{\xi}$ can then be based simply on minimizing the mismatch between the observed $\boldsymbol{x}(t)$ and its reconstruction $g(\boldsymbol{s}(t); \boldsymbol{\xi})$, or more generally on probabilistic modeling.

The models defined by Eqs. (8.1) or (8.2) have just one layer of latent variables which tries to represent everything there is to represent about the data. Such models have trouble letting go of any piece of information since this would increase the reconstruction error. Also, in many cases, an abstract invariant feature (such as "a face") cannot reduce the reconstruction error alone without plenty of accompanying details such as position, orientation, size, and so on. All of those details need to be represented alongside the relevant feature to show any benefit in reducing the reconstruction error. This means that latent variable models with a single layer of latent variables have trouble discarding information and focusing on abstract invariant features.

It is possible to fix the situation by introducing a hierarchy of latent variables:

$$p\left(\boldsymbol{s}^{(l)}(t)|\boldsymbol{s}^{(l+1)}(t), \boldsymbol{\xi}^{(l)}\right), \tag{8.3}$$

where the superscript (l) refers to variables on layer l. The observations can be taken into the equation by defining $\boldsymbol{s}^{(0)} := \boldsymbol{x}$. Now the latent variables on higher levels no longer need to represent everything. Lower levels can take care of representing details while higher levels can focus on selected features, abstract or not.

In such hierarchical models, higher-level latent variables can still represent just the mean of the lower-level variables,

$$\boldsymbol{s}^{(l)}(t) = g^{(l)}\left(\boldsymbol{s}^{(l+1)}(t); \boldsymbol{\xi}^{(l)}\right) + \boldsymbol{n}^{(l)}(t), \tag{8.4}$$

but more generally, the higher-level variables can represent any properties of the distribution, such as the variance [11]. For binary variables, sigmoid units are often used for representing the dependency (for a recent example of such a model, see [12]).

Exact inference, that is, computing the posterior probability of the unknown variables (latent variables and the parameters of the mappings), is typically mathematically intractable. Instead, approximate inference techniques such as variational Bayesian methods are employed (e.g., [11,12]). They amount to approximating the intractable exact posterior probability with a simpler tractable approximation. Learning then corresponds to iteratively minimizing the cost function with respect to the posterior approximation. For instance, in the case of continuous latent variables, the posterior could be approximated as Gaussian with a diagonal covariance. For each unknown variable, the mean and variance would then be estimated in the course learning. These posterior means and variances typically depend on the values on both lower and higher layers and inference would therefore proceed iteratively.

If hierarchical latent variable models were easy to define and learn, the problem would be solved. Unfortunately hierarchical models often require complex probabilistic methods to train them. They often involve approximations which compromise their performance [13] or are limited to restricted model structures which are mathematically tractable. Also, many training schemes require the latent variable values to be updated layer-wise by combining bottom-up information with top-down priors. This slows down the propagation of information in the network.

8.2.3 AUTOENCODERS AND DETERMINISTIC AND STOCHASTIC LATENT VARIABLES

Autoencoder networks resemble in many ways single-layer latent variable models. The key idea is that the inference process of mapping observations $\boldsymbol{x}(t)$ to the corresponding latent variables, now called hidden unit activations $\boldsymbol{h}(t)$, is modeled by an encoder network f and the mapping back to observations is modeled by a decoder network g:

$$\boldsymbol{h}(t) = f(\boldsymbol{x}(t); \boldsymbol{\xi}_f) \tag{8.5}$$

$$\hat{\boldsymbol{x}}(t) = g(\boldsymbol{h}(t); \boldsymbol{\xi}_g). \tag{8.6}$$

The mappings f and g are called encoder and decoder mappings, respectively. In connection to latent variable models, analogous mappings are called the recognition and reconstruction mappings.

Learning of autoencoders is based on minimizing the difference between the observation vector $\boldsymbol{x}(t)$ and its reconstruction $\hat{\boldsymbol{x}}(t)$, that is, minimizing the cost $\|\boldsymbol{x}(t) - \hat{\boldsymbol{x}}(t)\|^2$ with respect to the parameters $\boldsymbol{\xi}_f$ and $\boldsymbol{\xi}_g$. For the remainder of this chapter, all mappings f and g are assumed to have their own parameters but they are omitted for brevity.

Just like latent variable models, autoencoders can be stacked together:

$$\boldsymbol{h}^{(l)}(t) = f^{(l)}\left(\boldsymbol{h}^{(l-1)}(t)\right) \tag{8.7}$$

$$\hat{\boldsymbol{h}}^{(l-1)}(t) = g^{(l)}\left(\hat{\boldsymbol{h}}^{(l)}(t)\right). \tag{8.8}$$

As before, the observations are taken into the equation by defining $\boldsymbol{h}^{(0)} := \boldsymbol{x}$. Furthermore, now $\hat{\boldsymbol{h}}^{(L)} := \boldsymbol{h}^{(L)}$ for the last layer L, connecting the encoder and decoder paths.

Typically over the course of learning, new layers are added to the previously trained network. After adding and training the last layer, training can continue in a supervised manner using just the mappings $f^{(l)}$, which define a multi-layer feedforward network, and minimizing the squared distance between the actual outputs $\boldsymbol{h}^{(L)}$ and desired targets' outputs.

It is tempting to assume that the hierarchical version of the autoencoder in Eqs. (8.7) and (8.8) corresponds somehow to the hierarchical latent variable model in Eq. (8.4). Unfortunately this is not the case because the intermediate hidden layers $0 < l < L$ act as so-called deterministic variables while the hierarchical latent variable model requires so-called stochastic variables. The difference is that stochastic variables have independent representational capacity. No matter what the priors tell, stochastic latent variables $\boldsymbol{s}^{(l)}$ can overrule this and add their own bits of information to the reconstruction. By contrast, deterministic variables such as $\hat{h}^{(l)}$ add zero bits of information and, assuming the deterministic mappings $g^{(l)}$ are implemented as layered networks, correspond to the hidden layers of the mappings $g^{(l)}$ between the stochastic variables $\boldsymbol{s}^{(l)}$ in Eq. (8.3).

In order to fix this, we are going to take a cue from the inference structure of the hierarchical latent variable model in Eq. (8.3). The main difference to Eq. (8.8) is that inference of $\hat{\boldsymbol{s}}^{(l)}(t)$ combines information from bottom-up likelihood and top-down prior but Eq. (8.8) only depends on top-down information. In other words, $\hat{\boldsymbol{h}}(t)$ in Eq. (8.8) cannot add any new information to the representation because it does not receive that information from the bottom-up path. All we need to do is add a shortcut connection from the bottom-up encoder path to the modified top-down decoder path:

$$\hat{\boldsymbol{h}}^{(l-1)}(t) = g^{(l)}\left(\hat{\boldsymbol{h}}^{(l)}(t), \boldsymbol{h}^{(l-1)}(t)\right). \tag{8.9}$$

Now $\hat{\boldsymbol{h}}^{(l)}$ can recover information which is missing in $\hat{\boldsymbol{h}}^{(>l)}$. In other words, the higher layers do not need to represent all the details. Also, the mapping $g^{(l)}$ can learn to combine abstract information from higher levels, such as "face," with detailed information about position, orientation, size, and so on, from lower layers. This means that the higher layers can focus on representing abstract invariant features if they seem more relevant to the task at hand than the more detailed information.

Figure 8.1 shows roughly the inference structure of a hierarchical latent variable model and compares it with a standard autoencoder and the ladder network. Note that while $\hat{\boldsymbol{h}}^{(l)}(t)$ combines information both from bottom-up and top-down paths in the ladder network, $\boldsymbol{h}^{(l)}(t)$ does not. This direct path from inputs to the highest layer means that training signals from the highest layers can propagate directly through the network in the same way as in supervised learning. Gradient propagation already combines information from bottom-up activations and top-down gradients so there is no need for extra mixing of information.

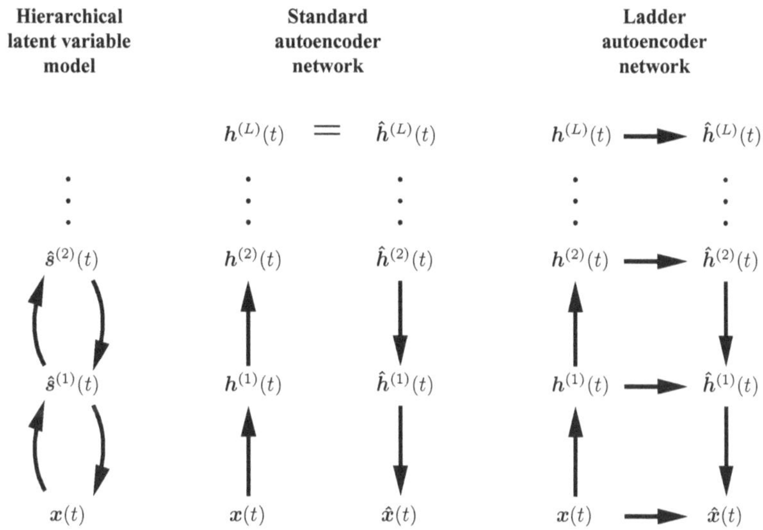

FIGURE 8.1

The inference structure of a hierarchical latent variable model is compared with the standard autoencoder and the proposed ladder network. The details of inference in latent variable models are often complex and the posterior distribution approximation is more complex than just the posterior mean $\hat{\boldsymbol{s}}^{(l)}(t)$, but overall the picture is approximately as shown in the left. Since all information in the standard autoencoder network has to go through the highest layer, it needs to represent all the details of the input $\boldsymbol{x}(t)$. Intermediate hidden layer activations $\hat{\boldsymbol{h}}^{(l)}(t)$ cannot independently represent information because they only receive information from the highest layer. In the ladder network, by contrast, lateral connections at each layer give a chance for each $\hat{\boldsymbol{h}}^{(l)}(t)$ to represent information independently from the higher layers. Also, abstract invariant representations at the higher levels can be interpreted in the context of detailed information without the higher levels having to represent all the details.

8.3 PARALLEL LEARNING ON EVERY LAYER

A general problem with deep models which have an error function only at the input layer (autoencoder) or at the output layer (supervised feedforward models) is that many parts of the network are far away from the source of training signals. In fact, if the ladder model shown in Figure 8.1 is trained in the same fashion as regular autoencoders, that is, by minimizing the difference between $\boldsymbol{x}(t)$ and $\hat{\boldsymbol{x}}(t)$, the problem only becomes worse. That is because each shortcut connection has a chance of contributing to the reconstruction $\hat{\boldsymbol{x}}(t)$, leaving a shrinking share of the error for the higher layers. Standard autoencoders force all training signals to pass through all levels in the hierarchy and even then learning through multiple layers of nonlinear functions is difficult and slow.

By contrast, hierarchical latent variable models have cost functions for all stochastic variables. Since the ladder network shares many properties with hierarchical latent variable models, it seems reasonable to try to find a way to introduce training signals at each level of the hierarchy of the ladder network. This section shows how this can be done by combining DSS framework [8] with training denoising functions to remove injected noise [9].

8.3.1 FROM NEURAL PCA TO DENOISING SOURCE SEPARATION

In order to develop a system where learning is distributed rather than guided by gradients propagating from a single error term, we shall turn our attention to competitive unsupervised learning.

The starting point for the algorithms we will study is the neural principal component analysis (PCA) learning rule by Oja [14] which can utilize second-order statistics of the input data to find principal component projections. When slight nonlinear modifications are made to the learning rule and input data are whitened, the method becomes sensitive to higher-order statistics and performs independent component analysis (ICA) [15].

The nonlinearity used in the algorithm can be interpreted as a contrast function which measures the non-Gaussianity of the source distribution. This is the interpretation originally given to the popular FastICA algorithm [16]. However, there is an alternative view: the nonlinearity can be interpreted as a denoising function. Hyvärinen [17] derived this as a maximum likelihood estimate but we are going to follow the derivation by Valpola and Pajunen [18] who showed that the nonlinearity can be interpreted as the expectation step of the expectation maximization algorithm. Overall, nonlinear PCA learning rule, combined with input whitening and orthogonalization of the projections, can be interpreted as an efficient approximation to the expectation maximization (EM) algorithm applied to a linear latent variable model tuned for ICA [18]. This interpretation led to the development of DSS framework [8].

The EM algorithm [19] is a method for optimizing the parametric mappings of latent variable models. It operates by alternating the E step (expectation of latent variables) and M step (maximization of likelihood of the parameters). The E step assumes the mapping fixed and updates the posterior distribution of $\boldsymbol{s}(t)$ for all t while the M step does the reverse, updating the mapping while assuming the posterior distribution of $\boldsymbol{s}(t)$ fixed.

The derivation by Valpola and Pajunen [18] assumed a linear reconstruction model

$$\hat{\boldsymbol{x}}(t) = g^{(0)}(\boldsymbol{s}(t)) = \boldsymbol{A}\boldsymbol{s}(t), \tag{8.10}$$

which means that the E step boils down to

$$\hat{\boldsymbol{s}}(t) = g^{(1)}(\boldsymbol{A}^{-1}\boldsymbol{x}(t)) = g^{(1)}(\boldsymbol{s}_0(t)), \tag{8.11}$$

where we have denoted $s_0(t) = A^{-1}x(t)$. The mapping $g^{(1)}$ depends on the prior distribution of $s(t)$ and the noise distribution p_n. When the noise has low variance σ_n^2, it can be approximated as

$$\hat{s}(t) = g^{(1)}(s_0(t)) \approx s_0(t) + \sigma_n^2 \left. \frac{\partial \log p_s(s(t))}{\partial s(t)} \right|_{s(t)=s_0(t)} . \tag{8.12}$$

The M step amounts to solving the regression problem in Eq. (8.10) with $\hat{s}(t)$ substituting $s(t)$, that is, minimizing the cost

$$C = \frac{1}{T} \sum_{t=1}^{T} \left\| x(t) - g^{(0)}\left(\hat{s}(t)\right) \right\|^2 . \tag{8.13}$$

The problem with the above EM algorithm for estimating the model is that its convergence scales with σ_n^2. Without noise, the algorithm stalls completely.

Valpola and Pajunen [18] showed that this can be remedied by considering the fixed point of the algorithm. At the same time, it is useful to parameterize the inverse mapping

$$s_0(t) = f^{(1)}(x(t)) = Wx(t) \tag{8.14}$$

and for M step minimize

$$C = \frac{1}{T} \sum_{t=1}^{T} \left\| s_0(t) - \hat{s}(t) \right\|^2 = \frac{1}{T} \sum_{t=1}^{T} \left\| f^{(1)}(x(t)) - \hat{s}(t) \right\|^2 \tag{8.15}$$

instead of Eq. (8.13). Equation (8.15) needs to be accompanied by constraints on the covariance of $s_0(t)$ because the trivial solution $W = \mathbf{0}$ and $\hat{s}(t) = \mathbf{0}$ yields $C = 0$.

The algorithm resulting from these assumptions is essentially the same as the nonlinear PCA learning rule. When the input data are whitened, the M step becomes simple matrix multiplication just as in the nonlinear PCA learning rule, amounting to essentially Hebbian learning. The nonlinearity $g^{(1)}(s_0(t))$ has the interpretation that it is the expected value of the latent variables given noisy observations, that is denoising.

What will be crucial for our topic, learning deep models, is that the cost function (Eq. 8.15) does not directly refer to the input $x(t)$ but only to the latent variable $s(t)$ and its denoised version. In a hierarchical model, this will mean that each layer contributes terms to the cost function, bringing the source of training signals close to the parameters on each layer.

Just as with the nonlinear PCA learning rule, there needs to be an additional constraint which implements competition between the latent variables because they could otherwise all converge to the same values. In the case of a linear model, the easiest approach is to require W to be orthogonal. This does not apply to nonlinear models. Instead, it is possible to require that the covariance matrix of the latent variables $s(t)$ is a unit matrix [20].

8.3.2 DENOISING AUTOENCODERS AND GENERATIVE STOCHASTIC NETWORKS

The denoising function used in Eq. (8.11) can be derived from the prior distribution p_s of the latent variables. There are many techniques for learning such distributions but a particularly useful technique that directly learns the denoising function was proposed by Vincent et al. [9] in connection with autoencoders. The idea is to corrupt the inputs fed into the autoencoder with noise and ask the network to reconstruct the original uncorrupted inputs. This forces the autoencoder to learn how to denoise the corrupted inputs.

Bengio and Thibodeau-Laufer [21] further showed that it is possible to sample from such models simply by iterating corruption and denoising. The distribution of the denoised samples converges to the original data distribution because during training, the denoising function learns to cancel the diffusion resulting from the corruption of input data. The diffusive forces are proportional to $-\frac{\partial \log p_x(\boldsymbol{x})}{\partial \boldsymbol{x}}$ and, on average, carry samples from areas of high density toward low densities. The denoising function learns to oppose this, with the same force but opposite sign.[c] When sampling starts with any given distribution, the combined steps of corruption and denoising produce an average flow of samples which only disappears when the diffusion flow caused by corruption exactly cancels the flow caused by denoising, that is, when the sample distribution follows the original training distribution. Bengio and Thibodeau-Laufer [21] suggested that sampling is more efficient in hierarchical models if corruption takes place not only on inputs but on all levels of the encoder path and called such networks generative stochastic networks (GSNs).

What is surprising is that from denoising functions it is even possible to derive probability estimates for the data. Note that the denoising function loses information about absolute probability and only conserves information about relative probabilities because the logarithm first turns multiplication into summation and the constant normalization term then disappears in differentiation. Such a representation bears similarity to energy-based probability models where only relative probabilities can be readily accessed. It turns out, however, that any model which can reconstruct missing data can be turned into a probability density estimator [22]. By using input erasure as corruption, the autoencoder can thus be used for deriving normalized probability estimates even if denoising function loses information about the normalization factor of the probability.

8.3.3 RECURSIVE DERIVATION OF THE LEARNING RULE

We are now ready to derive a learning rule with a distributed cost function for the ladder network. The basic idea is to apply a denoising autoencoder recursively. The starting point is the standard denoising autoencoder which minimizes the following cost C:

$$\tilde{\boldsymbol{x}}(t) = \text{corrupt}(\boldsymbol{x}(t)) \tag{8.16}$$

$$\hat{\boldsymbol{x}}(t) = g\left(\tilde{\boldsymbol{x}}(t)\right) \tag{8.17}$$

$$C = \frac{1}{T}\sum_{t=1}^{T} \left\|\boldsymbol{x}(t) - \hat{\boldsymbol{x}}(t)\right\|^2. \tag{8.18}$$

During learning the denoising function g learns to remove the noise which we are injecting to corrupt $\boldsymbol{x}(t)$. Now assume that the denoising function uses some internal variables $\boldsymbol{h}^{(1)}(t)$ for implementing a multi-layer mapping:

$$\tilde{\boldsymbol{h}}^{(1)}(t) = f^{(1)}\left(\tilde{\boldsymbol{x}}(t)\right) \tag{8.19}$$

$$\hat{\boldsymbol{x}}(t) = g\left(\tilde{\boldsymbol{h}}^{(1)}(t)\right). \tag{8.20}$$

Rather than giving all the responsibility of denoising to g, it is possible to learn first how to denoise $\boldsymbol{h}^{(1)}$ and then use that result for denoising $\boldsymbol{x}$:

$$\boldsymbol{h}^{(1)}(t) = f^{(1)}(\boldsymbol{x}(t)) \tag{8.21}$$

$$\tilde{\boldsymbol{h}}^{(1)}(t) = f^{(1)}\left(\tilde{\boldsymbol{x}}(t)\right) \tag{8.22}$$

$$\hat{\boldsymbol{h}}^{(1)}(t) = g^{(1)}\left(\tilde{\boldsymbol{h}}^{(1)}(t)\right) \tag{8.23}$$

$$\hat{\boldsymbol{x}}(t) = g^{(0)}\left(\hat{\boldsymbol{h}}^{(1)}(t)\right) \tag{8.24}$$

$$C^{(1)} = \frac{1}{T}\sum_{t=1}^{T} \left\|\boldsymbol{h}^{(1)}(t) - \hat{\boldsymbol{h}}^{(1)}(t)\right\|^2 \tag{8.25}$$

$$C^{(0)} = \frac{1}{T}\sum_{t=1}^{T} \left\|\boldsymbol{x}(t) - \hat{\boldsymbol{x}}(t)\right\|^2. \tag{8.26}$$

Training could alternate between training the mapping $g^{(1)}$ by minimizing $C^{(1)}$ and training all the mappings by minimizing $C^{(0)}$.

We can continue adding layers and also add the lateral connections of the ladder network:

$$\boldsymbol{h}^{(l)}(t) = f^{(l)}(\boldsymbol{h}^{(l-1)}(t)) \tag{8.27}$$

$$\tilde{\boldsymbol{h}}^{(l)}(t) = f^{(l)}\left(\tilde{\boldsymbol{h}}^{(l-1)}(t)\right) \tag{8.28}$$

$$\hat{\boldsymbol{h}}^{(l)}(t) = g^{(l)}\left(\tilde{\boldsymbol{h}}^{(l)}(t), \hat{\boldsymbol{h}}^{(l+1)}(t)\right) \tag{8.29}$$

$$C^{(l)} = \frac{1}{T}\sum_{t=1}^{T} \left\|\boldsymbol{h}^{(l)}(t) - \hat{\boldsymbol{h}}^{(l)}(t)\right\|^2. \tag{8.30}$$

As before, we assume that $\boldsymbol{h}^{(0)}$ refers to the observations $\boldsymbol{x}$.

This derivation suggests that the cost functions $C^{(l)}$ should only be used for learning the encoding mappings $f^{(>l)}$ of the layers above but not $f^{(\leq l)}$, those of the layers below, because $C^{(l)}$ is derived assuming $\boldsymbol{h}^{(l)}(t)$ is fixed. This is problematic because it means that there cannot be a single consistent cost function for the whole learning process and learning has to continuously alternate between learning different layers.

However, recall that in the DSS framework it is precisely the forward mapping $f^{(l)}$ which is updated using Eq. (8.15), practically the same equation as Eq. (8.30). This means that we can in fact minimize a single cost function

$$C = C^{(0)} + \sum_{l=1}^{L} \alpha_l C^{(l)}, \tag{8.31}$$

where the coefficients α_l determine the relative weights of the cost terms originating in different layers.

The DSS framework assumes that $\hat{\boldsymbol{h}}^{(l)}$ is constant and optimizes using gradients stemming from $\boldsymbol{h}^{(l)}$ while in the denoising autoencoder framework, the roles are reversed. This means that by combining the cost functions and learning everything by minimizing the cost with respect to all the parameters of the model, we are essentially making use of both types of learning.

Just like in hierarchical latent variable models, higher level priors offer guidance to lower-level forward mappings (cf. EM algorithm). Since the gradients propagate backward along the encoding path, this model is fully compatible with supervised learning: the standard supervised cost function can simply be added to the top-most layer L, measuring the distance between $\boldsymbol{h}^{(L)}(t)$ and the target output.

8.3.4 DECORRELATION TERM FOR THE COST FUNCTION

There is one final thing that we must take care of, the decorrelation term needed by DSS algorithms. Recall that Eq. (8.30) is minimized if $\boldsymbol{h}^{(l)}(t) = \hat{\boldsymbol{h}}^{(l)}(t) =$ constant. Minimization of Eq. (8.30) with respect to $\hat{\boldsymbol{h}}^{(l)}(t)$ actually typically promotes decorrelation because it amounts to regression and any extra information can be used to reduce the reconstruction error. Minimization of Eq. (8.30) with respect to $\boldsymbol{h}^{(l)}(t)$ promotes finding projections that can be predicted as well as be possible and, since mutual information is symmetric, therefore also help to predict other features as long as the entropy of the hidden unit activations is kept from collapsing by avoiding the trivial solution where $\boldsymbol{h}(t) =$ constant.

We are going to assume that the mappings $f^{(l)}$ and $g^{(l)}$ are sufficiently general that we can, without loss of generality, assume that the covariance matrix $\boldsymbol{\Sigma}^{(l)}$ of the hidden unit activations on layer l equals the unit matrix: $\boldsymbol{\Sigma}^{(l)} = \boldsymbol{I}$, where

$$\boldsymbol{\Sigma}^{(l)} = \frac{1}{T}\sum_{t=1}^{T} \boldsymbol{h}^{(l)}(t)\left[\boldsymbol{h}^{(l)}(t)\right]^{T}. \tag{8.32}$$

Here, we assumed that the average activation is zero, a constraint which we can also enforce without any loss of generality as long as the first stages of mappings $f^{(l)}$ and $g^{(l)}$ are affine transformations.

A very simple cost function to promote $\mathbf{\Sigma}^{(l)} \approx \boldsymbol{I}$ would be $\sum_{i,j}[\Sigma_{ij}^{(l)} - \delta_{ij}]^2$, where δ_{ij} is the Kronecker delta. In other words, this measures the sum of squares of the difference between $\mathbf{\Sigma}^{(l)}$ and $\boldsymbol{I}$. However, this cost function does not distinguish between too small and too large eigenvalues of $\mathbf{\Sigma}^{(l)}$ but from the viewpoint of keeping the DSS-style learning from collapsing the representation of $\boldsymbol{h}^{(l)}(t)$, only too small eigenvalues pose a problem. To analyze the situation, note that

$$\sum_{i,j}\left[\Sigma_{ij}^{(l)} - \delta(i,j)\right]^2 = \mathrm{tr}\left(\left[\mathbf{\Sigma}^{(l)} - \boldsymbol{I}\right]^2\right) = \sum_i \left(\lambda_i^{(l)} - 1\right)^2, \tag{8.33}$$

where $\lambda_i^{(l)}$ are the eigenvalues of $\mathbf{\Sigma}^{(l)}$. The first equality follows from the definition of the trace of a matrix and the second from the fact that the trace equals the sum of eigenvalues.

Since Eq. (8.33) is symmetric about $\lambda = 1$, it penalizes $\lambda = 0$ just as much as $\lambda = 2$ while the former is infinitely worse from the viewpoint of keeping $\boldsymbol{h}$ from collapsing.

A sound measure for the information content of a variable is the determinant of the covariance matrix because it measures the square of the (hyper)volume of the (hyper)cuboid whose sides have the length determined by the standard deviations of the distribution along its eigenvectors. Since the determinant of a matrix equals the product of its eigenvalues, the logarithm of the determinant equals the sum of the logarithms of the eigenvalues:

$$\log\det \mathbf{\Sigma}^{(l)} = \sum_i \log \lambda_i^{(l)} = \mathrm{tr}\left(\log \mathbf{\Sigma}^{(l)}\right). \tag{8.34}$$

The latter equality follows from the fact that any analytical function can be defined for square matrices so that it applies to the eigenvalues of the matrix. This is because $\left(\boldsymbol{E}\boldsymbol{\Lambda}\boldsymbol{E}^{-1}\right)^k = \boldsymbol{E}\boldsymbol{\Lambda}^k\boldsymbol{E}^{-1}$ and therefore any power series expansion of a matrix turns into the same power series expansion of the eigenvalues. Note that $\log \Sigma$ is the matrix logarithm, *not* the logarithm of the elements of the matrix.

Equation (8.34) is a measure which grows smaller when the information content diminishes but it can be turned into a sensible cost function which reaches its minimum value of 0 when $\lambda = 1$:

$$C_{\Sigma}^{(l)} = \sum_i \left(\lambda_i^{(l)} - \log \lambda_i^{(l)} - 1\right) = \mathrm{tr}\left(\mathbf{\Sigma}^{(l)} - \log \mathbf{\Sigma}^{(l)} - \boldsymbol{I}\right). \tag{8.35}$$

This cost penalizes $\lambda = 0$ infinitely and grows relatively modestly for $\lambda_i > 1$.

It is relatively simple to differentiate this cost with respect to $\mathbf{\Sigma}^{(l)}$ since, for any analytical function ϕ, it holds

$$\frac{\partial \mathrm{tr}\left(\phi(\mathbf{\Sigma})\right)}{\partial \mathbf{\Sigma}} = \phi'(\Sigma). \tag{8.36}$$

In our case $\phi(a) = a - \log a - 1$ and thus $\phi'(a) = 1 - a^{-1}$. We therefore have

$$\frac{\partial C_\Sigma^{(l)}}{\partial \boldsymbol{\Sigma}^{(l)}} = \boldsymbol{I} - [\boldsymbol{\Sigma}^{(l)}]^{-1}. \tag{8.37}$$

The rest of the formulas required for computing the gradients with the chain rule are straight-forward since Eq. (8.32) has a simple quadratic form.

Note that all twice differentiable cost functions that are minimized when $\lambda_i = 1$ have the same second-order behavior (up to scaling) close to the minimum so the simpler Eq. (8.33) works just as well if all λ_i are sufficiently close to 1. However, to avoid any potential problems, Eq. (8.35) was used in the experiments presented in this chapter.

Finally, as suggested earlier, we will add a simple term to the cost function to make sure that the hidden unit activations really have a zero mean:

$$\boldsymbol{\mu}^{(l)} = \frac{1}{T}\sum_{t=1}^{T} \boldsymbol{h}^{(l)}(t) \tag{8.38}$$

$$C_\mu^{(l)} = \left\| \boldsymbol{\mu}^{(l)} \right\|^2. \tag{8.39}$$

While DSS algorithms require decorrelation of the output representation, it is now also important to decorrelate (i.e., whiten) the inputs. Normally denoising autoencoders have a fixed input but now the cost functions on the higher layers can influence their input mappings and this creates a bias toward PCA-type solutions. This is because the amount of noise injected to $\boldsymbol{x}$ is relatively smaller for projections for which the variance of $\boldsymbol{x}$ is larger. The terms $C^{(\geq 1)}$ are therefore smaller if the network extracts mainly projections with larger variance, that is, PCA-type solutions. While PCA may be desirable in some cases, often it is not.

8.3.5 LEARNING RULE FOR THE LADDER NETWORK

We are now ready to collect together the recipe for learning the ladder network. Given (typically prewhitened) observations $\boldsymbol{h}^{(0)}(t) := \boldsymbol{x}(t)$, the cost function C is computed using the following formulas:

$$\boldsymbol{h}^{(l)}(t) = f^{(l)}\left(\boldsymbol{h}^{(l-1)}(t)\right) \quad \text{for } 1 \leq l \leq L \tag{8.40}$$

$$\tilde{\boldsymbol{h}}^{(0)}(t) = \text{corrupt}\left(\boldsymbol{h}^{(0)}(t)\right) \tag{8.41}$$

$$\tilde{\boldsymbol{h}}^{(l)}(t) = f^{(l)}\left(\tilde{\boldsymbol{h}}^{(l-1)}(t)\right) \quad \text{for } 1 \leq l \leq L \tag{8.42}$$

$$\hat{\boldsymbol{h}}^{(L)}(t) = g^{(L)}\left(\tilde{\boldsymbol{h}}^{(L)}(t)\right) \tag{8.43}$$

$$\hat{\boldsymbol{h}}^{(l)}(t) = g^{(l)}\left(\tilde{\boldsymbol{h}}^{(l)}(t), \hat{\boldsymbol{h}}^{(l+1)}(t)\right) \quad \text{for } 0 \leq l \leq L-1 \tag{8.44}$$

$$C^{(l)} = \frac{1}{T}\sum_{t=1}^{T}\left\|\boldsymbol{h}^{(l)}(t) - \hat{\boldsymbol{h}}^{(l)}(t)\right\|^2 \tag{8.45}$$

$$\boldsymbol{\Sigma}^{(l)} = \frac{1}{T}\sum_{t=1}^{T}\boldsymbol{h}^{(l)}(t)\left[\boldsymbol{h}^{(l)}(t)\right]^T \tag{8.46}$$

$$C_{\Sigma}^{(l)} = \mathrm{tr}\left(\boldsymbol{\Sigma}^{(l)} - \log\boldsymbol{\Sigma}^{(l)} - \boldsymbol{I}\right) \tag{8.47}$$

$$\boldsymbol{\mu}^{(l)} = \frac{1}{T}\sum_{t=1}^{T}\boldsymbol{h}^{(l)}(t) \tag{8.48}$$

$$C_{\mu}^{(l)} = \left\|\boldsymbol{\mu}^{(l)}\right\|^2 \tag{8.49}$$

$$C = C^{(0)} + \sum_{l=1}^{L}\alpha_l C^{(l)} + \beta_l C_{\Sigma}^{(l)} + \gamma_l C_{\mu}^{(l)}. \tag{8.50}$$

Learning the parameters of the mappings $f^{(l)}$ and $g^{(l)}$ is based on minimizing C. The simple solution is to apply gradient descent (stochastic or batch version) but basically any optimization method can be used; for example, nonlinear conjugate gradient or quasi-Newton methods. Whichever method is chosen, the existence of a single cost function to be minimized guarantees that learning converges as long as the minimization method performs properly.

Equation (8.42) could include corruption in the same manner as in GSN. However, to keep things simple, the experiments reported in this chapter only applied corruption to the input layer.

Figure 8.2 shows the computational diagram of the cost function of the ladder network. The two f paths going upward share their mappings $f^{(l)}$ and the only difference is that the inputs on the corrupted path are corrupted by noise. Each layer of the network adds its own term to the cost function, measuring how well the clean activations $\boldsymbol{h}^{(l)}(t)$ are reconstructed from the corrupted activations. During forward computations, information will flow from the observations toward the cost function terms along the arrows. During learning, gradients will flow from the cost function terms in the opposite direction. Training signals arriving along the clean f path correspond to DSS-style learning while the training signals in the denoising g path and corrupted f path correspond to the type of learning taking place in denoising autoencoders. The terms $C_{\Sigma}^{(l)}$ and $C_{\mu}^{(l)}$ which promote unit covariance and zero mean of the clean activations $\boldsymbol{h}^{(l)}(t)$, respectively, are not shown. They are required for keeping DSS-style learning from collapsing the representations and are functions of the clean f path only.

From the perspective of learning deep networks, it is important that any mapping, $f^{(l)}$ or $g^{(l)}$, is close to one of the cost function terms $C^{(l)}$. This means that learning is efficient even if propagating gradients through the mappings would not be efficient.

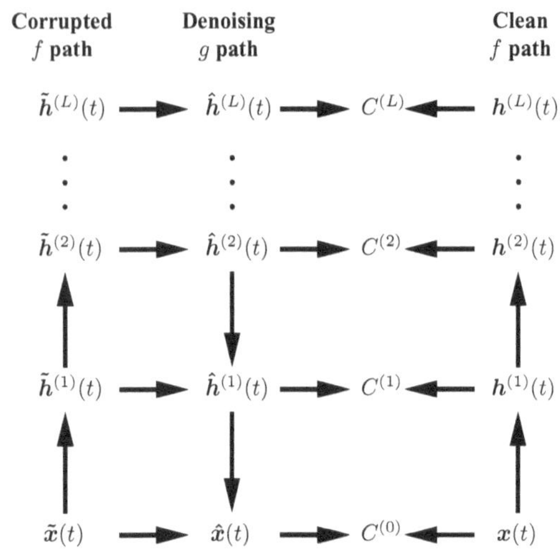

FIGURE 8.2

Ladder network's cost computations are illustrated. The clean f path shares exactly the same mappings $f^{(l)}$ as the corrupted f path. The only difference is that corruption noise is added in the corrupted path. The resulting corrupted activations are denoted by $\tilde{\boldsymbol{h}}^{(l)}(t)$. On each layer, the cost function has a term $C^{(l)}$, which measures the distance between the clean activations $\boldsymbol{h}^{(l)}(t)$ and their reconstructions $\hat{\boldsymbol{h}}^{(l)}(t)$. The terms $C_{\Sigma}^{(l)}$ and $C_{\mu}^{(l)}$ which measure how well the activations $\hat{\boldsymbol{h}}^{(l)}(t)$ are normalized are not shown.

8.4 EXPERIMENTS

This section presents a few simple experiments which demonstrate the key aspects of the ladder network:

- How denoising functions can represent probability distributions.
- How lateral connections relieve the pressure to represent every detail at the higher layers of the network and allow them to focus on abstract invariant features.
- How the cost function terms on higher layers speed up learning.

The three following sections will gradually develop a two-layered ladder network which can learn abstract invariant features. First, simple distributions are modeled. Then this is put into use in a linear ICA model with one hidden layer. Finally, a second layer is added which models the correlations between the variances of the first-layer activations. All the experiments used the set of learning rules described in Eqs. (8.40)–(8.50). The hyperparameters β_l were automatically adjusted to keep the smallest eigenvalue of $\boldsymbol{\Sigma}_l$ above 0.7. The hyperparameter γ_l was set to the same value as β_l.

8.4.1 REPRESENTING DISTRIBUTIONS WITH DENOISING FUNCTIONS

We will start by a simple experiment which elucidates the relation between the prior distribution of activations and their denoising functions, shown in Figure 8.3. Three different distributions were tested, super-Gaussian, sub-Gaussian, and Gaussian distributions. Each of them had a unit variance and zero mean. Each plot shows five different results overlaid on top of each other.

The super-Gaussian distribution was a Laplace distribution whose p.d.f. is

$$p(x) = \frac{1}{\sqrt{2}} \mathrm{e}^{-\sqrt{2}|x|}.$$

On a logarithmic scale, it behaves as $-\sqrt{2}|x|$ plus constant.

The sub-Gaussian distribution was generated by scaling a sinusoidal signal by $\sqrt{2}$ to obtain a variable with unit variance. The p.d.f. of this distribution is

$$p(x) = \frac{1}{\pi\sqrt{2 - x^2}}$$

for $|x| < \sqrt{2}$.

For this experiment, the model did not have any hidden layers ($L = 0$) and therefore there are no forward functions f, only one denoising function which was implemented as a single hidden neuron with tanh activation and a bypass connection:

$$\hat{x} = g(\tilde{x}) = \xi_1 \tilde{x} + \xi_2 \tanh(\xi_3 \tilde{x} + \xi_4) + \xi_5,$$

where ξ_i are scalar parameters. All parameters were optimized by minimizing $C^{(0)} = \sum_t \|\hat{x}(t) - x(t)\|^2$.

With small enough noise, the denoising functions should theoretically approach

$$\hat{x} \approx \tilde{x} + \sigma_{\mathrm{n}}^2 \frac{\partial \log p(x)}{x} \bigg|_{x=\tilde{x}},$$

where σ_{n}^2 is the variance of the noise used for corrupting $\tilde{x}$. With small corruption noise, this would then mean that the denoising function of the Laplacian input would be a sum of x and a scaled step function. This is not the case now since σ_{n}^2 is as large as the variance of the input. This tends to smoothen the denoising function.

As can be readily seen from Figure 8.3, the denoising function for a Gaussian observation is linear. Theoretically, the function should be

$$\hat{x} = \frac{\sigma_{\mathrm{x}}^2}{\sigma_{\mathrm{x}}^2 + \sigma_{\mathrm{n}}^2} \tilde{x},$$

where σ_{x}^2 is the variance of the observations. Since both variances equal one in these experiments, the theoretical optimum is $\hat{x} = \tilde{x}/2$. The estimated denoising function follows this very closely.

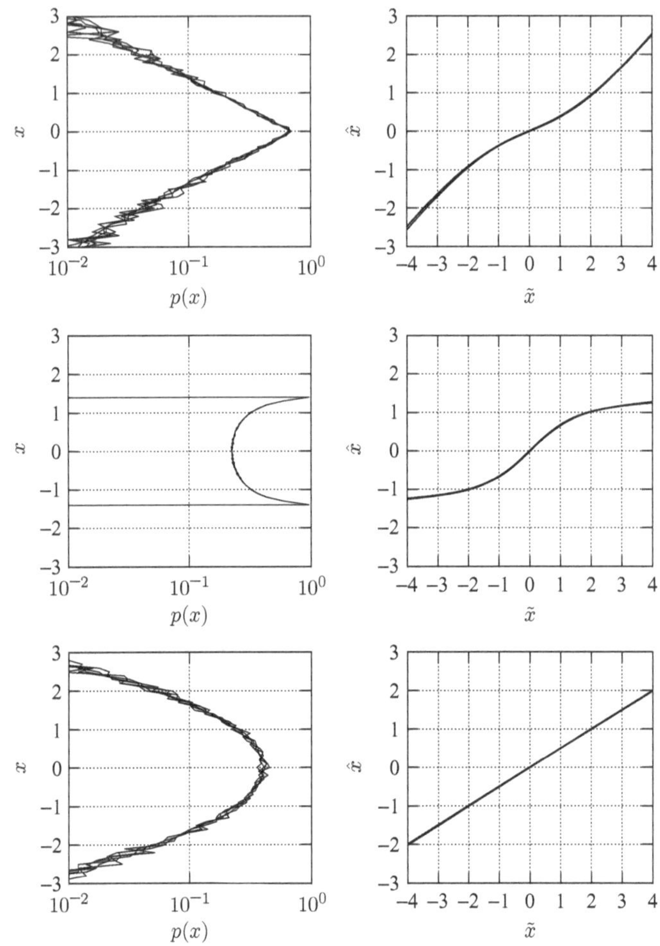

FIGURE 8.3

Illustration of the connection between the marginal distribution and denoising function. On the left, three different probability distributions are shown on a logarithmic scale. From top: super-Gaussian, sub-Gaussian, and Gaussian distributions. Each distribution has a unit variance and zero mean. On the right, denoising functions have been trained to remove corruptive noise with unit variance. In the Gaussian case, theoretically the optimal solution is $\hat{x} = \tilde{x}/2$. Each plot shows five different random samples plotted on top of each other. Note that in the plots showing the p.d.f. or the data, x is plotted on the vertical axis to help side-by-side comparison with the denoising function.

8.4.2 ICA MODEL

We will now move on to a simple linear ICA model which serves as an example of how statistical modeling will be translated into function approximation in this framework. It also gives some intuition on how the lateral connections help higher levels to focus on relevant features.

In linear ICA, these data are assumed to be a linear mixture of independent identically distributed sources. Unlike in PCA, the mixing is not restricted to be orthogonal because sources with non-Gaussian marginal distributions can be recovered. If more than one source has a Gaussian distribution, these sources cannot be expected to be recovered and will remain mixed. However, the subspace spanned by the Gaussian sources should be recoverable and separate from all the non-Gaussian sources.

One limitation is that unless there is some extra information available, the sources can only be recovered up to scaling and permutation. Scaling is usually fixed by assuming that source distributions have a unit variance. This will still leave permutation and sign of the sources ambiguous.

The dataset was generated by linearly mixing 10,000 samples from 15 sources into 15 observations. The elements of the mixing matrix were sampled from a zero-mean Gaussian distribution. The sources had the same distributions as in the previous example: five super-Gaussian, five sub-Gaussian, and five Gaussian sources. These data were *not* whitened because one purpose of the experiments was to demonstrate that normal autoencoders have a bias toward PCA solution even if the cost function terms $C^{(\geq 1)}$ are not used. In these experiments, α_l in Eq. (8.50) were set to zero.

8.4.2.1 Model structure

The model had one hidden layer ($L = 1$) and the only nonlinearity was on the hidden layer denoising which was a simplified version of the model used in the previous experiment. With essentially zero mean observations, there is no need for bias terms in the model. The mappings of the model are as follows:

$$f(\boldsymbol{x}) = \boldsymbol{W}\boldsymbol{x} \tag{8.51}$$

$$g_i^{(1)}(\boldsymbol{h}) = a_i h_i + b_i \tanh(h_i) \tag{8.52}$$

$$g^{(0)}(\boldsymbol{x}, \boldsymbol{h}) = \boldsymbol{A}\boldsymbol{h} + \boldsymbol{B}\boldsymbol{x}. \tag{8.53}$$

The parameters to be estimated are the three matrices $\boldsymbol{W}, \boldsymbol{A}$, and $\boldsymbol{B}$ and the two vectors $\boldsymbol{a}$ and $\boldsymbol{b}$, which were used for the denoising functions of individual hidden units.

Note that the underlying assumption of ICA models is that the sources are independent. This is incorporated in the model by making the denoising on the hidden layer unit-wise. Also, the lateral linear mapping with matrix $\boldsymbol{B}$ can model any covariance structure in the observation data. This means that the hidden layer should be able to focus on representing the non-Gaussian sources.

8.4.2.2 Results

Experiments verify that the model structure defined by Eqs. (8.51)–(8.53) is indeed able to recover the original sources that were used for generating the observed mixtures (up to permutation and scaling). This is apparent when studying the normalized loading matrix which measures how much the original sources contribute the values of the hidden units. The loading matrix is obtained from the product $\boldsymbol{W}\boldsymbol{A}_{\text{orig}}$, where $\boldsymbol{W}$ is the unmixing matrix learned by the model and $\boldsymbol{A}_{\text{orig}}$ is the original mixing matrix. Rows of this matrix measure how much contribution from the original source there is in each of the recovered hidden neuron activations. In the normalized loading matrix, the rows are scaled so that the squares sum up to one, that is, the row vectors have unit lengths. A successful unmixing is characterized by a single dominant loading in each row and can be measured by the average angle of each vector to the dominant source. In the experiments, a typical value was around 10° which corresponds to the contribution $\cos(10°) \approx 0.985$ from the dominant source.

The denoising mappings $g_i^{(1)}(\boldsymbol{h})$ for each source depend on the distribution of the source just as expected. In particular, the sign of the parameter b_i is determined by the super- or sub-Gaussianity of the sources. When there are more hidden units than non-Gaussian sources, the model will also represent Gaussian sources but the preference is for non-Gaussian sources. This is to be expected because the lateral mapping $\boldsymbol{B}$ at the lowest level of the network can already represent any Gaussian structure. In other words, the model performs just as expected.

What is more interesting is what happens if the lateral mapping $\boldsymbol{B}$ is missing. Since the reconstruction $\hat{\boldsymbol{x}}(t)$ can then only contain information which is present in the hidden units, the network has a strong pressure to conserve as much information as possible. Essentially the dominant mode of operation is then PCA: the network primarily extracts the subspace spanned by the eigenvectors of the data covariance matrix corresponding to the largest eigenvalues. The network can only secondarily align the representation along independent components. If the independent components do not happen to align with the principal subspace, PCA wins over ICA.

In one experiment, for instance, a network with $\boldsymbol{B}$ and 11 hidden units was able to retrieve the 10 non-Gaussian sources with loadings between 0.958 and 0.994, averaging 0.981. By contrast, the 10 best loadings in exactly the same setting but without $\boldsymbol{B}$ were between 0.499 and 0.824 and averaged 0.619 after the same number of iterations. This is not significantly different from random mappings which yield average loadings around 0.613 ± 0.026 (average $\pm$ std).

It turned out that the network was able to do better but it converged tremendously slowly, requiring about 100 times more iterations than with $\boldsymbol{B}$. Still, even after the network had seemingly converged, the 10 best loadings were between 0.645 and 0.956 and averaged 0.870. While this is clearly better than random, even the best loading was worse than the worst loading when the lateral connections $\boldsymbol{B}$ were used.

That this is due to the network's tendency to extract a principal subspace can be seen by analyzing how a large portion of the subspace spanned by $\boldsymbol{W}$ falls outside the subspace spanned by the 11 largest eigenvectors of the data covariance matrix. In the network with $\boldsymbol{B}$ this was 28% whereas in the network lacking $\boldsymbol{B}$ this was just 0.03%.

To be fair, it should be noted that prewhitening the inputs $\boldsymbol{x}(t)$ restores the autoencoder's ability to recover independent components just as it allows nonlinear PCA learning rule to perform ICA rather than principal subspace analysis. However, in more complex cases, it may not be as easy to normalize away the information that is not wanted.

8.4.3 HIERARCHICAL VARIANCE MODEL

We shall now move on to expanding the ICA model by adding a new layer to capture the nonlinear dependencies remaining in the hidden unit activations in the ICA model. This makes sense because it is usually impossible to produce truly statistically independent components simply by computing different linear projections of the observations. Even if the resulting feature activations lack linear correlations, there are normally higher-order dependencies between the features (for more discussion of such models, see [11,23]).

One typical example is that the variances of the activations are correlated because the underlying cause of a feature activation is likely to generate other activations, too. In order to find a network structure that could represent such correlated variances, let us recall that the optimal denoising of a Gaussian variable h with prior variance σ_{h} and Gaussian corruption noise σ_{n} is

$$\hat{h} = \frac{\sigma_{\mathrm{h}}^2}{\sigma_{\mathrm{h}}^2 + \sigma_{\mathrm{n}}^2}\tilde{h}.$$

This can be written as

$$\hat{h} = \frac{1}{1 + \sigma_{\mathrm{n}}^2/\sigma_{\mathrm{h}}^2}\tilde{h} = \mathrm{sigmoid}(\log \sigma_{\mathrm{h}}^2 - \log \sigma_{\mathrm{n}}^2)\tilde{h},$$

where the sigmoidal activation function is defined as

$$\mathrm{sigmoid}(x) = \frac{1}{1 + \mathrm{e}^{-x}}.$$

What this means is that information about the variance of h translates to a modulation of the connection strength in the denoising function mapping $\tilde{h}$ to $\hat{h}$.

The experiments in this section demonstrate that all we need to do to represent correlations in the variances in the hidden unit activations is add modulatory connections from the layer above and give enough flexibility for the forward mapping to the highest layer. The network will then find a way to use the representations of the highest layer. Moreover, lateral connections on the middle layer now play a crucial role because they allow the higher layers to focus on representing higher-order correlations. Unlike in the case with the ICA model in the previous section, this could not have been replaced by whitening the inputs. The experiments also demonstrate how the cost function terms on the higher layers speed up learning.

8.4.3.1 Data

The dataset of 10,000 samples was generated as a random linear mixture of sources s_i just as in the ICA experiment. This time, however, the sources were Gaussian with a changing variance. The variances of the sources were determined by higher-order variance sources, each of which was used by a group of four sources. There were four such groups, in other words, four higher-order variance sources which determined the variances of 16 sources. Such groups of dependent sources mean that the data follow the model used in independent subspace analysis [23].

The variance sources v_j were sampled from a Gaussian distribution and the variance σ_i^2 of the lower-level sources $i \in \mathbb{G}_j$ was obtained by computing e^{v_j}. The set $\mathbb{G}_j$ contain all the indices i for which the lower-level sources s_i are modulated by the variance source v_j. Since there were four nonoverlapping groups of four sources, $\mathbb{G}_1 = \{1, 2, 3, 4\}$, $\mathbb{G}_2 = \{5, 6, 7, 8\}$, and so on.

Note that although the sources are sampled from a Gaussian distribution, their marginal distribution is super-Gaussian since the variance is changing. In these experiments, the dataset was prewhitened.

8.4.3.2 Model structure

The linear mappings between the observations and the first layer were the same as with the ICA model but now a second nonlinear layer was added and denoising $g^{(1)}$ of the first layer was modified to make use of it:

$$
\begin{aligned}
f^{(1)}(\boldsymbol{x}) &= \mathbf{W}^{(1)}\boldsymbol{x} && (8.54)\\
f^{(2)}(\boldsymbol{h}^{(1)}) &= \mathrm{MLP}(\boldsymbol{h}^{(1)}) && (8.55)\\
g_i^{(2)}(\boldsymbol{h}^{(2)}) &= a_i h_i^{(2)} && (8.56)\\
g_i^{(1)}(\boldsymbol{h}^{(1)}, \boldsymbol{h}^{(2)}) &= \mathrm{sigmoid}(\boldsymbol{A}_i^{(2)}\boldsymbol{h}^{(2)} + b_i^{(1)})h_i^{(1)} && (8.57)\\
g^{(0)}(\boldsymbol{x}, \boldsymbol{h}^{(1)}) &= \boldsymbol{A}^{(0)}\boldsymbol{h}^{(1)} + \boldsymbol{B}\boldsymbol{x}. && (8.58)
\end{aligned}
$$

The multilayer perceptron (MLP) network used in the second-layer encoder mapping $f^{(2)}$ was

$$\mathrm{MLP}(\boldsymbol{h}) = \mathbf{W}^{(2b)}\psi(\mathbf{W}^{(2a)}\boldsymbol{h} + \boldsymbol{b}^{(2a)}) + \boldsymbol{b}^{(2b)}, \tag{8.59}$$

where the activation function $\psi(x) = \log(1 + e^x)$ operates on the elements of the vector separately. Note that we have included the bias term $\boldsymbol{b}^{(2b)}$ to make sure that the network can satisfy the constraint of having zero mean activations.

8.4.3.3 Results

Experiments verified that the network managed to separate individual source subspaces (Figure 8.4) and learned to model the correlations between the variances of different sources (Figure 8.5). The figures correspond to an experiment where the dimension of the first layer was 16 and second layer 10. The MLP network used for modeling $f^{(2)}$ had 50 hidden units. The figures show results after 1000 training for iterations. The results were relatively good already after 300 iterations and after 1000 iterations the network had practically converged.

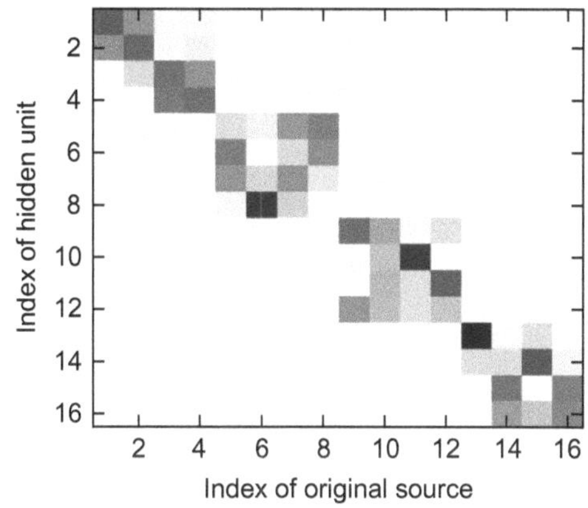

FIGURE 8.4

The squares of the normalized loading matrix from a hierarchical ladder network. Black corresponds to 1 and white to 0. As is evident from the plot, the four subspace $\mathbb{G}_j$ have been cleanly separated from each other but remain internally mixed. This is because the distribution of the sources within each subspace is spherically symmetric, which makes it impossible to determine the rotation within the subspace. This is reflected in the blocky structure of the matrix of loadings. The hidden units were ordered appropriately to reveal the block structure.

Figure 8.4 shows squares of the loadings, that is, element-wise squares of the matrix $\boldsymbol{W}^{(1)}\boldsymbol{A}_{\text{orig}}$, scaled such that the squares sum up to one for each hidden unit. Since these data were generated by modulating the variance of the Gaussian sources of each subspace, the data distribution was spherically symmetric within each subspace. This is reflected in the blocky appearance of the matrix of loadings as the network has settled on a random rotation within the subspace. Note that such rotation indeterminacy is a property of the input data. In other experiments, the model was fed with the ICA data from the previous section and readily separated the sources (results not shown here).

Figure 8.5 shows the learned denoising functions. Each plot shows how the term $\text{sigmoid}(\boldsymbol{A}_1^{(2)}\boldsymbol{h}^{(2)} + b_1^{(1)})$ belonging to the first hidden unit in Eq. (8.57) behaves as a function of one of the hidden neurons $h_i^{(1)}$. The sigmoid term is a function of all the hidden neuron activations through the second-layer hidden unit activations. Theoretically, the sigmoid term modulating the mapping from $\tilde{h}_1^{(1)}$ to $\hat{h}_1^{(1)}$ should be a function of the norm of the activation vector of subspace $\mathbb{G}_1$, that is, a function of $\sum_{i\in\mathbb{G}_1}[h_i^{(1)}]^2$. As is readily seen from the plots, the second-layer activations have apparently learned to represent such quadratic terms of first-layer activations because the sigmoid indeed appears to be a function of the norm of the activation vector of the first subspace. When ICA data were used (results not shown here), the sigmoid term learned to neglect all the other hidden units and developed to be a function of $h_1^{(1)}$ only (assuming we are looking at the sigmoid of hidden unit 1).

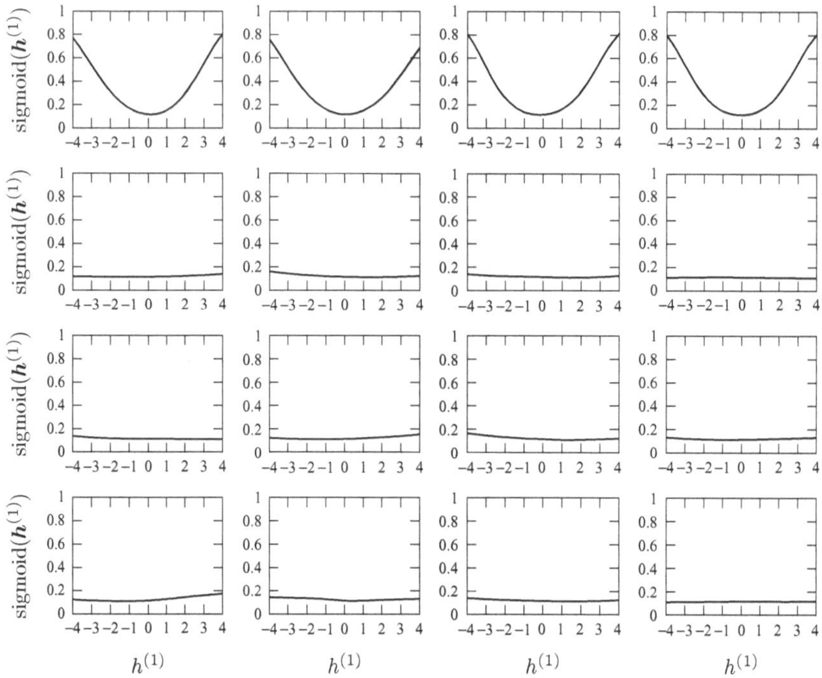

FIGURE 8.5

Illustration of one of the denoising functions learned by a hierarchical ladder network. Each plot shows how the term sigmoid$(\boldsymbol{A}_1^{(2)}\boldsymbol{h}^{(2)} + b_1^{(1)})$ modulating $\tilde{h}_1^{(1)}$ in Eq. (8.57) behaves as a function of one of the hidden neurons $h_i^{(1)}$. In each plot, $h_i^{(1)}$ for just one i takes nonzero values. The top row shows how sigmoid changes as a function of the hidden units, which belong to the same subspace $\mathbb{G}_1$ as $h_1^{(1)}$ whose sigmoid is shown. The rest of the plots correspond to hidden units from other groups $\mathbb{G}_{\neq 1}$.

The variance features developing on the second layer of the model can only improve the reconstruction of $\hat{\boldsymbol{x}}$ if they are combined with $\boldsymbol{h}^{(1)}$ because variance alone cannot say anything about the direction where the reconstruction should be changed. Without the shortcut connections in the ladder model, the highest layer with only 10 hidden units[d] could not have learned to represent the variance sources because there would not have been enough space to also represent the activations $\boldsymbol{h}^{(1)}$, which are also needed to make use of the variance sources. The higher layers can only let go of the details because they are recovered again when denoising proceeds from the highest layers toward the lowest.

Another important question was whether the terms of the cost function originating in the higher layers of the network really are useful for learning. To investigate this, 100 simulations were run with different datasets and random initializations of the network. It turned out that in this particular model, the higher layer cost function term $C^{(2)}$ was not important but $C^{(1)}$ could speed up learning considerably particularly during the early stages of learning. As expected, it was crucial to combine it with a proper decorrelation term $C_{\Sigma}^{(1)}$. The success of the model was measured by the value $C^{(0)}$ that was reached. Note that it is not *a priori* clear that adding other cost function terms could help reduce $C^{(0)}$. Nevertheless, this turned out to be the case. By iteration 100, the network consistently reached a lower value of $C^{(0)}$ than by iteration 200 when minimizing $C^{(0)}$ alone (as in standard denoising autoencoders). Subsequent learning continued approximately at the same pace which seems reasonable as denoising autoencoders should be able to optimize the model after it has been initialized close to a sensible solution.

Another interesting finding was that about one-third of the improvement seems to be attributable to the decorrelation term $C_{\Sigma}^{(1)}$. It was able to speed up learning alone without $C^{(1)}$ despite initializing the network with mappings which ensure that all the representations start out as decorrelated. Whereas the speedup of $C^{(1)}$ was most pronounced in the beginning of learning, the speedup offered by $C_{\Sigma}^{(1)}$ was more important during the middle phases of learning. Presumably this is because the representations start diverging from the uncorrelated initialization gradually.

At the optimum of the cost function $C^{(0)}$, the addition of any extra term can only make the situation worse from the viewpoint of minimizing $C^{(0)}$. It is therefore likely that optimally the weights α_l and β_l in Eq. (8.50) should be gradually decreased throughout training and could be set to zero for a final finetuning phase. For simplicity, all α_l were kept fixed in the simulations presented here.

8.5 DISCUSSION

The experiments verified that the ladder model with lateral shortcut connections and cost function terms at every level of the hierarchy is indeed able to learn abstract invariant features efficiently. Although the networks studied here only had a few layers (no more than six between $\tilde{x}$ and $\hat{x}$ no matter how they are counted), what is important is that the representations were abstract and invariant already on the second layer. In fact, the model with two layers, linear features on the first and variance features on the second, corresponds roughly to the architecture with simple and complex cells found by Hubel and Wiesel [1] (for a more detailed discussion see [23]). Promising as the results are, it is clearly necessary to conduct far larger experiments to verify that the ladder network really does support learning in much deeper hierarchies.

Similarly, it will be important to verify that the ladder network is indeed compatible with supervised learning and can therefore support useful semi-supervised learning. All the experiments reported in this chapter were unsupervised. However, all the results supported the notion that the shortcut connections of the ladder network allow it to discard information, making it a good fit with supervised learning.

One of the most appealing features of the approach taken here is that it replaces all probabilistic modeling with function approximation. In these experiments, an MLP network learned to extract higher-level sources that captured the dependencies in the first-level sources. This model corresponded to independent subspace analysis not because the model was tailored for it but because the input data had that structure. The forward mapping $f^{(2)}$ was very general, an MLP network. In these experiments, the denoising mapping $g^{(1)}$ had a somewhat more limited structure mainly to simplify analysis of the results. It will be interesting to study whether $g^{(l)}$ can also be replaced by a more general mapping.

Another important avenue for research will be to take advantage of all the machinery developed for GSN and related methods, such as sampling, calculating probability densities, and making use of multiple rounds of corruption and denoising during learning [21,22,24]. Particularly, the ability to sample from the model should be very useful. In order to make full use of these possibilities, the corruption procedure should be extended. Now, simple Gaussian noise was added to the inputs. Noise could be added at every layer to better support sampling [21] and could also involve masking out some elements of the input vector completely [22]. If different types of corruption are needed at different times, it might be possible to extend the denoising functions to handle different types of corruption (information about the corruption strategy could be provided as side information to the denoising functions) or it might be possible to relearn just the denoising functions $g^{(l)}$ while keeping the previously learned forward mappings $f^{(l)}$ fixed.

A crucial aspect of the ladder network is that it captures the essential features of the inference structure of hierarchical latent variable models. Along the same line, it should be possible to extend the model to support even more complex inferences, such as those that take place in Kalman filters.

The model studied by Yli-Krekola [25] takes even one step further: it implements a dynamical biasing process which gives rise to an emergent attention-like selection of information in a similar fashion as in the model suggested by Deco and Rolls [26]. The model studied by Yli-Krekola [25] is derived from the DSS framework and already has a structure which is reminiscent of the ladder architecture presented here. The model is otherwise very elegant but is prone to overfit its lateral connections and exaggerate the feedback loops between units. By using the same tricks as here, injecting noise for the benefit of learning lateral and top-down denoising, it might be possible to learn the lateral connections reliably.

8.6 CONCLUSIONS

In this chapter, a ladder network structure was proposed for autoencoder networks. The network's lateral shortcut connections give each layer the same representational capacity as stochastic latent variables have in hierarchical latent variable models. This allows the higher levels of the network to discard information and focus on representing more abstract invariant features.

In order to support efficient unsupervised learning in deep ladder networks, a new type of cost function was proposed. The key aspect is that each layer of the network contributes its own terms to the cost function. This means that every mapping in the network receives training signals directly from some term which measures local reconstruction errors. In addition to the immediate training information, the network also propagates gradient information throughout the network. This means that it is also possible to add terms which correspond to supervised learning.

The price to pay is that each higher-level cost function needs to be matched with a decorrelation term which prevents the representation from collapsing. This is analogous to the competition used in unsupervised competitive learning. Additionally, it is often useful to decorrelate the inputs because otherwise the network is biased toward finding a PCA solution.

Preliminary experiments verified that the network was able to learn abstract invariant features and that the extra terms in the cost function speed up learning. The experiments support the notion that the model scales to very deep models and works well together with supervised learning but much larger experiments are still required to verify these claims.

ACKNOWLEDGMENTS

I would like to thank Tapani Raiko and Antti Rasmus for useful discussions. Antti Rasmus has been running experiments in parallel to this work and his input on how different versions performed has been invaluable. Jürgen Schmidhuber, Kyunghyun Cho, and Miquel Perelló Nieto have made available collections of citations, which have saved plenty of my time when preparing the manuscript.

Last but certainly not least, I would like to thank Erkki Oja for creating the environment where the ideas which underly the work presented here have been able to develop. Erkki has always supported my research. His example has shown how it is possible to follow intuition in designing unsupervised learning algorithms but he has also always emphasized the importance of rigorous analysis of the convergence and other properties of the resulting algorithms. Without this combination and his pioneering work in neural PCA, nonlinear PCA learning rule, and ICA, none of the research reported here would have gotten very far.

NOTES

a. Unsupervised learning aims at representing structure in the input data, often by means of features. The resulting features can be used as input for classification tasks or as initialization for further supervised learning.

b. Consider, for example, the ImageNet classification problem that Krizhevsky et al. [7] tackled. With 1000 target classes, each label carries less than 10 bits of information. Compare this with the amount of information contained in the 256×256 RGB images used as input. It is impossible to say exactly how many bits of information each image carries but certainly several orders of magnitude more than 10 bits.

c. Notice the similarity to the denoising in Eq. (8.12).

d. In principle, the highest layer only needs four hidden units to represent the four variance sources. The network was indeed able to learn such a compact representation but learning tended to be slower than with 10 hidden units.

REFERENCES

[1] D.H. Hubel, T. Wiesel, Receptive fields, binocular interaction, and functional architecture in the cat's visual cortex, J. Physiol. Lond. 160 (1962) 106-154.
[2] K. Fukushima, Neural network model for a mechanism of pattern recognition unaffected by shift in position – neocognitron, Trans. IECE J62-A (10) (1979) 658-665.
[3] J. Schmidhuber, Deep learning in neural networks: an overview. Neural Networks 61 (2015) 85-117.
[4] G. Hinton, R. Salakhutdinov, Reducing the dimensionality of data with neural networks, Science 313 (5786) (2006) 504-507.
[5] G.E. Hinton, S. Osindero, Y.-W. Teh, A fast learning algorithm for deep belief nets, Neural Comput. 18 (7) (2006) 1527-1554.
[6] D.C. Ciresan, U. Meier, L.M. Gambardella, J. Schmidhuber, Deep big simple neural nets for handwritten digit recognition, Neural Comput. 22 (2010) 3207-3220.
[7] A. Krizhevsky, I. Sutskever, G.E. Hinton, ImageNet classification with deep convolutional neural networks, in: Advances in Neural Information Processing Systems (NIPS 2012), 2012, pp. 1106-1114.
[8] J. Särelä, H. Valpola, Denoising source separation, J. Mach. Learn. Res. 6 (2005) 233-272.
[9] P. Vincent, L. Hugo, Y. Bengio, P.-A. Manzagol, Extracting and composing robust features with denoising autoencoders, in: Proceedings of the 25th International Conference on Machine Learning (ICML '08), Helsinki, Finland, 2008, pp. 1096-1103.
[10] C. Bishop, Latent variable models, in: M. Jordan (Ed.), Learning in Graphical Models. MIT Press, Cambridge, MA, 1999, pp. 371-403.
[11] H. Valpola, M. Harva, J. Karhunen, Hierarchical models of variance sources, Signal Process. 84 (2) (2004) 267-282.
[12] K. Gregor, I. Danihelka, A. Mnih, C. Blundell, D. Wierstra, Deep autoregressive networks, in: Proceedings of the 31st International Conference on Machine Learning, ICML'14, Beijing, China, 2014, pp. 1242-1250.
[13] A. Ilin, H. Valpola, On the effect of the form of the posterior approximation in variational learning of ICA models, in: Proc. of the 4th Int. Symp. on Independent Component Analysis and Blind Signal Separation (ICA2003), Nara, Japan, 2003, pp. 915-920.
[14] E. Oja, A simplified neuron model as a principal component analyzer, J. Math. Biol. 15 (1982) 267-273.
[15] E. Oja, The nonlinear PCA learning rule in independent component analysis, Neurocomputing 17 (1) (1997) 25-46.
[16] A. Hyvärinen, E. Oja, A fast fixed-point algorithm for independent component analysis, Neural Comput. 9(7) (1997) 1483-1492.
[17] A. Hyvärinen, Independent component analysis in the presence of Gaussian noise by maximizing joint likelihood, Neurocomputing 22 (1-3) (1998) 49-67.
[18] H. Valpola, P. Pajunen, Fast algorithms for Bayesian independent component analysis, in: Proc. Int. Workshop on Independent Component Analysis and Blind Signal Separation (ICA2000), Helsinki, Finland, 2000, pp. 233-237.
[19] A.P. Dempster, N.M. Laird, D.B. Rubin, Maximum likelihood from incomplete data via the EM algorithm, J. R. Stat. Soc. Ser. B 39 (1) (1977) 1-38.

[20] M.S.C. Almeida, H. Valpola, J. Särelä, Separation of nonlinear image mixtures by denoising source separation, in: Proceedings of the 6th International Conference on Independent Component Analysis and Blind Signal Separation, ICA 2006, Charleston, SC, USA, 2006, pp. 8-15.

[21] Y. Bengio, É. Thibodeau-Laufer, G. Alain, J. Yosinski, Deep generative stochastic networks trainable by backprop, in: Proceedings of the 31st International Conference on Machine Learning, ICML'14, Beijing, China, 2014, pp. 226-234.

[22] B. Uria, I. Murray, H. Larochelle, A deep and tractable density estimator, in: Proceedings of the 31st International Conference on Machine Learning, ICML'14, Beijing, China, 2014, pp. 467-475.

[23] A. Hyvärinen, P. Hoyer, Emergence of phase and shift invariant features by decomposition of natural images into independent feature subspaces, Neural Comput. 12 (7) (2000) 1705-1720.

[24] T. Raiko, L. Yao, K. Cho, Y. Bengio, Iterative neural autoregressive distribution estimator (NADE-k), in: Advances in Neural Information Processing Systems 27 (NIPS 2014), 2014, pp. 325-333.

[25] A. Yli-Krekola, A bio-inspired computational model of covert attention and learning (MA thesis), Helsinki University of Technology, Finland, 2007.

[26] G. Deco, E.T. Rolls, A neurodynamical cortical model of visual attention and invariant object recognition, Vision Res. 44 (2004) 621-642.

PART

Applications

II

CHAPTER

Two decades of local binary patterns: A survey

9

Matti Pietikäinen and Guoying Zhao

Center for Machine Vision Research, Department of Computer Science and Engineering, University of Oulu, Finland

9.1 INTRODUCTION

Texture is an important characteristic of many types of images. It can be seen in images ranging from multispectral remotely sensed data to microscopic images. Texture can play a key role in a wide variety of applications of computer vision and image analysis. Therefore, the analysis of textures has been a topic of intensive research since the 1960s. Most of the proposed methods have not been, however, capable to perform well enough for real-world textures.

In recent years, very discriminative and computationally efficient local texture descriptors have been developed, such as local binary patterns (LBPs), which have led to significant progress in applying texture methods to different problems and applications. The focus of research has broadened from 2-D textures to 3-D textures and spatiotemporal (dynamic) textures. Due to this progress, the division between texture descriptors and more generic image or video descriptors has been disappearing.

The LBP operator can be seen as a unifying approach to the traditionally divergent statistical and structural models of texture analysis. Perhaps the most important property of the LBP operator is its robustness to monotonic grayscale changes caused, for example, by illumination variations. Another important property is its computational simplicity, which makes it possible to analyze images in challenging real-time settings.

The original LBP was invented two decades ago, published first at ICPR 1994 and then in *Pattern Recognition* [1], but at that time, one could not imagine what a great success it would be today. In the 1990s, LBP did not receive interest in the scientific community, because it was regarded as an ad hoc method. However, the promising power of LBP was already evident, because it performed much better in texture classification and segmentation tasks than the state of the art at that time [2]. Due to its computational simplicity and good performance, LBP was also successfully used in some applications such as industrial inspection. The theoretical foundations of LBP became much more clear after our research on multidimensional distributions of signed gray-level differences, carried out jointly with Prof. Erkki Oja and Dr. Kimmo Valkealahti [3].

E. Bingham, S. Kaski, J. Laaksonen and J. Lampinen (Eds) Advances in Independent Component Analysis and Learning Machines.
http://dx.doi.org/10.1016/B978-0-12-802806-3.00009-9

To a large extent, the scientific community accepted LBP after its generalized version was published in *IEEE PAMI* journal [4], and further it was shown to be highly successful in face recognition, published first at ECCV 2004 and then in *IEEE PAMI* [5]. In 2014, the ECCV paper was awarded with the prestigious Koenderink Prize for Fundamental Contributions in Computer Vision. Different types of applications of LBP to motion analysis have been proposed after the spatiotemporal LBP was introduced, also in *IEEE PAMI* [6]. All these papers are highly cited, reflecting the increasing popularity of LBP. Due to this success of LBP, a book describing the basic methods and surveying different variants and applications was published [7]. Another survey on LBP and its applications in face analysis appeared in Ref. [8]. An edited book on some LBP variants and applications was published in 2013 [9].

After these surveys, which covered the progress until 2010, the interest on LBP has been growing further. LBP is no longer just a simple texture operator, but it forms the foundation for a new direction of research dealing with local binary image and video descriptors. Many different variants of LBP have been proposed to improve its robustness and to increase its discriminative power and applicability to different types of problems. A large number of new papers have been published in leading journals and conferences. Due to its discriminative power and computational simplicity, the LBP operator has become highly popular in various applications, including facial image analysis, biometrics, medical image analysis, motion and activity analysis, and content-based retrieval from image and video databases. Figure 9.1 depicts the number of citations in Web of Science and Google Scholar to the landmark LBP paper published in 2002 [4], showing a clear increase especially after LBP was successfully adopted for face recognition in 2006 and its spatiotemporal version was proposed in 2007.

Based on all these, the publication of this survey is very timely, covering the progress until early 2014. It complements and extends our preliminary survey presented in Sections 2.9 and 3.5 of Ref. [7], by focusing on the most important recent developments, new types of variants, and future challenges.

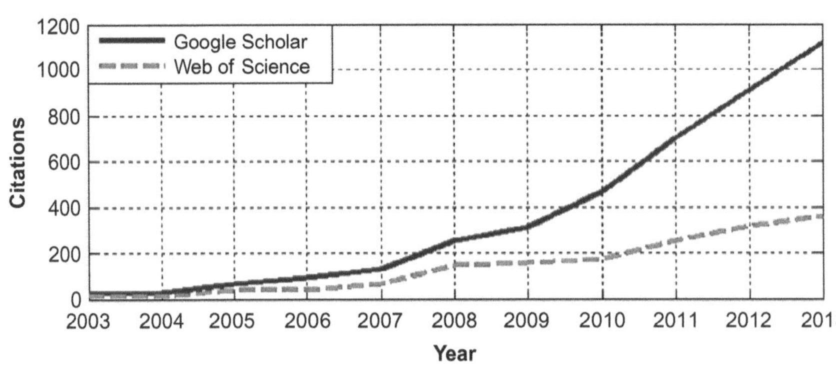

FIGURE 9.1

Annual citations to the PAMI 2002 paper on LBP.

9.2 AN OVERVIEW OF BASIC LBP OPERATORS

The LBP is based on the assumption that texture has locally two complementary aspects, a pattern and its strength. The pixels of an image are labeled by thresholding the neighborhood of each pixel, and the result is considered as a binary number. The distribution of the LBP labels computed over a region is then used for texture description.

The original LBP operator shown in Figure 9.2 works in a 3 × 3 neighborhood, using the center value as a threshold [1]. The thresholded values are multiplied with weights of the corresponding pixels, and by summing up the result an LBP code is obtained. The contrast measure *C* is obtained by subtracting the average gray level of the pixels below the threshold from that of those above (or equal to) the center pixel. If all neighbors of the center pixel have the same value (1 or 0), the value of *C* is set to 0.

The distributions of LBP codes or 2-D distributions of LBP and *C* are used as features in texture recognition. The LBP operator was extended to use neighborhoods of different sizes in Ref. [4]. Using a circular neighborhood and bilinearly interpolating values at noninteger pixel coordinates allow any radius and number of pixels in the neighborhood. The multiscale LBP (MLBP) is obtained by concatenating histograms produced by operators at different radii. The grayscale variance of the local neighborhood can be used as the complementary contrast measure.

Ojala et al. [4] found that some of the LBP patterns occur more frequently than others, representing, for example, edges, curves, line ends, flat areas, and spots. Based on this observation, the so-called uniform patterns were defined to reduce the number of patterns. Uniform patterns have been widely used and were necessary, for example, to reduce the length of the feature vector in face description.

In their recent paper, Guo et al. [10] proposed a completed modeling of the LBP operator and developed an associated completed LBP (CLBP) scheme for texture classification. The local differences are decomposed into two complementary components: the signs (like LBP) and the magnitudes (Figure 9.3). The magnitude component provides an effective alternative for the complementary contrast measure of LBP. In addition to this texture-related information, they also include information about image intensity in their representation.

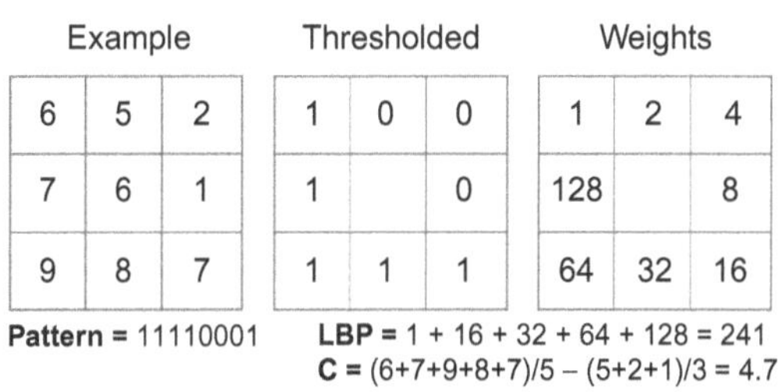

FIGURE 9.2

The original LBP.

9	12	34
10	25	28
99	64	56

(a)

−16	−13	9
−15		3
74	39	31

(b)

−1	−1	1
−1		1
1	1	1

(c)

16	13	9
15		3
74	39	31

(d)

−1	−1	−1
−1		−1
1	1	1

(e)

FIGURE 9.3

CLBP: (a) 3 × 3 sample block, (b) the local differences, (c) the sign, (d) magnitude components, and (e) thresholded binary value of (d) against the average magnitude in whole image region.

Dynamic texture descriptors provide a very effective new tool for motion analysis. Zhao and Pietikäinen [6] extended the original LBP to a spatiotemporal representation for dynamic texture analysis. For this purpose, the so-called volume local binary pattern (VLBP) operator was proposed, in which dynamic texture is considered as a set of volumes in the (X, Y, T) space where X and Y denote the spatial coordinates and T denotes the frame index (time). The neighborhood of each pixel is thus defined in three-dimensional space. The VLBP combines motion and appearance together to describe dynamic textures. To make VLBP computationally simple and easy to extend, an operator based on co-occurrences of LBPs on three orthogonal planes (LBP-TOPs) considers three orthogonal planes – *XY*, *XT*, and *YT* – and concatenates LBP co-occurrence statistics in these three directions as shown in Figure 9.4. The circular neighborhoods are generalized to elliptical sampling to fit to the space-time statistics.

The first application problems to which spatiotemporal LBP was adopted were facial expression recognition, face and gender recognition, lip-reading, and action recognition [7].

Facial images are composed local texture patterns, and based on this observation, Ahonen et al. [5] introduced a method for combining local and global information for face (object) description. The face image is divided into several regions from which the LBP feature distributions are extracted and concatenated into an enhanced feature vector to be used as a face descriptor.

This approach has emerged as one of the key paradigms for facial image analysis and has also been adopted to other application domains. The length of the feature vector is an important design parameter in this description. Ahonen et al. used uniform patterns to reduce the dimensionality. Figure 9.5 illustrates the use of LBP for face description [5].

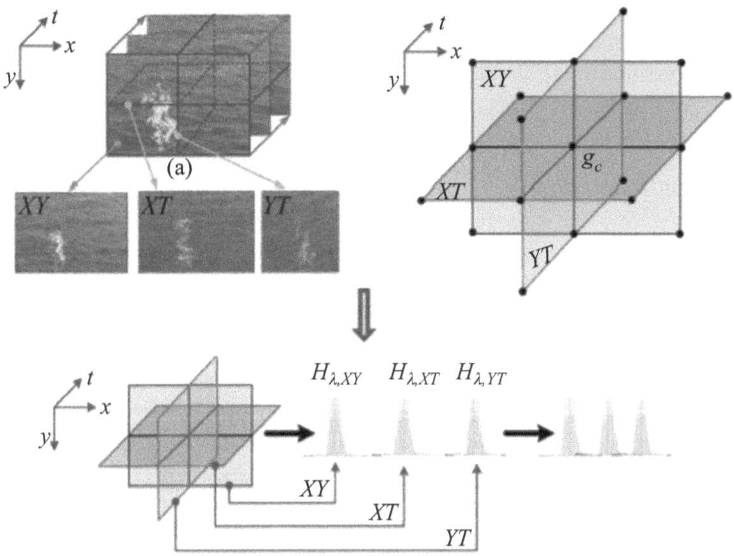

FIGURE 9.4

LBP-TOP representation for spatiotemporal description.

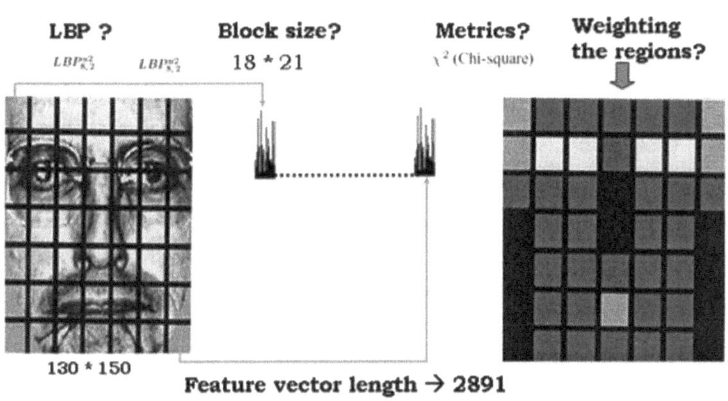

FIGURE 9.5

Face description using LBP.

9.3 LBP VARIANTS IN THE SPATIAL DOMAIN

9.3.1 DIFFERENT TYPES OF VARIANTS

Due to its flexibility, the LBP method can be easily modified to make it suitable for the needs of different types of problems. Several extensions and modifications of LBP have been proposed with an aim to increase its robustness and discriminative power. In this section, different variants are divided into such categories that describe their roles in feature extraction. Some of the variants could belong to more than one category, but only the most obvious categories are considered here. The choice of a proper method for a given application depends on many factors, such as the discriminative power, computational efficiency, robustness to lighting and other variations, and the imaging system used.

9.3.1.1 Preprocessing

In many applications, it is useful to preprocess the input image prior to LBP feature extraction. Especially, multiscale Gabor filtering and edge detection have been used for this purpose. Gabor filtering has been widely used before LBP computation in face recognition. A motivation for this is that methods based on Gabor filtering and LBP provide complementary information: LBP captures small and fine details, while Gabor filters encode appearance information over a broader range of scales. For example, Zhang et al. [11] proposed the extraction of LBP features from images obtained by filtering a facial image with 40 Gabor filters of different scales and orientations. The extracted features are called local Gabor binary patterns (LGBPs). A downside of the method is the high dimensionality of the LGBP representation.

Tan and Triggs [12] developed a very effective preprocessing chain for compensating illumination variations in face images. It is composed of gamma-correction, difference of Gaussian filtering, masking (optional), and equalization of variation. This approach has been very successful in LBP-based face recognition under varying illumination conditions. When using it for the basic LBP, the last step can be omitted due to LBP's invariance to monotonic grayscale changes.

In many recent studies, edge detection has been used prior to LBP computation to enhance the gradient information. Gradient images are more insensitive to lighting variations than the original images. Perhaps the first one was Yao and Chen [13] proposing local edge patterns (LEPs) to be used with color features for color texture retrieval. In LEPs, the Sobel edge detection and thresholding are used to find strong edges, and then LBP-like computation is used to derive the LEPs.

Li et al. [14] presented an approach based on capturing the intrinsic structural information of face appearances with multiscale heat kernel matrices. Heat kernels perform well in characterizing the topological structural information of face appearance. Histograms of LBPs computed for nonoverlapping blocks are then used for face description.

Also, other types of preprocessing have been applied with LBP, including wavelets and momentograms, as described in Sections 9.3.1.4 and 9.3.1.9.

9.3.1.2 Neighborhood topology and sampling

One important factor that makes the LBP approach so flexible to different types of problems is that the topology of the neighborhood from which the LBP features are computed can be different, depending on the needs of the given application.

The extraction of LBP features is usually done in a circular or square neighborhood. A circular neighborhood is important especially for rotation-invariant operators. However, in some applications, such as face recognition, rotation invariance is not required, but anisotropic information may be important. To exploit this,

Liao and Chung [15] proposed an elliptical binary pattern for face recognition. Nanni et al. [16] investigated the use of different neighborhood topologies (circle, ellipse, parabola, hyperbola, and Archimedean spiral) and encodings in their research on LBP variants for medical image analysis. An operator using quinary encoding in an elliptic neighborhood (EQP) provided the best performance. He et al. [17] used rotation-invariant LBP with elliptic sampling in four directions together with circular sampling to get anisotropic and isotropic information. A multistructure LBP (Ms-LBP) was achieved by applying this operator at different layers of an image pyramid. A downside of the method is that quite large image samples are needed for extracted macrostructures. Wolf et al. [18] considered different ways of using bit strings to encode the similarities between patches of pixels, which could capture complementary information to pixel-based descriptors. They proposed a three-patch LBP (TPLBP) and four-patch LBP (FPLBP). For each pixel in TPLBP, for example, a $w \times w$ patch centered at the pixel and S additional patches distributed uniformly in a ring of radius r around it are considered. Then, the values for pairs of patches located on the circle at a specified distance apart are compared with those of the central patch. The value of a single bit is set according to which of the two patches is more similar to the central patch. The code produced will have S bits per pixel. In FPLBP, two rings centered on the pixel were used instead of one ring in TPLBP.

Orjuela et al. [19] presented geometric local textural patterns (GLTPs), which are based on exploring intensity changes on oriented neighborhoods. An oriented neighborhood describes a particular geometry composed of points on circles with different radii around the center pixel. A digital representation of the points on the oriented neighborhood defines a GLTP code. The simple case called geometric local binary pattern (GLBP) is based on Boolean comparisons.

Wang et al. [20] proposed a sampling structure based on combining pixel and patch to mimic the retinal sampling grid (pixel to patch [PTP]). Also, a neighboring intensity relationship (NIR) operator was proposed to complement LBP texture information by exploring grayscale properties between neighborhoods. Two rotation-invariant descriptors were also proposed: the local intensity relationship pattern (LNIRP) based on the NIR operator and LNIRP_PTP using the PTP sampling structure. The method is computationally simple, has small feature dimensionality, and is training-free. Ylioinas et al. [21] introduced a dense sampling approach through the form of upsampling to extract more stable and discriminative texture patterns in local regions. Experiments on face recognition, texture classification, and age group estimation problems on various challenging benchmark databases demonstrate the efficiency of the proposed scheme.

9.3.1.3 Thresholding and encoding

A drawback of the LBP method, as well as of all local descriptors that apply vector quantization, is that they are not robust in the sense that a small change in the input image would always cause a small change in the output. LBP may not work properly for noisy images or on flat image areas of constant gray level. This is due to the thresholding scheme of the operator.

Instead of using the value of the center pixel for thresholding in the local neighborhood, other techniques have also been considered. Hafiane et al. [22] proposed median binary pattern operator by thresholding the local pixel values, including the center pixel, against the median within the neighborhood. The so-called improved LBP, on the other hand, compares the values of the neighboring pixels against the mean gray level of the local neighborhood [23].

Tan and Triggs [12] proposed a three-level operator called local ternary patterns (LTPs) (using one threshold T), for example, to deal with problems on near constant image areas. In ternary encoding, the difference between the center pixel and a neighboring pixel is encoded by three values (1, 0, and −1) according to a threshold T. The ternary pattern is divided into two binary patterns taking into account its positive and negative components. The histograms from these components computed over a region are then concatenated. Figure 9.6 depicts an example of splitting a ternary code into positive and negative codes.

In order to make the LBP more robust against negligible changes in pixel values, the thresholding scheme of the operator was modified in Ref. [24] by adding a small constant a in local difference computations. The bigger the value of $|a|$, the bigger the allowed changes in pixel values without affecting the thresholding results. In order to retain the discriminative power of the LBP operator, a relatively small value should be used. An advantage of this modified LBP compared with the three-valued LBPs described above is that the feature vector length remains the same as in the ordinary LBP. Heikkilä et al. [25] proposed center-symmetric LBP (CS-LBP) to reduce the feature vector length of the original LBP. The center-symmetric pairs of pixels in the neighborhood are compared, instead of comparing all neighbors with the center pixel. Therefore, the CS-LBP produces only 16 patterns instead of 256 for the basic LBP. An interest region descriptor based on CS-LBP compared favorably to the SIFT operator in image matching and categorization experiments, especially for images with lighting variations. A similar thresholding approach as in Ref. [24] was used to improve the robustness of the CS-LBP descriptor.

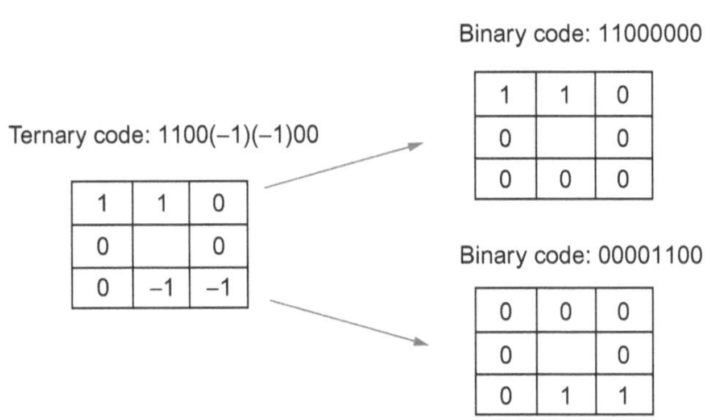

FIGURE 9.6

Local ternary pattern (LTP) operator.

Nanni et al. [16] studied the effects of different encodings of the local grayscale differences, using binary (B), ternary (T), and quinary (Q) encodings. In binary coding, the difference between a neighboring pixel and the center pixel is encoded by two values (0 and 1) as in LBP; in ternary encoding, it is encoded by three values as in LTP; and in quinary encoding, it is encoded by five values ($-2, -1, 0, 1$, and 2) according to two thresholds (T_1 and T_2). A quinary code can be split into four binary LBP codes.

A downside of the methods using one or two thresholds is that the methods are not strictly invariant to local monotonic gray-level changes as the original LBP. The feature vector lengths of these operators are also longer.

A soft three-valued LBP using fuzzy membership functions was proposed by Ahonen and Pietikäinen [26] to improve the robustness. In soft LBP, one pixel typically contributes to more than one bin in the histogram. An extensive study on generalized fuzzy LBPs was carried out by Katsigiannis et al. [27]. A disadvantage of the fuzzy methods is their increased computational cost.

Liao et al. [28] noticed that adding the small threshold (T) in LTP is not invariant under scaling of intensity values. The intensity scale-invariant property of a local comparison operator is very important, for example, in background modeling, because lighting variations, either global or local, often cause sudden changes of grayscale intensities of neighboring pixels simultaneously, which would approximately be a scale transform with a constant factor. Therefore, a scale-invariant local ternary pattern operator was developed for dealing with the grayscale intensity changes in complex background. They also proposed a pattern kernel density estimation technique to effectively model the probability distribution of local patterns in the pixel process. Ylioinas et al. [29] adopted a related method for image appearance description. Compared to the standard LBP histogram statistics where one labeled pixel always contributes to one bin of the histogram, the proposed method exploits a kernel-like similarity function to determine weighted votes contributing several possible pattern types in the statistic. As a result, the method yields a more reliable estimate of the underlying LBP distribution of the given image, providing improved performance especially for small-sized samples.

Trefny and Matas [30] proposed two encoding schemes, which are complementary to the standard LBPs and also invariant to monotonic intensity transformations. The binary value transition coded LBP is composed of neighbor pixel comparisons in a clockwise direction for all pixels except the central pixel (CI), encoding the relation between neighboring pixels. Direction coded LBP is related to the CS-LBP operator but uses also center pixel information for encoding. Intensity variation along each of the four basic directions is coded into two bits. The first bit encodes whether the center pixel is an extreme point and the second bit encodes whether the difference of border pixels compared to the center pixel grows or falls.

Liu et al. [31] extracted two different and complementary types of features, pixel intensities and differences, from local patches. The intensity-based features consider the intensities of the CI and its neighbors (NI), while for the difference-based feature, two components are computed: the radial difference (RD) and the angular difference

(AD). Two intensity-based descriptors CI-LBP and NI-LBP and two difference-based descriptors RD-LBP and AD-LBP were proposed.

Inspired by LBP, higher-order local derivative patterns (LDPs) were proposed by Zhang et al. [32], with applications in face recognition. The basic LBP represents the first-order circular derivative pattern of images, a micropattern generated by the concatenation of the binary gradient directions as was shown in Ref. [33]. The higher-order derivative patterns extracted by LDP will provide more detailed information but may also be more sensitive to noise than in LBP. The length of the feature vector is also four times the length of LBP, and the issue of rotation invariance was not addressed. To address these problems, Guo et al. [34] proposed local directional derivative patterns (LDDPs), a special case of which is the LBP. Similar to LBP, rotation invariance can be easily defined also for LDDP. Best recognition accuracy can be obtained by combining first- and second-order LDDPs.

Hussain and Triggs [35] tackled the problem of earlier LBP variants based on hand-specific codings, which limits them to small spatial support and coarse gray-level comparisons. They proposed local quantized patterns (LQPs), using vector quantization to code larger or deeper patterns. Precomputed lookup tables are used to make coding very fast. Their approach outperformed HOG, LBP, and LTP in challenging object detection and texture classification problems. Later, Huang et al. [36] introduced CLQP to include also magnitude (like in CLBP) and orientation information. A better and much faster way for initializing the vector quantization was also proposed.

Jiang et al. [37] proposed gradient LBPs for human detection. After LBP is calculated for each pixel, a 56-bin histogram is built with the co-occurrence of width value and angle value of LBPs. Width is the number of value 1s in the binary code of this pixel. Eight direction codes with 0-7 are defined in the direction of eight NIs. Angle value is the direction code of the middle pixel in 1 area of its binary code. Magnitude of gradient is used as weight to this 56-bin histogram.

9.3.1.4 Multiscale analysis

From a signal processing point of view, the sparse sampling used by the MLBP operator [4] may not result in an adequate representation of the signal, resulting in aliasing effects. Due to this, some low-pass filtering would be needed to make the operator more robust. From the statistical point of view, however, even sparse sampling is acceptable provided that the number of samples is large enough. The sparse sampling is commonly used, for example, with the methods based on grayscale difference or co-occurrence statistics. Mäenpää and Pietikäinen [38] proposed to use Gaussian low-pass filters for collecting texture information from a larger area than the original single pixel. The filters and sampling position were designed to cope with the neighborhood while minimizing the redundant information. With this approach, the radii of the LBP operators used in the multiscale version grow exponentially.

Another extension of MLBP operator is the multiscale block local binary pattern (MB-LBP) [39], which has gained popularity especially in facial image analysis. The key idea of MB-LBP is to compare average pixel values within small blocks instead of comparing pixel values. Instead of the fixed uniform pattern mapping, MB-LBP was proposed to be used with a mapping that is dynamically learned from training data. In this mapping, the N most often occurring MB-LBP patterns receive labels $0.N-1$, and all the remaining patterns share a single label. The number of labels, and consequently the length of the MB-LBP histogram, is a parameter the user can set.

A straightforward way for multiscale analysis is to utilize a pyramid of the input image computed at different resolutions and then concatenate LBP distributions computed from different levels of the pyramid. In their research on contextual analysis of textured scene images, Turtinen and Pietikäinen [40] combined this kind of idea with the original MLBP approach: image patches at three different scales were resized to the same size, and then LBP features were computed using LBPs with three different radii.

He et al. [41] developed a pyramid-based Ms-LBP for texture classification. It is obtained by executing the LBP on different layers of image pyramid, allowing to extract both micro- and macrostructures from textures. Five templates are used for creating the pyramid. The first one is a 2-D Gaussian function used to smooth the image. Other four anisotropic filters are used to create anisotropic subimages of the pyramid in four directions. Qian et al. [42] also extended LBP to pyramid transform domain. Different ways of sampling in pyramid construction are considered, including no sampling, partial sampling, and spatial pyramid sampling, of which the last one leads to the smallest computational cost.

Song and Li [43] combined wavelets and LBP. They built up the image description using a hierarchical framework based on low-dimensional wave LBP features, which not only extracts multiscale-oriented features and local image patterns but also captures multilevel (pixel, patch, and image) features. A very competitive performance was demonstrated in experiments.

Liu et al. [44] proposed a computationally simple approach to multiscale analysis with their binary rotation-invariant and noise-tolerant texture descriptor. Points are sampled in a circular neighborhood, but the number of bins in a single scale LBP histogram is kept constant and small by averaging over several contiguous pixels in the circle before binarization. This allows to encode a large number of scales and also reduces the effects of noise. Both sign and magnitude components, like in CLBP, are considered.

9.3.1.5 Handling rotation and scale variations

LBPs have been used for rotation-invariant texture recognition since the late 1990s [45]. In the most widely used version proposed in Ref. [4], the neighboring n binary bits around a pixel are clockwise-rotated n times. A maximal number of the most significant bits are used to express this pixel.

Guo et al. [46] developed an adaptive LBP (ALBP) by incorporating the directional statistical information for rotation-invariant classification. The directional statistical features, specifically the mean and standard deviation of the local absolute difference, are extracted and used to improve the LBP classification efficiency. In addition, the least square estimation is used to adaptively minimize the local difference for more stable directional statistical features. Garcia-Olalla et al. [47] extended the ALBP method to an approach called adaptive local binary pattern with oriented standard deviation, which adds an oriented standard deviation term to the LBP operator instead of using this information in the matching. Excellent results were obtained when assessing boar sperm vitality.

In Ref. [48], LBP variance (LBPV) was proposed as a rotation-invariant descriptor. For LBPV, there are three stages:

1. using the local contrast (grayscale variance) as an adaptive weight to adjust the contribution of the LBP code in histogram calculation,
2. learning the principal directions: the extracted LBPV features are used to estimate the principal orientations, and then the features are aligned to the principal orientations, and
3. determining the nondominant patterns and thus by reducing them, feature dimension reduction was achieved.

Zhang et al. [49] proposed monogenic LBP, which integrates the traditional rotation-invariant LBP operator with two other rotation-invariant measures: the local phase and local surface type computed by the first- and second-order Riesz transforms, respectively. The local phase corresponds to a qualitative measure of local structure (step, peak, etc.), whereas the monogenic curvature tensor extracts local surface type information. Zhao et al. [50] extracted grayscale difference information but totally abandoned structural information by proposing a completed local binary count operator. Their results suggest that microstructure is not always needed for rotation-invariant classification.

Zhao et al. [51] proposed an approach to compute rotation-invariant features from histograms of local, noninvariant patterns. They applied this approach to both static and dynamic LBP descriptors. For static textures, they presented local binary pattern histogram Fourier (LBP-HF) features. LBP-HF is computed from discrete Fourier transforms of LBP histograms. The approach can also be generalized to embed any uniform features into this framework, and combining supplementary information, for example, sign and magnitude components of LBP, together can improve the description ability. Moreover, two variants of rotation-invariant LBP-HF descriptors were proposed for LBP-TOP, which is not rotation-invariant. Experiments showed that LBP-HF and its extensions perform very well in rotation-invariant texture classification. They are also robust with respect to changes in viewpoint, outperforming recent methods proposed for view-invariant recognition of dynamic textures.

Li et al. [52] considered scale- and rotation-invariant LBP. A circular neighboring set of a pixel is defined as a self-adaptive texton. The optimal scale of each pixel is adaptively selected based on the scale space derived by the Laplacian of Gaussian and used to determine the radius of the scale-adaptive texton. Different pixels have different optimal scales, resulting in scale invariance. The subuniform patterns of each uniform pattern are defined to improve the discrimination, and the circular shift LBP histogram is computed to obtain rotation invariance.

9.3.1.6 Considering co-occurrences

Recent studies show that encoding co-occurrences of LBPs can significantly improve the performance. Co-occurrences can be considered in different ways, within the same LBP operator, between adjacent operators, or at region level.

Qi et al. [53] introduced a pairwise rotation-invariant co-occurrence LBP (PRI-CoLBP), which incorporates two types of context: spatial co-occurrence and orientation co-occurrence. The method aims to preserve the relative angle between the orientations of individual features. The relative angle provides information about local curvature. For each co-occurrence pattern, the gradient magnitude of the two points is used to weight the copattern. Excellent results using a three-scale PRI-CoLBP are reported for many different datasets. The length of the feature vector (590 for the single scale operator and 6×590 when extracting six co-occurrence patterns in two angles and three scales) may limit the applicability of this and also many other co-occurrence methods in certain applications, such as face recognition. Qi et al. [54] also proposed a rotation-invariant multiscale joint encoding of LBP (MSJ-LBP) operator. In the original MLBP, each scale is encoded into histograms separately, and then the histograms are concatenated. This ignores the correlation between different scales. MSJ-LBP encodes jointly the LBPs of two scales around the center point to capture their correlation. For each point at two chosen scales, its joint MS-LBP pattern (one of the 590 patterns) and its gradient magnitude are computed. The gradient magnitude is used to weight the given joint pattern.

Nosaka et al. [55] proposed a co-occurrence among adjacent LBP (CoALBP) operator. The co-occurrence of adjacent LBPs is defined as an index of how often their combination occurs in the whole image (or region). Co-occurrence is measured with an autocorrelation matrix generated from multiple LBPs. Later, they extended it to rotation-invariant co-occurrence among adjacent LBP operator, which is enabled by introducing the concept of rotation equivalence class to CoALBP [56]. Louis and Plataniotis [57] considered co-occurrences of rotation-invariant LBPs (CoLBPs) at region level. Basic rotation-invariant LBPs are computed for all possible scales in the examined scanning window. Then, instead of computing histograms, multiple instances of rotation-invariant LBPs are selected using the sequential forward selection algorithm.

9.3.1.7 Handling color

LBP has also been widely applied to color images. To describe color and texture jointly, opponent color LBP (OCLBP) was defined in Ref. [58]. In OCLBP, the operator is used on each color channel independently and then for pairs of color channels so that the center pixel is taken from one channel and the neighboring pixels from the other. Opposing pairs, such as R-G and G-R, are highly redundant, so either of them can be used in the analysis. In total, six histograms (out of nine) are utilized (R, G, B, R-G, R-B, and G-B), making the descriptor six times longer than the monochrome LBP histogram. The OCLBP descriptor fares well in comparison with other color texture descriptors. It has been later used successfully, for example, for face recognition. However, the authors of this paper did not recommend joint color and texture description as in their experiments: "all joint color texture descriptors and all methods of combining color and texture on a higher level are outperformed by either color or gray-scale texture alone."

A popular way is to apply the ordinary LBP to different color channels separately. Instead of the original R, G, and B channels, other more discriminative and invariant color features derived from them can be used for LBP feature extraction as well. Along this line, Zhu et al. [59] proposed multiscale color LBPs for visual object class recognition. Six operators were defined by applying MLBP on different types of channels and then concatenating the results. From these, the hue LBP (computed from the hue channel of the HSV color space), opponent LBP (computed over all three channels of the opponent color space), and an opponent LBP (computed over two channels of the normalized opponent space) provided good performance in experiments. Later, Zhu et al. [60] extended their approach to image region description using orthogonal combination of LBPs (OC-LBPs) and six new local descriptors based on OC-LBP enhanced with color information. OC-LBP is obtained by combining the histograms of $P/4$ different 4-orthogonal neighbor operators. The dimension of the descriptor is $4 \times P$, which is linear with the number of neighboring pixels in comparison to 2P for the original LBP. Color OC-LBP descriptor is obtained by concatenating OC-LBP histograms computed over selected color channels. Experiments on image matching, object recognition, and scene classification were used to show the effectiveness of the approach.

Qi et al. [61] presented an approach to encode cross channel texture correlation for color texture classification. The texture correlation between different RGB channels is first empirically studied using LBP as texture descriptor and Shannon's information entropy as correlation measurement. For color texture description, pairwise color channels are jointly encoded to obtain cross channel LBPs (CCLBPs). A multiscale CCLBP is also developed. Excellent results were reported.

Banerji et al. [62] present new image descriptors based on color, texture, shape, and wavelets for object and scene image classification. A three-dimensional LBP descriptor is proposed for encoding both color information and texture information and H-descriptor to integrate the 3-D LBP and the HOG of its wavelet transform, to encode color, texture, shape, and local information. The H-descriptor is comparatively assessed on seven well-known color spaces. A new H-fusion is also presented by fusing the principal component analysis (PCA) features of the H-descriptors in the seven color spaces. Experimental results on the Caltech 256 object categories, the UIUC sports event, and the MIT scene datasets demonstrate excellent performance.

9.3.1.8 Handling noise

Sensitivity to noise in images is one of the key problems of the original LBP. Kylberg and Sintorn [63] evaluated the noise robustness of eight LBP variants on five different datasets. None of the descriptors was generally more noise robust for all datasets and noise levels. However, LTP was often among the best performing descriptors, and improved LBP [23] often performed slightly better than the original LBP. Median LBP [22] and local quinary pattern [16] performed worst.

Chen et al. [64] proposed a simple robust version of LBP (i.e., RLBP) by changing the coding of bits of LBP, which could otherwise be changed by noise. Experimental results on texture datasets demonstrated that the RLBP outperforms many widely used descriptors and other variants of LBP, especially when noise is added in the images. Experimental results in face recognition also provided very promising results. Ren et al. [65] introduced a noise-resistant LBP aiming to preserve the local image structures in the presence of noise. A small pixel difference is vulnerable to noise, and thus it is first encoded as uncertain, and then its value is determined based on the other bits of LBP code to form a code of local uniform pattern. They also proposed extended noise-resistant LBP to capture line patterns. Experiments with added Gaussian and uniform noise on various datasets demonstrate the efficiency of their approach. Zhao et al. [66] proposed a completed robust local binary pattern, in which each center pixel is replaced with the average of the neighborhood, as in improved LBP. To make the operator more robust and stable, a weighted local gray level is also introduced to replace the value of the center pixel. The reported results for the UIUC database are clearly worse than those of Chen et al. [64].

9.3.1.9 Combining local and global information

LBP reflects the correlation among pixels within a local area (e.g., 3×3 area for $LBP_{8,1}$), which mainly represents the local information. Recently, there have been many works combining more global or more local information with LBP, for getting a more discriminative description from different feature levels.

Liao and Chung [67] extracted first dominant patterns, the so-called advanced LBPs from images, and labeled their locations with "1" and "0" otherwise. The gray-level aura matrix was used to extract the spatial information from each binary image. Guo et al. [68] investigated rotation-invariant image description with a linear model-based descriptor named MiC, which is suited to modeling microscopic configuration of images. To explore multichannel discriminative information on both the microscopic configuration and local structures, the feature extraction process is formulated as an unsupervised framework. It consists of (1) the configuration model to encode image microscopic configuration and (2) local patterns to describe local structural information. In this way, images are represented by a novel feature: local configuration pattern. Khellah [69] proposed a method that computes global rotation-invariant features from estimated dominant neighborhood structure and combines them with local LBP features, providing very good performance also in experiments with additive Gaussian noise.

Covariance matrices (CovMs) capture correlation among elementary features of pixels over an image region. Ordinary LBP features cannot be used as elementary features, since they are not numerical variables in Euclidean spaces. To address this problem, Hong et al. [70] developed a powerful descriptor, named COV-LBP. Firstly, a variant of LBPs in Euclidean spaces, named the LBP difference (LBPD) feature, was proposed. LBPD reflects how far one LBP lies from the LBP mean of a given image region. Secondly, applying LBPD together with some other features provided a bank of discriminative features for CovMs, providing very good performance in experiments. Papagostas et al. [71] introduced moment-based LBPs. Their approach consists of two steps: the momentogram construction and the application of LBP on it. As a result, an enhanced LBP histogram is obtained, which is invariant under common geometric transformations (translation, rotation, and scaling), enclosing local and global information.

9.3.1.10 Complementary descriptors

A current trend in the development of new effective local image and video descriptors is to combine the strengths of complementary descriptors. From the beginning, the LBP operator was designed as a complementary measure of local image contrast. An interesting alternative for putting the local contrast into the one-dimensional LBP histogram was proposed by Guo et al. [48] in their LBPV method. As presented earlier, Guo et al. [10] also introduced the CLBP operator, in which the magnitude component is used as an improved contrast measure. Magnitude LBP and LBPV both contain supplementary information to LBP. They were embedded to the histogram Fourier framework [51] and concatenated to LBP-HF features as complementary descriptors to improve the description power for dealing with rotation variations.

In addition to applying LBP to Gabor-filtered face images, the joint use of LBP and Gabor and LBP and local phase quantization (LPQ) has provided excellent results in face recognition [12,72]. The HOG-LBP, combining LBP with the histogram of oriented gradients operator [73], has performed very well in human detection with partial occlusion handling [74]. Combining ideas from Haar and LBP features have given excellent results in accurate and illumination-invariant face detection [75]. A CS-LBP method for combining the strengths of SIFT and LBP in interest region description has also been developed [25].

The Weber law descriptor (WLD) is based on the fact that human perception of a pattern depends not only on the change of a stimulus (such as sound and lighting) but also on the original intensity of the stimulus [76]. WLD consists of two components: differential excitation and orientation. The differential excitation component is a function of the ratio between two terms: one is the relative intensity differences of a current pixel against its neighbors and the other is the intensity of the current pixel. The orientation component is the gradient orientation of the current pixel. For a given image, the two components are used to construct a concatenated WLD histogram. The joint use of LBP and the excitation component of WLD descriptor, together with the histogram of optic flow in dynamic texture segmentation, was considered by Chen et al. [77]. This indicates that the excitation component could be useful in replacing the contrast measure of LBP in other problems. The method of Liu et al. [78] also uses differential excitation of WLD together with LBP. The former is improved by using Laplacian of Gaussian.

Jun et al. [79] proposed local gradient patterns and binary histograms of oriented gradients and a hybrid feature that combines several local transform features using AdaBoost. Excellent results were reported in face and human detection experiments. In Nguyen et al. [80], contour templates representing object shape are used to derive a set of ("edge-like") key points at which local appearance features are extracted.

Both spatial information and orientation information are used in their computation. At each keypoint, nonredundant local binary pattern (NR-LBP) is computed. An object descriptor is formed by concatenating NR-LBP features from all keypoints. Very good performance is obtained for MIT and INRIA datasets. A downside of the method is that computation is quite time-consuming. The paper by Ma et al. [81] demonstrates that orientation information is critical in human detection. They proposed an oriented local binary pattern (OLBP) feature, which integrates pixel intensity difference information with texture orientation information to capture a salient object feature. Also, a set of edge orientation histograms and OLBP-based intrablock and interblock features is proposed to describe cell-level and block-level information. Experiments on INRIA and Caltech datasets demonstrate that the approach has a competitive performance and higher speed than existing detectors.

9.3.2 FEATURE SELECTION AND LEARNING

It has been shown by many studies that the dimensionality of the LBP distribution can be effectively decreased by reducing the number of neighboring pixels or by selecting a subset of bins available. In many cases, a properly chosen subset of LBP patterns can perform better than the whole set of patterns.

9.3.2.1 Rule-based selection

Already, the early studies on LBP indicated that some problems considering only four neighbors of the center pixel (i.e., 16 bins) can provide almost as good results as eight neighbors (256 bins). Mäenpää et al. [82] showed that a major part of the discriminative power lies in a small properly selected subset of patterns. In addition to the uniform patterns, they also considered a method based on beam search in which, starting from one, the size of the pattern set is iteratively increased up to a specified dimension D, and the best B pattern sets produced so far are always considered.

Smith and Windeatt [83] used the fast correlation-based filtering (FCBF) algorithm to select the most discriminative LBP patterns. FCBF operates by repeatedly choosing the feature that is most correlated with a given class (e.g., person identity in the case of face recognition), excluding those features already chosen or rejected and rejecting any features that are more correlated with it than with the class. As a measure of correlation, the information-theoretic concept of symmetric uncertainty is used. When applied to the LBP features, FCBF reduced their number from 107,000 to 120.

Liao et al. [84] introduced dominant local binary patterns that make use of the most frequently occurred patterns of LBP to improve the recognition accuracy compared to the original uniform patterns. The method has also rotation-invariant characteristics. To obtain discriminative patterns, Guo et al. [85] presented a learning model that is formulated into a three-layered model. It estimates the optimal pattern subset of interest by simultaneously considering the robustness, discriminative power, and representation capability of features. This model is generalized and can be integrated with existing LBP variants such as conventional LBP, rotation-invariant patterns, local patterns with anisotropic structure, CLBP, and LTP to derive new image features for texture classification.

9.3.2.2 Boosting

Boosting has become a very popular approach for feature selection. It has been widely adopted for LBP feature selection in various tasks, for example, 3-D face recognition and face detection. AdaBoost is commonly used for selecting optimal LBP settings (such as the size and the location of local regions and the number of neighboring pixels) or for selecting the most discriminative bins of an LBP histogram. For instance, Zhang et al. [86] used AdaBoost learning for selecting an optimal set for local regions and their weights for face recognition. Since then, many related approaches have been used at region level for LBP-based face analysis. Shan and Gritti [87], on the other hand, used AdaBoost for learning discriminative LBP histogram bins, with an application to facial expression recognition.

Heng et al. [88] presented a "shrink boost" method for selecting features from multiple LBP histograms. Motivation for it was that feature selection from sparse and high-dimension features using conventional greedy-based boosting leads to poor generalization. The shrink boost method solves the sparse regularization problem with two iterative steps. A "boosting" step first uses weighted training samples to learn a full high-dimensional classifier on all features, avoiding overfitting to few features, and improves generalization. Then, a "shrinkage" step shrinks least discriminative classifier dimension to zero to remove the redundant features. In well-known (INRIA human detection, DaimlerChrysler pedestrian detection, and bird detection) object detection problems, they used "shrink boost" to select sparse features from concatenated LBP histograms of multiple quantization and image channels to learn classifier of additive lookup tables, obtaining improved generalization even under limited training samples.

9.3.2.3 Subspace learning

Another approach for deriving compact and discriminative LBP-based feature vectors consists of applying subspace methods for learning and projecting the LBP features from the original high-dimensional space into a lower-dimensional space. For instance, Chan et al. [72] used linear discriminant analysis (LDA) to project high-dimensional MLBP features into a discriminant space, yielding very good results. To deal with the small sample size problem of LDA, Shan et al. [89] constructed ensemble of piecewise Fisher discriminant analysis classifiers, each of which is designed based on one segment of the high-dimensional histogram of LGBP features. Their approach was shown to be more effective than applying LDA to high-dimensional holistic feature vectors.

Tan and Triggs [12] combined Gabor wavelets and LBP features and projected them to PCA space. Then, the kernel discriminative common vectors are applied to extract discriminant nonlinear compact features for face recognition. Zhao et al. [90] applied Laplacian PCA (LPCA) for LBP feature selection and pointed out the superiority of LPCA over PCA and KPCA for feature selection. Hussain and Triggs [91] exploited the complementarity of three sets of features, namely, HOG, LBP, and

LTP, and adopted partial least squares dimensionality reduction for selecting the most discriminative features, yielding fast and efficient visual object detector.

Nanni et al. [92] explored the use of random subspace, known to work well with noise and correlated features, to train features based also on nonuniform patterns. The approach fuses classifiers (SVM) trained considering the uniform patterns and random subspace classifiers trained considering only the nonuniform patterns.

9.3.2.4 Other methods

Ren et al. [93] found that most existing approaches rely on a predefined LBP structure to extract features and that those structures can be generalized as the patterns constructed from the binarized pixel differences in a local neighborhood. Instead of using a predefined structure, they learn binarized pixel-difference patterns (BPPs), casting the BPP structure discovery as a feature selection problem, which is solved via incremental minimal-redundancy-maximal-relevance algorithms. The method outperformed existing methods (e.g., CENTRIST [94]) in two scene recognition problems.

From the observation that LBP is equivalent to the application of a fixed binary decision tree, Maturana et al. [95] proposed a new method for learning discriminative LBP-like patterns from training data using decision tree induction algorithms. For each local image region, a binary decision tree is constructed from training data, thus obtaining an adaptive tree whose main branches are specially tuned to encode discriminative patterns in each region. Among the drawbacks of the proposed decision tree LBP is the high cost of constructing and storage of the decision trees especially when large pixel neighborhoods are used.

9.3.3 OTHER METHODS INSPIRED BY LBP

LBP has also inspired the development of new effective local image descriptors related to LBP.

The LPQ descriptor is based on quantizing the Fourier transform phase in local neighborhoods [96]. The phase can be shown to be a blur-invariant property under certain commonly fulfilled conditions. In texture analysis, histograms of LPQ labels computed within local regions are used as a texture descriptor. Generation of the labels and their histograms is similar to the LBP method. Extensions of LPQ to multiple scales [72], spatiotemporal domain [97], and color images have also been developed. The LPQ descriptor has recently received considerable interest in blur-invariant face recognition [72,96].

Lategahn et al. [98] developed a framework that filters a texture region by a set of filters and subsequently estimates the joint probability density functions by Gaussian mixture models (GMMs). Using the oriented difference filters of the LBP method [33], they showed that this method avoids the quantization errors of LBP, obtaining better results than with the basic LBP. Additional performance improvement of the GMM-based density estimator was obtained when the elementary filters were replaced by wavelet frame transform filter banks.

Vu and Caplier [99] proposed a feature descriptor called pattern of oriented edge magnitudes (POEMs) for face recognition and image matching. First, image gradient is computed, and then a histogram of orientations is accumulated and assigned to the CI of a cell (a spatial region around the current pixel). The gradient magnitude is used to weight the contribution of each pixel. Finally, the accumulated magnitudes are encoded using the LBP operator. Descriptors based on these principles performed very well in face recognition and image matching experiments.

Sharma et al. [100] employed local higher-order statistics (LHSs) of local nonbinarized patterns for image description. The LHS requires neither any user-specified quantization of the space of pixel patterns nor any heuristics for discarding low occupancy volumes of the space. Experiments with texture and face databases, with an SVM classifier, demonstrate very good performance. Zhang et al. [101] proposed local energy pattern (LeP) for texture classification using self-adaptive quantization thresholds. The method generates local feature vectors obtained by rectifying the responses of the 2-D Gaussian-like second derivative filters and then utilizes N-nary coding quantization instead of binary one. LEP was also extended to spatiotemporal analysis.

In Ref. [102], texture images are first decomposed by the shearlet transform, followed by construction of local energy features. These are then quantized and encoded to be rotation-invariant. The energy histograms accumulated over all decomposition levels reflect the different energy distributions. The method extracts more directional features like orientations and is robust with respect to noise. Experiments show very promising performance, especially with additive Gaussian noise.

Inspired by LBP, Maani et al. [103] introduced a method in which local frequency components are computed by applying 1-D Fourier transform on a neighboring function defined on a circle of radius R at each pixel. They observed that the low-frequency components are the major constituents of the circular functions and can effectively represent textures. Three sets of features were extracted from the low-frequency components, two based on the phase and one based on the magnitude. The method has advantages of a very good performance, relatively small number of features, and robustness to noise.

Wu and Rehg [94] proposed CENTRIST, a holistic *cen*sus *tr*ansform h*ist*ogram-based visual descriptor for recognizing places and scene categories. LBP-like census transform [104] histograms are used to encode structural properties within an image and suppress detailed textural information. Constraints between neighboring pixels are utilized to capture the structural characteristic within a small image patch. In larger scales, spatial hierarchy of CENTRIST is used to catch rough geometric information. A downside of the method is that it is not rotation-invariant.

Crosier and Griffin [105] proposed basic image features (BIFs) for texture classification. BIF engineers, like LBP, a dataset-independent dictionary of local features over which textures are represented statistically. Zeroth-, first-, and second-order Gaussian derivative filters are used for local description. Inspired by methods like LBP that produce binary codes, Kannala and Rahtu [106] proposed binarized statistical image features (BSIFs). BSIF computes a binary code for each pixel by linearly projecting local image patches onto a subspace, whose basis vectors are learned from natural images via independent component analysis.

LBP has also given inspiration to recent interest on binary local feature descriptors, including BRIEF [107], ORB, and BRISK. The binary descriptors provide a comparable matching performance with the widely used interest region descriptors such as SIFT and SURF but have very fast extraction times and very low memory requirements needed, for example, in emerging applications using mobile devices with limited computational capabilities. Comparative evaluations of these descriptors can be found in Refs. [108,109].

9.4 SPATIOTEMPORAL AND OTHER DOMAINS

9.4.1 VARIANTS OF SPATIOTEMPORAL LBP

The high success of spatiotemporal LBP methods in various computer vision problems and applications has led to many other teams investigating the approach, and several extensions and modifications of spatiotemporal LBP have been proposed to increase its robustness and discriminative power.

Zhao and Pietikäinen [110] extended LBP-TOP to multiscale spatiotemporal space, with an application to facial expression recognition. AdaBoost was used to learn the principal appearance and motion, for selecting the most important expression-related features for all the classes or between every pair of expressions. Rotation-invariant variants of LBP-TOP based on histogram Fourier features were proposed by Zhao et al. [51].

LTP-TOP was developed by Nanni et al. [111]. The encoding function was modified for considering both the ternary patterns and the three orthogonal planes. Their experiments on a 10-class Weizmann dataset obtained very good results. WLD has also been extended to the spatiotemporal domain in the same way as LBP-TOP, yielding WLD-TOP for supplementing LBP-TOP in dynamic texture segmentation [77]. Mattivi and Shao [112] proposed Extended Gradient LBP-TOP for action recognition. Two modifications were made on the basis of LBP-TOP. Firstly, the computation of LBP was extended to nine slices, three for each axis. Therefore, on the *XY* dimension, there is the original *XY* plane (centered in the middle of the cuboid) plus two other *XY* planes located at 1/4 and 3/4 of the cuboid's length. The same is done for *XT* and *YT* dimensions. Secondly, computation of LBP operator on gradient images was introduced. The gradient image contains information about the rapidity of pixel intensity changes along a specific direction, has large magnitude values at edges, and can further increment LBP operator's performances, since LBP encodes local primitives such as curved edges, spots, and flat areas. For each cuboid, the brightness gradient is calculated along x, y, and t directions, and the resulting three cuboids containing specific gradient information are summed in absolute values. Before computing the image gradients, the cuboid is slightly smoothed with a Gaussian filter in order to reduce noise. The extended LBP-TOP is then performed on the gradient cuboid. Experiments on a KTH human action dataset showed the effectiveness of the method.

For dealing with 2-D face recognition, Lei et al. [113] proposed effective LBP operator on three orthogonal planes of Gabor volume (E-GV-LBP). Firstly, the Gabor face images are formulated as a third-order Gabor volume. Then, LBP operator is applied on three orthogonal planes of Gabor volume, respectively, named GV-LBP-TOP in short. In this way, the neighboring changes both in spatial space and during different types of Gabor faces can be encoded. Moreover, in order to reduce the computational complexity, an effective GVLBP (E-GV-LBP) descriptor was developed that describes the neighboring changes according to the central point in spatial, scale, and orientation domains simultaneously for face representation.

Visual information from captured video is important for speaker identification under noisy conditions. Combination of LBP dynamic texture and EdgeMap structural features was proposed to take both motion and appearance into account [114], providing the description ability for spatiotemporal development in speech. Spatiotemporal dynamic texture features of LBPs extracted from localized mouth regions are used for describing motion information in utterances, which can capture the spatial and temporal transition characteristics. Structural edge map features are extracted from the image frames for representing appearance characteristics. Combination of dynamic texture and structural features takes both motion and appearance together into account, providing the description ability for spatiotemporal development in speech. In the experiments on BANCA and XM2VTS databases, the proposed method obtained promising recognition results compared to the other features.

Goswami et al. [115] proposed a novel approach to ordinal contrast measurement called local ordinal contrast patterns (LOCPs). Instead of computing the ordinal contrast with respect to any fixed value such as that at the center pixel or the average intensity value, it computes the pairwise ordinal contrasts for the chain of pixels representing the circular neighborhoods starting from the center pixel. Then, it was extended for dynamic texture analysis by extracting the LOCP in three orthonormal planes to generate LOCP-TOP. Together with LDA, its performance of mouth-region biometrics in the XM2VTS database received good results.

Spatial representation of LPQ was extended to a dynamic texture descriptor called the volume local phase quantization by Päivärinta et al. [97]. The local Fourier transform is computed by 1-D convolutions for each dimension in a 3-D volume. The data achieved are compressed to a smaller dimension before a scalar quantization procedure. Finally, a histogram of all code words from dynamic texture is formed.

Huang et al. [116] proposed to use spatiotemporal monogenic binary patterns to describe the appearance and motion information of the dynamic sequences. Firstly, they used monogenic signals analysis to extract the magnitude, the real picture, and the imaginary picture of the orientation of each frame, since the magnitude can provide much appearance information and the orientation can provide complementary information. Secondly, the phase-quadrant encoding method and the local bit exclusive operator are utilized to encode the real and imaginary pictures from orientation in three orthogonal planes, and the LBP operator is used to capture the texture and motion information from the magnitude through three orthogonal planes. Finally, both the concatenation method and multiple kernel learning method are exploited to handle the feature fusion. The experimental results on the extended Cohn-Kanade and Oulu-CASIA facial expression databases demonstrated a state-of-the-art performance.

Ruiz-Hernandez and Pietikäinen [117] proposed a method for encoding LBPs using a reparametrization (RP) of the second-local order Gaussian jet. The information provided by RP generates robust and reliable histograms and is thus suitable for different facial analysis tasks. The proposed method has two main processes: the RP process that is used to compute needed parameters in a video sequence and the encoding process that combines the textural information provided by the LBP and the robustness of the RP. They showed that this approach can be used for recognizing facial microexpressions from videos, obtaining competitive performance on two different datasets.

Video texture synthesis is the process of providing a continuous and infinitely varying stream of frames. Guo et al. [118] proposed a variant of LBP-TOP called multiframe LBP-TOP to find the most appropriate matching pairs of frames for video texture synthesis. To achieve seamless synthesis results, a diffeomorphic growth model was applied to matching frames identified. The proposed approach has potential in other applications, for example, motion interpolation for videos and dynamic texture retrieval.

Inspired by spatiotemporal LBP, the Haar-like features were extended to represent the dynamic characteristic of facial expression [119]. The dynamic Haar-like features were built by two steps: (1) thousands of Haar-like features are extracted in each frame, and (2) the same Haar-like features in the consecutive frames are combined as dynamic features, as Figure 9.7 shows.

With the increasing amount of surveillance data, segmentation of moving objects in the compressed domain is receiving growing attention. Yang et al. [120] proposed a method in which the motion vectors are first accumulated and filtered to have

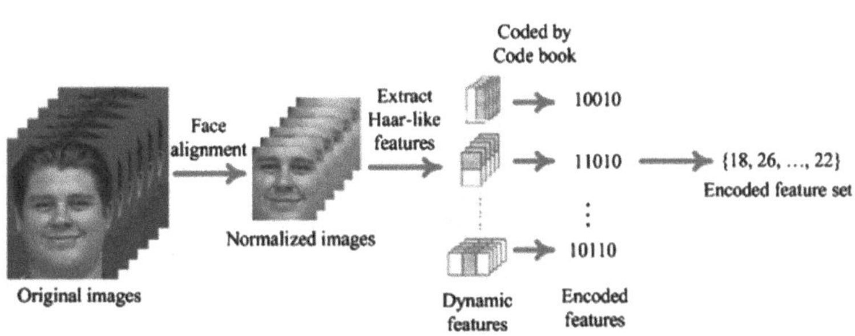

FIGURE 9.7

Dynamic Haar-like features.

reliable motion information. Then, spatiotemporal LBP features of motion vectors are extracted to obtain coarse and initial object regions in the H.264 compression domain. Final region refinement is done according to the distribution of DCT coefficients.

A local spatiotemporal directional descriptor was proposed for speaker identification by analyzing mouth movements [121]. The movements of mouth regions are described using LBPs and intensity contrast from six directions in three orthogonal planes. In addition, besides sign features, magnitude information encoded as weight for the bins with the same sign value was developed to improve the discriminative ability. Moreover, decorrelation is exploited to remove the redundancy of features. Experimental results on the challenging XM2VTS database showed the effectiveness.

9.4.2 LBP IN OTHER DOMAINS

9.4.2.1 Analysis of depth and 4-D images

Research on LBP variants for depth (range) images has focused on applications in face analysis. In most of the early papers, the depth values were first interpolated onto a regular grid, and then a more or less ordinary 2-D LBP approach was applied to the intensity (depth) values [122].

Huynh et al. [123] improved these by considering orientation differences in LBP computations, obtaining eight $LBP_{8,1}$-based orientations of the depth differences in a local 3×3 neighborhood. Eight depth difference histograms are obtained for the given region and concatenated to form an oriented histogram of depth differences. Sandbach et al. [124] extended the LBP methodology to 3-D face data by utilizing surface orientation information to detect facial action units (AUs). Local normal binary patterns employ the normals of the triangular polygons that form the 3-D face mesh to encode the shape of the mesh at each point. This is related to the gradient of a 2-D image, providing a richer source of local shape than the intensity alone. Bayramoglu et al. [125] also used surface orientation information in their method for facial AU detection. To get a short feature vector without degrading the performance, they used the CS-LBP approach with 3-D surface orientation information in computing the CS-3-D LBP operator. In Ref. [126], with the help of extracted keypoints, the 3-D face is divided into a mesh. LBP is computed using eight neighboring vertices of a current vertex. The depth and normal information of each vertex are extracted separately and encoded by LBP. The depth information describes the concavo-convex attribute, and the normal information indicates the face curvature variety.

Due to the progress of depth sensor technology, the analysis of 4-D images (i.e., depth images in motion) is an emerging research topic, for example, in facial expression recognition. Fang et al. [127] adopted the spatiotemporal LBP-TOP operator for 4-D facial expression recognition. A robust approach for registering consecutive frames of 4-D data is first applied, and the resulting "geometry" images are considered as frames in 2-D videos. LBP-TOP descriptors are computed on the difference between a frame and the first frame of the sequence, using the idea of the so-called flow image representing the deformation vectors in a subsequent frame w.r.t.

the first frame. Promising results are obtained for the BU-4DFE facial expression database. Reale et al. [128] used LBP-TOP as a comparative method in their paper on the 4-D spatiotemporal "Nebula" feature. The LBP-TOP approach, applied to 2-D and depth images, was in many ways similar to that of [127], but they did not use their alignment method and their tests on a flow image. The results for the Nebula feature were the best, but also, LBP-TOP performed reasonably well considering its simple way of implementation in this study.

9.4.2.2 Analysis of 3-D volume images

Extension of LBP to 3-D volume images is challenging. A circle in 2-D translates to a sphere in 3-D, and equidistant sampling on a sphere is not so trivial. The notion of ordering is also lost in 3-D due to the dimensionality [129].

Fehr and Burkhardt [130] extended the original LBP from 2-D images to 3-D volume data, achieving a full rotation invariance. For each LBP computation, the correlation between the values of all points on the neighborhood sphere with radius R and the weight factor that is a volume representation in an arbitrary but fixed order binomial factor is performed in the spherical harmonic domain. Rotation invariance is obtained from the computation of the minimum over all angles. Another method for rotationally invariant 3-D LBP using spectral harmonic decomposition was proposed by Banerjee et al. [129]. Unlike the original approach, the invariance is constructed implicitly, without considering all possible combinations of the pattern. Experiments were carried out with phantom data and clinical liver CTA data.

Morgado et al. [131] proposed an approach able to closely replicate in 3-D and without any approximation both uniformity and rotation invariance concepts originally proposed for the 2-D setting. Feature selection is done using correlation coefficients to quantify the relevance of each individual feature, and the SVM is used as the learning machine. The experiments demonstrate that the proposed method is able to enhance the diagnostic system and that the texture of the FDG-PET scans contains distinctive information about the presence of both Alzheimer's disease and mild cognitive impairment. Burner et al. [132] considered sign, contrast, and intensity information in computing texture bags at multiple scales for 3-D volumes. Instead of using histograms to match entire images, they search for regions that have similar local appearance. This is necessary to cope with the variability encountered in medical imaging data and the comparably subtle effects of disease. Based on the similarity of the texture structure of local regions, images are being ranked.

9.4.2.3 LBP in 1-D signal analysis

Analysis of 1-D signals is an emerging and potentially very important application area for LBP. For example, speech systems such as hearing aids require fast and inexpensive signal processing. Chatlani and Soraghan [133] proposed a straightforward simplification of the ordinary LBP to 1-D signals, with a preliminary application to simple signal segmentation and voice activity detection (VAD) to estimate periods of speech and nonspeech. Later, Zhu et al. [134] used LBP-based VAD in HMM-based speech recognition. Speech is first denoised by adaptive empirical model decomposition and processed with LBP-based VAD, in which 1-D LBP is used to find start and end points of voiced speech segment so that it is distinguished from noise, unvoiced, or mute segments. In another application example, 1-D LBP is applied for bone texture classification by Houam et al. [135]. Global texture information is characterized by image projections, and local information is extracted from these using 1-D LBP.

Another interesting direction for 1-D signal analysis is to first derive a 2-D representation of the signal and then apply the spatial domain LBP to this representation. Lazic and Aarabi [136] successfully applied 2-D LBP for detecting spoken terms from visual spectrogram representation derived from the audio signal. A similar approach was used by Costa et al. [137] for classifying music genre from visual spectrograms derived from the audio signal. Esfahanian et al. [138], on the other hand, borrowing principles of facial image analysis [5], divided the visual spectrograms first into nonoverlapping regions and then used concatenated histograms to classify dolphin calls.

9.5 FUTURE CHALLENGES

Due to the considerable effort on spatial domain variants, future research should focus more on new challenges.

A great majority of the research on dense texture descriptors has been based on the assumption that there are no significant view or scale variations in the scene or objects to be analyzed. In practice, these variations are very common and may include self-occlusion when an object is imaged from different views. An obvious but impractical solution would be to train the recognition system with samples viewed from a large number of different positions and then derive a set of representative models for each class [7]. Moore and Bowden [139] showed that a larger-sized operator (Gabor LBP or MLBP) works better than small-sized operators in multiview facial expression recognition. It would very be useful to develop new approaches that are designed for handling effectively problems of view and scale variation.

For a large number of applications, an ability to analyze small sample sizes at high speed is vital, including face analysis, interest region description, segmentation, background subtraction, and tracking. This means that a compact region description with a short feature vector is needed. Many of the proposed descriptors would fail in this respect. It is important to evaluate the performance of a new descriptor also with smaller sample sizes, as was done, for example, in Ref. [29] by cropping 41×41 patches from the original 200×200 pixel CUReT textures.

There is also a need for more research on developing approaches that are robust to lighting variations. The use of Gabor filtering [11] or a preprocessing by Tan and Triggs [12] is an example of the state of the art for face recognition, but this approach also tends to smooth small details from the images. Some of the recent works have successfully used image gradient orientation computation prior to feature extraction, as gradient orientations are known to be less sensitive to lighting variations than the original images.

Most of the past research has focused on improving the robustness of LBP for some specific tasks, like preprocessing prior to LBP computation (e.g., Gabor filtering and gradient computation), using different methods for encoding (e.g., LTP), reducing the number of neighbors in multiscale analysis (e.g., LQP), and processing LBP codes in different ways (e.g., LBP-HF). A future direction would be to consider different stages of feature computation together to optimize the performance. A good example of this direction is by Lei et al. [140], who obtained excellent results in face recognition by combining preprocessing by learning-based linear filtering, soft sampling to find the best set of neighbors for LBP computation, and clustering to find the optimal way for encoding.

Combining different types of complementary local operators is another way to go ahead, as, for example, the recent works on human detection and face recognition show. Approaches combining the strengths of LBP and HOG (or SIFT), for example, have led to increased performance.

Research on spatiotemporal LBP variants has not been as active as in the spatial domain. One would expect to see much more research in this area. Different effective ways of obtaining information in the temporal domain combined with novel ideas proposed for analyzing images in the spatial domain could be one way to progress. The robustness of spatiotemporal operators to different variations is largely unexplored. One example of this kind of work is by Zhao et al. [51], who demonstrated that their spatiotemporal LBP-HF features are robust with respect to view variations.

The use of LBP for analyzing 1-D signals has not been much studied, but has potential for a large number of novel applications, as the recent examples presented above demonstrate. In these works, quite primitive LBP-based analysis was used both for 1-D signal analysis and for 2-D analysis of spectrograms derived from 1-D signals. Borrowing ideas from the recent developments in spatial domain LBP variants could lead to much more powerful methods. The results on analyzing 2-D spectrograms also suggest that LBP variants could be very useful in a wide variety of applications of exploratory data analysis, in which relations between neighboring data elements should be considered.

With the introduction of Kinect and emerging more accurate sensors for depth sensing, the interest on processing depth images (3-D) and depth image sequences (4-D) has been growing, with applications, for example, in action and gesture recognition, face recognition, and recognition of facial AUs. Some LBP variants for depth images have been developed especially for face analysis, but research on 4-D data is largely unexplored. There is also much space for new developments in 3-D volume image analysis widely used in medicine. Simple 3-D variants of ordinary LBP have provided very promising improvements, but much more could be possible with more sophisticated solutions.

The databases used for comparing texture descriptors should be reconsidered. A good practice in some recent papers is to use both texture and face databases and datasets from other application domains. For 2-D faces, there are many good options available like Labeled Faces in the Wild and Face Recognition Grand Challenge (FRGC v2.0), representing different challenges. A problem with most of the used texture databases is that the best methods currently obtain over 95% accuracy. KTH-TIPS-2 represents a more difficult problem in this respect. It would be valuable to have extensive benchmarks with challenging databases for the most promising variants, as was done for texture retrieval in Ref. [141].

9.6 CONCLUSIONS

Due to its advantages, that is, flexibility, invariance to monotonic gray-level changes, and computational simplicity, LBP is a very powerful descriptor to represent local structures in images. A large number of variants have been designed aiming to obtain improved performance and/or robustness in one or more aspects of the original LBP. We have divided the extensions and modifications into 10 categories and introduced some representative variants in each of them.

A number of new effective local image descriptors, including LPQ, POEMs, and LEPs, have also been inspired by LBP. They provide the ability, for example, to deal with blur or noise and can be jointly used with LBP to complement each other.

Even though it is commonly agreed that multiscale analysis and joint use of complementary descriptors can improve the performance, a downside is the large dimensionality of the produced feature vector. To obtain a small set of the most discriminative LBP-based features, different feature selection and learning strategies were also discussed. Extensions of LBP to spatiotemporal domain extend the applicability of LBP from static images to dynamic video sequences. Different variants of the original spatiotemporal LBP were introduced.

Moreover, the future challenges of LBP were discussed. Among these are an improved robustness to view, scale, and lighting variations, optimization of different stages of feature computation, and combinations of different descriptors. There is also a need for further research on problems such as spatiotemporal LBPs and analysis of 1-D signals, depth, 3-D volume images, and 4-D depth image sequences.

REFERENCES

[1] T. Ojala, M. Pietikäinen, D. Harwood, A comparative study of texture measures with classification based on feature distributions, Pattern Recogn. 29 (1) (1996) 51–59.

[2] T. Ojala, M. Pietikäinen, Unsupervised texture segmentation using feature distributions, Pattern Recogn. 32 (1999) 477–486.

[3] T. Ojala, K. Valkealahti, E. Oja, M. Pietikäinen, Texture discrimination with multidimensional distributions of signed gray-level differences, Pattern Recogn. 34 (3) (2001) 727–739.

[4] T. Ojala, M. Pietikäinen, T. Mäenpää, Multiresolution gray-scale and rotation invariant texture classification with local binary patterns, IEEE Trans. Pattern Anal. Mach. Intell. 24 (7) (2002) 971–987.

[5] T. Ahonen, A. Hadid, M. Pietikäinen, Face description with local binary patterns: application to face recognition, IEEE Trans Pattern Anal. Mach. Intell. 28 (12) (2006) 2037–2041.

[6] G. Zhao, M. Pietikäinen, Dynamic texture recognition using local binary patterns with an application to facial expressions, IEEE Trans. Pattern Anal. Mach. Intell. 29 (6) (2007) 915–928.

[7] M. Pietikäinen, A. Hadid, G. Zhao, T. Ahonen, Computer Vision Using Local Binary Patterns, Springer, Berlin, Heidelberg, 2011.

[8] D. Huang, C. Shan, A. Mohsen, Y. H. Wang, L. Chen, Local binary patterns and its applications on facial image analysis: a survey, IEEE Trans. Syst. Man Cybern. C Appl. Rev. 41 (6) (2011) 765–781.

[9] S. Brahnam, L. C. Jain, L. Nanni, A. Lumini (Eds.), Local Binary Patterns: New Variants and Applications, Springer, Berlin, Heidelberg, 2013.

[10] Z. Guo, L. Zhang, D. Zhang, A completed modeling of local binary pattern operator for texture classification, IEEE Trans. Image Process. 19 (2010) 1657–1663.

[11] W. Zhang, S. Shan, W. Gao, X. Chen, H. Zhang, Local Gabor binary pattern histogram sequence (LGBPHS): a non-statistical model for face representation and recognition, in: Proceedings of IEEE International Conference on Computer Vision, vol. 1, 2005, pp. 786–791.

[12] X. Tan, B. Triggs, Enhanced local texture feature sets for face recognition under difficult lighting conditions, IEEE Trans. Image Process. 19 (6) (2010) 1635–1650.

[13] C. H. Yao, S. Y. Chen, Retrieval of translated, rotated and scaled color textures, Pattern Recogn. 36 (4) (2003) 913–929.

[14] X. Li, W. Hu, Z. Zhang, H. Wang, Heat kernel based local binary pattern for face representation, IEEE Signal Process. Lett. 17 (2010) 308–311.

[15] S. Liao, A. C. S. Chung, Face recognition by using elongated local binary patterns with average maximum distance gradient magnitude, in: Proceedings of Asian Conference on Computer Vision, 2007, pp. 672–679.

[16] L. Nanni, A. Lumini, S. Brahnam, Local binary patterns variants as texture descriptors for medical image analysis, Artif. Intell. Med. 49 (2010) 117–125.

[17] Y. He, N. Sang, C. Gao, Multi-structure local binary patterns for texture classification, Pattern Anal. Appl. 16 (4) (2012) 595–607.

[18] L. Wolf, T. Hassner, Y. Taigman, Effective unconstrained face recognition by combining multiple descriptors and learned background statistics, IEEE Trans. Pattern Anal. Mach. Intell. 33 (10) (2011) 1978–1990.

[19] S. A. Orjuela, J. P. Yanez Puentes, P. Philips, The geometric local texture patterns (GLTP), in: S. Brahnam, L. C. Jain, L. Nanni, A. Lumini (Eds.), Local Binary Patterns: New Variants and Applications, Springer, Berlin, Heidelberg, 2013, pp. 85–112.

[20] K. Wang, C.-E. Bichot, C. Chu, B. Li, Pixel to patch sampling structure and local neighboring intensity relationship patterns for texture classification, IEEE Signal Process. Lett. 20 (9) (2013) 853–856.

[21] J. Ylioinas, A. Hadid, Y. Guo, M. Pietikäinen, Efficient image appearance description using dense sampling based local binary patterns, in: Proceedings of Asian Conference on Computer Vision, 2013, pp. 375–388.

[22] A. Hafiane, G. Seetharam, B. Zavidovique, Median binary pattern for texture classification, in: Proceedings of International Conference on Image Analysis and Recognition, 2007, pp. 387–398.
[23] H. Jin, Q. Liu, H. Lu, X. Tong, Face detection using improved LBP under Bayesian framework, in: Proceedings of International Conference on Image and Graphics, 2004, pp. 306–309.
[24] M. Heikkilä, M. Pietikäinen, A texture-based method for modeling the background and detecting moving objects, IEEE Trans. Pattern Anal. Mach. Intell. 28 (4) (2006) 657–662.
[25] M. Heikkilä, M. Pietikäinen, C. Schmid, Description of interest regions with local binary patterns, Pattern Recogn. 42 (3) (2009) 425–436.
[26] T. Ahonen, M. Pietikäinen, Soft histograms for local binary patterns, in: Proceedings of Finnish Signal Processing Symposium, 2007, 4 p.
[27] S. Katsigiannis, E. Keramidas, D. Maroulis, FLBP: fuzzy local binary patterns, in: S. Brahnam, L. C. Jain, L. Nanni, A. Lumini (Eds.), Local Binary Patterns: New Variants and Applications, Springer, Berlin, Heidelberg, 2013, pp. 149–175.
[28] S. Liao, G. Zhao, V. Kellokumpu, M. Pietikäinen, S. Z. Li, Modeling pixel process with scale invariant local patterns for background subtraction in complex scenes, in: Proceedings of IEEE Conference on Computer Vision and Pattern Recognition, 2010, 8 p.
[29] J. Ylioinas, X. Hong, M. Pietikäinen, Constructing local binary pattern statistics by soft voting, in: Proceedings of Scandinavian Conference on Image Analysis, 2013, pp. 119–130.
[30] J. Trefny, J. Matas, Extended set of local binary pattern for rapid object detection, in: Proceedings of Computer Vision Winter Workshop, 2010.
[31] L. Liu, L. Zhao, Y. Long, G. Kuang, P. Fieguth, Extended local binary patterns for texture classification, Image Vision Comput. 30 (2012) 86–99.
[32] B. Zhang, Y. Gao, S. Zhao, J. Liu, Local derivative pattern versus local binary pattern: face recognition with high-order local pattern descriptor, IEEE Trans. Image Process. 19 (2) (2010) 533–544.
[33] T. Ahonen, M. Pietikäinen, Image description using joint distribution of filter bank responses, Pattern Recogn. Lett. 34 (2009) 368–376.
[34] Z. Guo, Q. Li, J. You, D. Zhang, W. Liu, Local directional derivative pattern for rotation invariant texture classification, Neural Comput. Appl. 8 (21) (2012) 1893–1904.
[35] S. ul Hussain, B. Triggs, Visual recognition using local quantized patterns, in: Proceedings of European Conference on Computer Vision, Part II, 2012, pp. 716–729.
[36] X. Huang, G. Zhao, X. Hong, M. Pietikäinen, W. Zheng, Texture description with completed local quantized patterns, in: Proceedings of Scandinavian Conference on Image Analysis, 2013, pp. 1–10.
[37] N. Jiang, J. Xu, S. Goto, Pedestrian detection using gradient local binary patterns, IEICE Trans. Fund. E95-A (8) (2012) 1280–1287.
[38] T. Mäenpää, M. Pietikäinen, Multi-scale binary patterns for texture analysis, in: Proceedings of Scandinavian Conference on Image Analysis, 2003, pp. 885–892.
[39] S. Liao, X. Zhu, Z. Lei, L. Zhang, S. Z. Li, Learning multi-scale block local binary patterns for face recognition, in: Proceedings of International Conference on Biometrics, 2007, pp. 828–837.
[40] M. Turtinen, M. Pietikäinen, Contextual analysis of textured scene images, in: Proceedings of British Machine Vision Conference, 2006, pp. 849–858.

[41] Y. He, N. Sang, C. Gao, Pyramid-based multi-structure local binary pattern for texture classification, in: Proceedings of Asian Conference on Computer Vision, vol. 3, 2010, pp. 1435–1446.

[42] X. Qian, X. Hua, P. Chen, L. Ke, PLBP: an effective local binary patterns texture descriptor with pyramid representation, Pattern Recogn. 44 (10/11) (2011) 2502–2515.

[43] T. Song, H. Li, WaveLBP based features for image classification, Pattern Recognit. Lett. 34 (2013) 1323–1328.

[44] L. Liu, Y. Long, P. Fieguth, S. Lao, G. Zhao, BRINT: binary rotation invariant and noise tolerant texture classification, IEEE Trans. Image Process. 23 (7) (2014) 3071–3084.

[45] M. Pietikäinen, T. Ojala, Z. Xu, Rotation-invariant texture classification using feature distributions, Pattern Recogn. 33 (2000) 43–52.

[46] Z. H. Guo, L. Zhang, D. Zhang, Rotation invariant texture classification using adaptive LBP with directional statistical features, in: Proceedings of International Conference on Image Processing, 2010, pp. 285–288.

[47] O. Garcia-Olalla, E. Alegre, R. Fernandez-Robles, M. T. Garcia-Ordas, Vitality assessment of boar sperm using an adaptive LBP based on oriented deviation, in: Proceedings of ACCV 2012 Workshops, 2013, pp. 61–72.

[48] Z. H. Guo, L. Zhang, D. Zhang, Rotation invariant texture classification using LBP variance (LBPV) with global matching, Pattern Recogn. 43 (3) (2010) 706–719.

[49] L. Zhang, D. Zhang, Z. Guo, D. Zhang, Monogenic-LBP: a new approach for rotation invariant texture classification, in: Proceedings of International Conference on Image Processing, 2010, pp. 2677–2770.

[50] Y. Zhao, D.-S. Huang, W. Jia, Completed local binary count for rotation invariant texture classification, IEEE Trans. Image Process. 21 (10) (2012) 4492–4497.

[51] G. Zhao, T. Ahonen, J. Matas, M. Pietikäinen, Rotation-invariant image and video description with local binary pattern features, IEEE Trans. Image Process. 21 (4) (2012) 1465–1477.

[52] Z. Li, G. Liu, Y. Yang, J. You, Scale- and rotation-invariant local binary pattern using scale-adaptive texton and subuniform-based circular shift, IEEE Trans. Image Process. 21 (4) (2012) 2130–2140.

[53] X. Qi, R. Xiao, C.-G. Li, Y. Qiao, J. Guo, X. Tang, Pairwise rotation invariant co-occurrence local binary pattern, IEEE Trans. Pattern Anal. Mach. Intell. 36 (11) (2014) 2199–2213.

[54] X. Qi, Y. Qiao, C.-G. Li, J. Guo, Multi-scale joint encoding of local binary patterns for texture and material classification, in: Proceedings of British Machine Vision Conference, 2013.

[55] R. Nosaka, Y. Ohkava, K. Fukui, Feature extraction based on co-occurrence of adjacent local binary patterns, in: Proceedings of PSIVT, Part II, 2011, pp. 82–91.

[56] R. Nosaka, C. H. Suryanto, K. Fukui, Rotation-invariant co-occurrence among adjacent LBPs, in: Proceedings of ACCV Workshops, 2013, pp. 15–25.

[57] W. Louis, K. N. Plataniotis, Co-occurrence of local binary patterns features for frontal face detection in surveillance applications, EURASIP J. Image Video Process. (2011), doi:10.1155/2011/745487.

[58] T. Mäenpää, M. Pietikäinen, Classification with color and texture: jointly or separately? Pattern Recogn. 37 (2004) 1629–1640.

[59] C. Zhu, C. E. Bichot, L. Chen, Multi-scale color local binary patterns for visual object classes recognition, in: Proceedings of International Conference on Pattern Recognition, 2010, pp. 3065–3068.

[60] C. Zhu, C.-E. Bichot, L. Chen, Image region description using orthogonal combination of local binary patterns enhanced with color information, Pattern Recogn. 46 (2013) 1949–1963.
[61] X. Qi, Y. Qiao, C.-G. Li, J. Guo, Exploring cross-channel texture correlation for color texture classification, in: Proceedings of British Machine Vision Conference, 2013.
[62] S. Banerji, A. Sinha, C. Liu, New image descriptors based on color, texture, shape, and wavelets for object and scene image classification, Neurocomputing 117 (2013) 173–185.
[63] G. Kylberg, I.-M. Sintorn, Evaluation of noise robustness for local binary pattern descriptors in texture classification, EURASIP J. Image Video Process. 17 (2013), 1-20.
[64] J. Chen, V. Kellokumpu, G. Zhao, M. Pietikäinen, RLBP: robust local binary pattern, in: Proceedings of British Machine Vision Conference, 2013.
[65] J. Ren, X. Jiang, J. Yuan, Noise-resistant local binary pattern with an embedded error-correction mechanism, IEEE Trans. Image Process. 22 (10) (2013) 4049–4060.
[66] Y. Zhao, W. Jia, R.-X. Hu, H. Min, Completed robust local binary pattern for texture classification, Neurocomputing 106 (2013) 68–76.
[67] S. Liao, A. C. S. Chung, Texture classification by using advanced local binary patterns and spatial distribution of dominant patterns, in: Proceedings of IEEE International Conference on Acoustics, Speech and Signal Processing, vol. 1, 2007, pp. 1221–1224.
[68] Y. Guo, G. Zhao, M. Pietikäinen, Texture classification using a linear configuration model based descriptor, in: Proceedings of British Machine Vision Conference, 2011.
[69] F. M. Khellah, Texture classification using dominant neighborhood structure, IEEE Trans. Image Process. 20 (11) (2011) 3270–3279.
[70] X. Hong, G. Zhao, M. Pietikäinen, X. Chen, Combining LBP difference and feature correlation for texture description, IEEE Trans. Image Process. 23 (6) (2014) 2557–2568.
[71] G. A. Papagostas, E. D. E. Koulouriotis, E. Karakasis, V. D. Tourassis, Moment-based local binary patterns: a novel descriptor for invariant pattern recognition applications, Neurocomputing 99 (2013) 358–371.
[72] C. H. Chan, M. A. Tahir, J. Kittler, M. Pietikäinen, Multiscale local phase quantization for robust component-based face recognition using kernel fusion of multiple descriptors, IEEE Trans. Pattern Anal. Mach. Intell. 35 (5) (2013) 1164–1177.
[73] N. Dalal, B. Triggs, Histograms of oriented gradients for human detection, in: Proceedings of IEEE International Conference on Computer Vision and Pattern Recognition, vol. 2, 2005, pp. 886–893.
[74] X. Wang, T. X. Han, S. Yan, An HOG-LBP human detector with partial occlusion handling, in: Proceedings of IEEE International Conference on Computer Vision, 2009, pp. 32–39.
[75] S. Yan, S. Shan, X. Chen, W. Gao, Locally assembled binary (LAB) feature with feature-centric cascade for fast and accurate face detection, in: Proceedings of IEEE International Conference on Computer Vision, 2008, 8 p.
[76] J. Chen, S. Shan, C. He, G. Zhao, M. Pietikäinen, X. Chen, W. Gao, WLD: a robust local image descriptor, IEEE Trans. Pattern Anal. Mach. Intell. 32 (9) (2010) 1705–1720.
[77] J. Chen, G. Zhao, M. Salo, E. Rahtu, M. Pietikäinen, Automatic dynamic texture segmentation using local descriptors and optical flow, IEEE Trans. Image Process. 22 (1) (2013) 326–339.
[78] F. Liu, Z. Tang, J. Tang, WLBP: Weber local binary pattern for local image description, Neurocomputing 120 (2013) 325–335.

[79] B. Jun, I. Choi, D. Kim, Local transform features and hybridization for accurate face and human detection, IEEE Trans. Pattern Anal. Mach. Intell. 35 (8) (2013) 1423–1436.
[80] D. T. Nguyen, P. O. Ogunbona, W. Li, A novel shape-based non-redundant local binary pattern descriptor for object detection, Pattern Recogn. 46 (2013) 1485–1500.
[81] Y. Ma, L. Deng, X. Chen, N. Guo, Integrating orientation cue with EOH-OLBP based multi-level features for human detection, IEEE Trans. Circ. Syst. Video Tech. 23 (10) (2013) 1755–1766.
[82] T. Mäenpää, T. Ojala, M. Pietikäinen, M. Soriano, Robust texture classification by subsets of local binary patterns, in: Proceedings of International Conference on Pattern Recognition, 2000, pp. 947–950.
[83] R. S. Smith, T. Windeatt, Facial expression detection using filtered local binary pattern features with ECOC classifiers and Platt scaling, in: Proceedings of JMLR Workshop on Applications of Pattern Analysis, vol. 11, 2010, pp. 111–118.
[84] S. Liao, M. W. K. Law, A. C. S. Chung, Dominant local binary patterns for texture classification, IEEE Trans. Image Process. 18 (5) (2009) 1107–1118.
[85] Y. Guo, G. Zhao, M. Pietikäinen, Discriminative features for texture description, Pattern Recogn. 45 (10) (2012) 3834–3843.
[86] G. Zhang, X. Huang, S. Li, Y. Wang, X. Wu, Boosting local binary pattern (LBP)-based face recognition, in: Proceedings of Advances in Biometric Authentication, 2005, pp. 179–186.
[87] C. Shan, T. Gritti, Learning discriminative LBP-histogram bins for facial expression recognition, in: Proceedings of British Machine Vision Conference, 2008, 10 p.
[88] C. K. Heng, S. Yokomitsu, Y. Matsumoto, H. Tamura, Shrink boost for selecting multi-LBP histogram features in object detection, in: Proceedings of International Conference on Computer Vision and Pattern Recognition, 2012, pp. 3250–3257.
[89] S. Shan, W. Zhang, Y. Su, X. Chen, W. Gao, Ensemble of piecewise FDA based on spatial histogram of local (Gabor) binary patterns for face recognition, in: Proceedings of International Conference on Pattern Recognition, vol. 4, 2006, pp. 606–609.
[90] D. Zhao, Z. Lin, X. Tang, Laplacian PCA and its applications, in: Proceedings of International Conference on Computer Vision, 2007, 8 p.
[91] S. ul Hussain, B. Triggs, Feature sets and dimensionality reduction for visual object detection, in: Proceedings of British Machine Vision Conference, 2010, 10 p.
[92] L. Nanni, S. Brahnam, A. Lumini, A simple method for improving local binary patterns by considering non-uniform patterns, Pattern Recogn. 45 (2012) 3844–3851.
[93] J. Ren, X. Jiang, J. Yuan, Learning binarized pixel-difference pattern for scene recognition, in: Proceedings of International Conference on Image Processing, 2013, 5 p.
[94] J. Wu, J. H. Rehg, CENTRIST: a visual descriptor for scene categorization, IEEE Trans. Pattern Anal. Mach. Intell. 33 (8) (2011) 1489–1501.
[95] D. Maturna, A. Soto, D. Mery, Face recognition with decision tree-based local binary patterns, in: Proceedings of Asian Conference on Computer Vision, 2010.
[96] E. Rahtu, J. Heikkilä, V. Ojansivu, T. Ahonen, Local phase quantization for blur-insensitive image analysis, Image Vis. Comput. 30 (8) (2012) 501–512.
[97] J. Päivärinta, E. Rahtu, J. Heikkilä, Volume local phase quantization for blur-insensitive dynamic texture classification, in: Proceedings of Scandinavian Conference on Image Analysis, 2011, pp. 360–369.
[98] H. Lategahn, S. Gross, S. Stehle, T. Aach, Texture classification by modeling joint distributions of local patterns with Gaussian mixtures, IEEE Trans. Image Process. 19 (6) (2010) 1548–1557.

[99] N.-S. Vu, A. Caplier, Enhanced patterns of oriented edge magnitudes for face recognition and image matching, IEEE Trans. Image Process. 21 (3) (2012) 1352–1365.
[100] G. Sharma, S. ul Hussain, F. Jurien, Local higher-order statistics (LHS) for texture categorization and facial analysis, in: Proceedings of European Conference on Computer Vision, Part VII, 2012, pp. 1–12.
[101] J. Zhang, J. Liang, H. Zhao, Local energy pattern for texture classification using self-adaptive quantization thresholds, IEEE Trans. Image Process. 22 (1) (2013) 31–42.
[102] J. He, H. Ji, X. Yang, Rotation invariant texture descriptor using local shearlet-based energy histograms, IEEE Signal Process. Lett. 30 (9) (2013) 905–908.
[103] R. Maani, S. Kalra, Y.-H. Yang, Rotation invariant local frequency descriptors for texture classification, IEEE Trans. Image Process. 22 (6) (2013) 2409–2419.
[104] R. Zabih, J. Woodfill, Non-parametric local transforms for computing visual correspondences, in: Proceedings of the Third European Conference on Computer Vision, Stockholm, vol. 2, May 1994, pp. 151–158.
[105] M. Crosier, L. D. Griffin, Using basic image features for texture classification, Int. J. Comput. Vision 88 (2010) 447–460.
[106] J. Kannala, E. Rahtu, BSIF: binarized statistical image features, in: Proceedings of International Conference on Pattern Recognition, 2012, pp. 1363–1366.
[107] M. Calonder, V. Lepetit, M. Özuysal, T. Trzcinski, C. Strecha, P. Fua, BRIEF: computing a local binary descriptor very fast, IEEE Trans. Pattern Anal. Mach. Intell. 34 (7) (2012) 1281–1298.
[108] J. Heinly, E. Dunn, J.-M. Frahm, Comparative evaluation of binary features, in: Proceedings of European Conference on Computer Vision, Part II, 2012, pp. 759–773.
[109] O. Miksik, K. Mikolajczyk, Evaluation of local detectors and descriptors for fast feature matching, in: Proceedings of International Conference on Pattern Recognition, 2012, pp. 2681–2684.
[110] G. Zhao, M. Pietikäinen, Boosted multi-resolution spatiotemporal descriptors for facial expression recognition, Pattern Recognit. Lett. 30 (12) (2009) 1117–1127.
[111] L. Nanni, S. Brahnam, A. Lumini, Local ternary patterns from three orthogonal planes for human action classification, Expert Syst. Appl. 38 (2011) 5125–5128.
[112] R. Mattivi, L. Shao, Human action recognition using LBP-TOP as sparse spatio-temporal feature descriptor, in: Proceedings of International Conference on Computer Analysis of Images and Patterns, 2009, pp. 740–747.
[113] Z. Lei, S. Liao, M. Pietikäinen, S. Z. Li, Face recognition by exploring information jointly in space, scale and orientation, IEEE Trans. Image Process. 20 (1) (2011) 247–256.
[114] G. Zhao, X. Huang, Y. Gizatdinova, M. Pietikäinen, Combining dynamic texture and structural features for speaker identification, in: Proceedings of ACM Multimedia Workshop Multimedia in Forensics, Security and Intelligence, 2010, pp. 93–98.
[115] B. Goswami, C. H. Chan, J. Kittler, B. Christmas, Local ordinal contrast pattern histograms for spatio-temporal, lip-based speaker authentication, in: Proceedings of IEEE International Conference on Biometrics: Theory, Applications and Systems, 2010, 6 p.
[116] X. Huang, G. Zhao, M. Pietikäinen, W. Zheng, Spatiotemporal local monogenic binary patterns for facial expression recognition, IEEE Signal Process. Lett. 19 (5) (2012) 243–246.

[117] J. Ruiz-Hernandez, M. Pietikäinen, Encoding local binary patterns using the re-parametrization of the second order Gaussian jet, in: Proceedings of IEEE International Conference on Automatic Face and Gesture Recognition, 2013, 6 p.
[118] Y. Guo, G. Zhao, Z. Zhou, M. Pietikäinen, Video texture synthesis with multi-frame LBP-TOP and diffeomorphic growth model, IEEE Trans. Image Process. 22 (10) (2013) 3879–3891.
[119] P. Yang, Q. Liu, D. Metaxas, Boosting encoded dynamic features for facial expression recognition, Pattern Recognit. Lett. 30 (2) (2009) 132–139.
[120] J. Yang, S. Wang, Z. Lei, Y. Zhao, S. Z. Li, Spatio-temporal LBP based moving object segmentation in compressed domain, in: Proceedings of IEEE International Conference on Advanced Video and Signal-Based Surveillance, 2012, pp. 252–257.
[121] G. Zhao, M. Pietikäinen, Visual speaker identification with spatiotemporal directional features, in: Proceedings of International Conference on Image Analysis and Recognition, 2013, pp. 1–10.
[122] Y. Huang, Y. Wang, T. Tan, Combining statistics of geometrical and correlative features for 3D face recognition, in: Proceedings of British Machine Vision Conference, 2006, pp. 879–888.
[123] T. Huynh, R. Min, J.-L. Dugelay, An efficient LBP-based descriptor for facial depth images applied to gender recognition using RGB-D face data, in: Proceedings of ACCV Workshops, 2013, pp. 133–145.
[124] G. Sandbach, S. Zafeiriou, M. Pantic, Local normal binary patterns for 3D facial action unit detection, in: Proceedings of IEEE International Conference on Image Processing, 2012.
[125] N. Bayramoglu, G. Zhao, M. Pietikäinen, CS-3DLBP and geometry based person independent 3D facial action unit detection, in: Proceedings of IEEE/IAPR International Conference on Biometrics, 2013.
[126] H. Tang, B. Yin, Y. Sun, Y. Hu, 3D face recognition using local binary patterns, Signal Proc. 93 (2013) 2190–2198.
[127] T. Fang, X. Zhao, S. K. Shah, I. Kakadiaris, 4D facial expression recognition, in: Proceedings of ICCV Workshops, 2011, pp. 1594–1601.
[128] M. Reale, X. Zhang, L. Yin, Nebula feature: a space-time feature for posed and spontaneous 4D facial behavior analysis, in: Proceedings of IEEE International Conference on Automatic Face and Gesture Recognition, 2013, 8 p.
[129] J. Banerjee, A. Moelker, W. J. Niessen, T. van Walsum, 3D LBP-based rotationally invariant region description, in: Proceedings of ACCV 2012 Workshops, 2013, pp. 26–37.
[130] J. Fehr, H. Burkhardt, 3D rotation invariant local binary patterns, in: Proceedings of International Conference on Pattern Recognition, 2008, 4 p.
[131] P. Morgado, M. Silveira, J. S. Marques, Diagnosis of Alzheimers disease using 3D local binary patterns, Comput. Methods Biomech. Biomed. Eng Imaging Vis. 1 (1) (2013) 2–12.
[132] A. Burner, R. Donner, M. Mauerhoefer, M. Holzer, F. Kainberger, G. Langs, Texture bags: anomaly retrieval in medical images based on local 3D-texture similarity, in: Proceedings of MICCAI, 2011.
[133] N. Chatlani, J. J. Soraghan, Local binary patterns for 1-D signal processing, in: Proceedings of European Signal Processing Conference, 2010, pp. 95–99.

[134] Q. Zhu, N. Chatlani, J. S. Soraghan, 1-D local binary patterns based VAD used in HMM-based improved speech recognition, in: Proceedings of European Signal Processing Conference, 2012, pp. 1633–1637.

[135] L. Houam, A. Hafiane, A. Boukrouche, E. Lespessailles, R. Jenanna, One dimensional local binary pattern for bone texture classification, Pattern Anal. Appl. 17 (1) (2012) 179–193.

[136] N. Lazic, P. Aarabi, Spoken term detection using visual spectrogram matching, in: Proceedings of IEEE International Symposium on Multimedia, 2008, pp. 637–642.

[137] Y. M. G. Costa, L. S. Oliveira, A. L. Koerich, F. Gouyon, J. G. Martins, Music genre classification using LBP textural features, Signal Proc. 92 (2012) 2723–2737.

[138] M. Esfahanian, H. Zhuang, N. Erdol, Using local binary patterns as features for classification of dolphin calls, JASA Express Lett. 134 (1) (2013) 105–111.

[139] S. Moore, R. Bowden, Local binary patterns for multi-view facial expression recognition, Comput. Vis. Image Understand. 115 (5) (2011) 541–558.

[140] Z. Lei, M. Pietikäinen, S. Z. Li, Learning discriminant face descriptor, IEEE Trans. Pattern Anal. Mach. Intell. 36 (2) (2014) 289–302.

[141] N. P. Doshi, G. Schaefer, A comprehensive benchmark of local binary pattern algorithms for texture retrieval, in: Proceedings of International Conference on Pattern Recognition, 2012, pp. 2760–2763.

CHAPTER

Subspace approach in spectral color science

10

Jussi Parkkinen[1,2], Hannu Laamanen[3] and Markku Hauta-Kasari[1,3]

[1] *School of Computing, University of Eastern Finland, Joensuu, Finland,* [2] *School of Engineering, Monash University Malaysia, Jalan Lagoon Selatan, Selangor, Malaysia,* [3] *Institute of Photonics, University of Eastern Finland, Joensuu, Finland*

10.1 INTRODUCTION

Color is an important element of our everyday life. We use color to select fruits, we decide our clothes partly based on the color and color gives us a feeling of pleasure in art works. Humans recognize colors through their visual system, which includes the retina as the detector layer and the nerve structure for response manipulation and for creating the color sensation [1].

Most of the standard color coordinate systems and color representations are based on the model of the human eye. In the human retina, there are different types of light sensitive cells [1]. Cone type cells contribute mainly to the color sensation. There are three types of cone cells in the retina and they have different wavelength sensitivities.

Throughout history, color science has been a multidisciplinary field of research with contributions from early Greek philosophers through great figures of the eighteenth century like Isaac Newton and Johan Wolfgang von Goethe to present day scientists in disciplines varying from Mathematics and Physics to Psychology and Arts. Throughout these years, there has also been discussion about what color is. This is an essential question, if we like to build a measurement system and develop some computational methods for color analysis and management. The traditional thinking of color is defined through human vision. This approach has problems from the measuring point of view. Human color vision is difficult to measure, since the understanding of color is finally formed in the human brain. The signal, however, which causes the color sensation, can be measured. It is the light that enters the human eye. The signal, which forms the color sensation, is electromagnetic radiation on the wavelength range visible to the human eye, which is between 380 and 780 nm. There is another problem, if we define color through the human color vision, the color vision of other animals and many artificial color vision systems is not included. In this chapter, we consider color as the signal, electromagnetic spectrum between the wavelengths λ_1 and λ_2, which reaches the detection system, biological or artificial, and causes detection, which can be considered as a color. In order to call the detection

E. Bingham, S. Kaski, J. Laaksonen and J. Lampinen (Eds) Advances in Independent Component Analysis and Learning Machines.
http://dx.doi.org/10.1016/B978-0-12-802806-3.00010-5

of the signal a color, we need to measure the intensity of this signal at least at two distinct wavelengths. Otherwise, we would measure only brightness of the light or in case of imaging, a gray-level image. In this chapter, we treat the color signal as an n-dimensional vector originated from a continuous intensity function of the above-mentioned spectrum.

The first widely accepted attempt to represent color as a combination of spectral components of light was given by Newton [2]. He spread sunlight using a prism into a spectrum, where he counted seven component colors: violet, indigo, blue, green, yellow, orange, and red. Then the color science was developed based on the model of human vision and three different component-based color representations became standards in color science. Currently color camera and color display technology is mostly based on the idea of three-color components. Three-component representation has many problems in color science, for example metamerism. This is because there is not enough information about the color in a three-component representation of a color signal. Therefore, a spectrum is needed as a basis for accurate color management.

Analysis of color information using eigenvectors of covariance or correlation matrix from a set of color spectra started slowly in the 1950s and 1960s. Morris and Morrissey [3] calculated eigenvectors of spectral densities of Kodak Echtachrome film. Simonds [4] describes a method to analyze photographic films by characteristic vectors of spectral response curves. Judd et al. [5] analyzed daylight data using eigenvectors of covariance matrix of a set of 622 daylight spectra. Cohen [6] was the first one who calculated characteristic spectra of standard Munsell color chips spectra. He used 150 Munsell chips for linear component analysis by the centroid method. Ohta [7] used principal component analysis (PCA) to find component spectra of linear expansion of spectral density distributions of the dye mixtures. As can be seen from the references, the early days of using the spectral information for color analysis were mostly done in big photographic film companies. The size of data and the accuracy of results were restricted by the power of computing machines.

A modern computation intensive era for color spectral analysis started in the mid-1980s. Then also the modeling of color space in the spectral domain started to evolve. Earlier the spectral approach was just targeted to the analysis of color spectra. Maloney and Wandell [8] used linear expansion of color spectra into modeling of color constancy and Maloney [9] described a way to represent color spectral data by a small number of parameters using a linear model of the spectral color space.

Steps toward a more comprehensive understanding of the structure of the spectral color space and to manage colors in the spectral space instead of three-dimensional standard color space were taken [10,11]. In these chapters, the idea of dividing the spectral color space into subspaces of colors was initiated. The idea originated from Erkki Oja's studies on subspace methods [12].

In the 1980s, the rise of optical computing was predominant. Caulfield [13] proposed a method for color spectrum classification using an optical computing technique. Soon after that Jaaskelainen et al. [14] proposed a realization of learning subspace method by optical computing. A realization of the subspace method by optical computing using liquid crystal spatial light modulator was presented in [15].

In this chapter we use the term "spectral image" for an image where each pixel is represented by an n-dimensional spectrum. In the literature, the terms "multispectral image" and "hyperspectral image" are used. However, we think that spectral image follows the same logic as "black-and-white image," "gray-level image," or "color image." In all those cases, there is one black-and-white, gray-level, or color value for each pixel. As in spectral image, there is one spectrum in each pixel. Multispectral image does not follow this convention. There are not that many spectra in each pixel as "multi" would suggest.

10.2 PRINCIPAL COMPONENT ANALYSIS

During the last few decades PCA [16,17] has become a standard tool in spectral color science. Oja has linked the PCA also to neural networks [18,19], which gained much interest in data analysis, including spectral color science, in the late 1980s and 1990s. Later, independent component analysis (ICA) became a popular data analysis tool [20]. Here, we introduce the mathematical formulation of the PCA and later introduce the use of ICA in spectral color analysis.

Let us consider a random vector $\mathbf{x} = [x_1, x_2, \ldots, x_N]^T$ having expectation $\mathrm{E}\{\mathbf{x}\} = \mu$. In PCA, we calculate the eigenvalues λ and eigenvectors $\mathbf{v}$ of the covariance matrix $\mathbf{C}$ given by

$$\mathbf{C} = \mathrm{E}\{(\mathbf{x} - \mu)(\mathbf{x} - \mu)^T\}. \tag{10.1}$$

In a similar method for the analysis of continuous stochastic processes, called the Karhunen-Loéve (KL) expansion [21], the covariance matrix is replaced by the autocovariance function.

Spectral color data consist of radiance or reflectance spectra measured with constant intervals over the range of the electromagnetic spectrum. In the case of human vision, it covers the wavelength range from 380 to 780 nm. When N sample points of each spectrum are arranged into the shape of a column vector, a spectral data set consisting of M spectra can be represented as an $N \times M$ matrix $\mathbf{X}$. Instead of calculating a covariance matrix by using centered data, where the mean vector is removed, it is more common to use a correlation matrix in color-related applications, because the subtraction of the mean vector would remove important brightness information.

The $N \times N$ correlation matrix is calculated as

$$\mathbf{R} = \mathbf{X}\mathbf{X}^T. \tag{10.2}$$

The eigenvalues and eigenvectors of the correlation matrix can be solved from the matrix equation

$$\mathbf{RV} = \mathbf{VD}, \tag{10.3}$$

where the diagonal elements of the matrix $\mathbf{D}$ contain eigenvalues and the columns of the matrix $\mathbf{V}$ are the corresponding eigenvectors in the same order. The form of the eigenvectors is characteristic to the spectral data set used in the calculations.

The principal components, elements of the matrix **Y** below, are obtained by projecting color spectra onto the eigenvectors

$$\mathbf{Y} = \mathbf{V}^T\mathbf{X}. \tag{10.4}$$

In principle, each column of the matrix **Y** contains the same amount of information as the corresponding column of the matrix **X**. The relative information content of each component can be estimated by the relative size of the corresponding eigenvalue.

The advantage of the PCA is that it compresses all essential information into the first few components corresponding to the largest eigenvalues. The other components carry mainly noise. Thus, original data can be compressed by selecting only the first P components to represent it and by discarding all other components. This results in the following approximation,

$$\widehat{\mathbf{X}} = \mathbf{V_P}\mathbf{V_P}^T\mathbf{X}, \tag{10.5}$$

where the columns of the matrix $\mathbf{V_P}$ consist of the first P eigenvectors. The quality of the spectral reconstruction can be best estimated by the length of the difference vector between the reconstructed $\hat{\mathbf{x}}_i$ and original spectrum $\mathbf{x}_i$,

$$\delta = \|\hat{\mathbf{x}}_i - \mathbf{x}_i\|. \tag{10.6}$$

Another possibility would be to use the goodness-of-fit coefficient (GFC) suggested by Romero et al. [22].

Figure 10.1 shows five eigenvectors corresponding to the five largest eigenvalues of the correlation matrix **R** using Munsell colors [23,24] as the spectral color data set. These data are measured by 1 nm spectral resolution. An interesting feature is that when the spectral data set consists of a unique set of optimal color spectra, representing the most saturated object colors, the correlation matrix becomes a circulant Toeplitz matrix and its eigenvectors are discrete representations of the trigonometric functions [25].

When necessary, important or interesting wavelength regions can be weighted by multiplying spectral data set **X** with an $N \times N$ diagonal weight matrix **U**, for example, for improving color reproduction accuracy in the compression of spectral color data [26]. When weighted spectral data are compressed and reconstructed, an approximation for the original unweighted spectral data can be obtained by multiplying the reconstruction $\widehat{\mathbf{UX}}$ from the left side with a diagonal matrix **W**, whose diagonal elements are the reciprocal values of the corresponding diagonal elements of the weight matrix **U**.

PCA has been used as a basis also for spectral color space analysis and defining of new basis vectors of the spectral color space. Lenz and Bui [27] studied the structure of spectral color space based on the fact that the spectra are positive functions and are located in the positive octant of the space. They used PCA in their analysis and also showed that in practice the first eigenvector of a set of spectra is close to the mean vector of the set, although mathematically they are not equal. In [28], Piche calculated the PCA eigenvectors for the spectral data set and from them a set of positive basis vectors has been computed. The all positive basis is important in some practical issues. Instead of analyzing the spectral color set as a whole, the color space

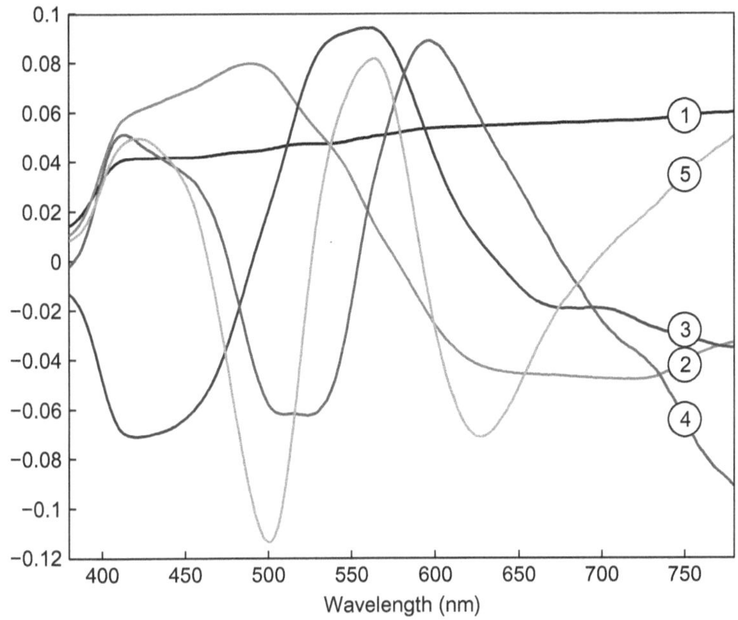

FIGURE 10.1

First five eigenvectors calculated for the Munsell dataset.

can be divided into subspaces and each subspace is defined by PCA eigenvectors. This approach has been used by Zhang and Xu [29], who divided the colors into subgroups and PCA is run for each subgroup separately. This method has been used to improve the spectral reconstruction accuracy. An early attempt to create the subspaces was to use the average learning subspace method (ALSM) [11], a classification method based on PCA [12]. Another approach to improve the reconstruction accuracy is to use the nonlinear PCA. This approach has been taken by Barakzehi et al. [30], where nonlinear PCA has been realized by a neural network structure and this approach improves the spectral reconstruction accuracy in fluorescent samples, when compared to the plain PCA reconstruction.

10.3 INDEPENDENT COMPONENT ANALYSIS

Here, we give a short and in many ways incomplete summary of ICA. A comprehensive description of the method can be found from the book written by Hyvärinen et al. [20].

In the ICA method, the main purpose is to find mutually statistically independent source signals s_i from observed mixtures x_i. Independent source signals are called independent components (ICs) and they are estimated by finding a separation matrix $\mathbf{W}$, so that the estimates for the ICs, $\hat{\mathbf{s}} = \mathbf{W}\mathbf{x}$, are as independent as possible. Matrix $\mathbf{x} = [x_1, x_2, \ldots, x_n]^T$ represents a set composed of n observables.

Few preprocessing steps are required before the analysis of the observed data. In the ICA model, signals s_i and their linear combinations x_i are assumed to be zero mean signals, thus the data should be centered by subtracting the mean values. Another assumption is that observed data are white, thus before the analysis matrix **x**, including the observed mixtures x_i, is multiplied by a whitening matrix to remove possible correlations and scale variances to unity. Whitening also ensures that the rows of the separation matrix are orthogonal. In the ICA model, the numbers of ICs and observed mixtures should be equal. When necessary, the dimension of the matrix **x** can be reduced by using PCA.

Values of the ICs are expected to follow non-Gaussian distributions. The central limit theorem [31] implies that "the sum of independent random variables has a distribution that is closer to Gaussian than any of the original random variables" [20], thus non-Gaussianity indicates statistical independence. Kurtosis measures peakedness of the probability distribution and can be used as a quantitative measure of non-Gaussianity of a zero mean random variable x and is defined as

$$\mathrm{kurt}(x) = \mathrm{E}\{x^4\} - 3[\mathrm{E}\{x^2\}]^2. \tag{10.7}$$

In the case of PCA, the first eigenvector indicates the direction where the variance of the data is largest. The second eigenvector gives the direction of maximum variance, subject to being uncorrelated to the first eigenvector, etc. A rather similar approach can be used in the case of ICA: let us mark an arbitrary linear combination of whitened observables by **wx**, where **w** is a row vector. One row of the separation matrix can be solved by maximizing absolute value of kurtosis,

$$\max|\mathrm{kurt}(\mathbf{wx})| = |\mathrm{kurt}(\mathbf{w}_1\mathbf{x})| \tag{10.8}$$

and other rows can be solved by setting

$$\max|\mathrm{kurt}(\mathbf{wx})| = |\mathrm{kurt}(\mathbf{w}_j\mathbf{x})| \tag{10.9}$$

and requiring that

$$\mathbf{w}_i\mathbf{w}_j^T = 0 \quad \text{with all } 1 \le i < j. \tag{10.10}$$

An alternative measure for non-Gaussianity is negentropy, a quantity representing entropy, which is more stable against possible outliers and thus a better choice for practical applications [20]. In practice, **W** can be solved by using, for example, the gradient descent method or Newton's method having better convergence properties (FastICA algorithm) [20]. After ICs are solved and normalized to unity, they form an orthonormal linear basis that can be used like an eigenvector basis.

10.4 COMPARISON OF METHODS

Table 10.1 compares the reconstruction accuracy of PCA, weighted PCA and ICA. Spectra of the Munsell color chips are median filtered before the analysis to remove measurement noise. Color reproduction accuracy of the weighted PCA seems to be superior compared to the other two methods, but its spectral reconstruction accuracy

Table 10.1 Reflectance Spectra of the Munsell Color Chips Reconstructed by Using Three Different Methods and a Different Number of Components

	PCA		Weighted PCA		ICA	
Components	$\overline{\Delta E_{00}}$	$\overline{\delta}$	$\overline{\Delta E_{00}}$	$\overline{\delta}$	$\overline{\Delta E_{00}}$	$\overline{\delta}$
1	16.8788	1.5169	16.4430	1.7318	12.6301	1.0557
2	12.3050	0.8390	12.0952	1.0920	4.7273	0.7059
3	3.1816	0.4628	0.7497	0.5908	3.0654	0.4088
4	1.1636	0.3366	0.5866	0.5052	0.8169	0.2770
5	1.0443	0.2675	0.3246	0.4228	0.7362	0.2008
6	0.7079	0.1892	0.2476	0.3832	0.4228	0.1563
7	0.3260	0.1439	0.2108	0.3655	0.3335	0.1224
10	0.1321	0.0678	0.0998	0.2789	0.0813	0.0571
15	0.0356	0.0300	0.0188	0.1743	0.0295	0.0269
20	0.0055	0.0157	0.0055	0.1259	0.0045	0.0146

Notes: *Average color differences are evaluated by using the CIE ΔE_{00} color difference formula and CIE D65 standard source and average spectral reconstruction errors are calculated by Eq.* (10.6).

is the worst. At first glance ICA seems to give slightly better results than PCA, but actually this small difference is caused by ICA's requirement to use zero mean data and subtracted mean values can be considered to form one additional component. Analyzing PCA and ICA from a natural scene color information representation point of view can be found in [32]. During recent years, also in spectral data and image analysis, interest toward Kernel methods has increased [33].

10.5 SPECTRAL COLOR APPLICATIONS

In the human visual system, there are three types of light sensitive cells, which form the color representation in the human brain. The simple model for the sensor-level response of these cells can be given as

$$\alpha_i = \int l(\lambda) r(\lambda) s_i(\lambda)\, d\lambda, \qquad (10.11)$$

where α_i is the response of the cone, whose wavelength sensitivity is $s_i(\lambda)$, $l(\lambda)$ is the light source spectrum, and $r(\lambda)$ is the object reflectance function [34]. Let us define $f(\lambda) = l(\lambda)r(\lambda)$ as the color signal reaching the eye. When taking the color signal and cone sensitivities as continuous functions, they can be considered as elements in a Hilbert space. In this Hilbert space, the integral in Eq. (10.11) represents an inner product between $f(\lambda)$ and $s_i(\lambda)$. In this framework, we can interpret the cone response space as a three-dimensional subspace of the spectral color space, which the spectrum $f(\lambda)$ belongs to.

Models of the human color vision include the opponent-color representation of the color sensation [34]. According to this model, the color is represented by three components: lightness, red-green opponency component, and blue-yellow opponency component. There are studies on the relation of PCA and ICA eigenvectors to the human vision color opponency [32].

In addition to modeling the human color vision by the subspaces, spectral decomposition methods have also been used to analyze eye images. Eye fundus imaging is a common tool in studying eye and retina in clinical routines. During recent years the imaging technique has developed enabling use of spectral imaging of the eye fundus [35].

The use of PCA eigenvectors in practical implementations was considered at a time when computation power was not very high for eigenvector analysis. In the 1980s, the use of PCA and KL transform in spectral color and image analysis started to become a practice. At the same time optical computing was receiving increased attention. An idea is to use PCA eigenvectors as filters and do part of spectral image processing by optical computation. A problem in the optical implementation, however, was that eigenvectors other than the first also had negative values. This problem was studied in [36], where a transform into all positive basis vectors was realized. These basis vectors could then be implemented by optical filters. A spectral imaging system where inner products of Eq. (10.5) were computed by optical processing is given in [37].

In the case of spectral images, principal components can be used to visualize possible hidden details, otherwise indistinguishable with the naked eye, and it is also possible to form the true-color representations for the principal components and their combinations [25,38]. When the spectral data set is a representative sample of the whole color space including variety of different hues, as in the case of the Munsell colors, in a rough approximation the first component can be considered to consist of achromatic brightness information and the second and third components to correspond to chromatic opponent-color channels [38]. A proper weighting of spectral data can be used to strengthen this property [26].

Use of PCA, modified PCAs and ICA in spectral data analysis ranging from UV to far IR and spectral image analysis include quite a variety of applications. Examples of these are in geology [39], food science [40], archeology [41], dermatology [42], skin color analysis [43], studying cartilage conditions [44], aerosolic science [45], climatology [46], spectral image compression in remote sensing [47], arts [48], ceramic tiles analysis [49], and analysis of daylight [50]. PCA in spectral imaging has been also implemented in a parallel processing environment in GPU for real-time analysis [51].

10.6 CONCLUSIONS

The development in spectral imaging systems is very active. The spectral range of cameras is expanding, and they are becoming smaller and more affordable. This means that spectral imaging is gaining popularity in many fields of application. Hence, there is a need for development of spectral data and image analysis methods. Furthermore, color is understood more as a spectral signal instead of only three-dimensional coordinate values. Spectral understanding is necessary because many applications are using the electromagnetic spectrum outside the human visual range, in the UV and IR regions. In this case the only useful approach is the spectral approach.

The development in applications without wavelength limits also links to theoretical studies about colors. The study of colors as entities in a high dimensional spectral space and the structure and essence of this space are interesting topics, which require active research for the benefit of color science, and also for practical solutions in demanding engineering problems. PCA, ICA, and related methods developed, for example, by Erkki Oja still form a solid basis for this research.

REFERENCES

[1] D. Purves, et al., Neuroscience, fifth ed., Sinauer Associates, Inc., Sunderland, MA, USA, 2012.

[2] I. Newton, Opticks or, a Treatise of the Reflections, Refractions, Inflections, and Colours of Light, 1704. Available as Google eBook, http://books.google.fi/books/about/Opticks.html?id=GnAFAAAAQAAJ&redir_esc=y (valid November 12, 2014).

[3] R.H. Morris, J.H. Morrissey, An objective method for determination of equivalent neutral densities of color film images. II. Determination of primary equivalent neutral densities, J. Opt. Soc. Am. 44 (7) (1954) 530-534.

[4] J.L. Simmonds, Application of characteristic vector analysis to photographic and optical response data, J. Opt. Soc. Am. 53 (8) (1963) 968-974.

[5] D.B. Judd, et al., Spectral distribution of typical daylight as a function of correlated color temperature, J. Opt. Soc. Am. 54 (8) (1963) 1031-1040.

[6] J. Cohen, Dependency of the spectral reflectance curves of the Munsell color chips, Psychonom. Sci. 1 (1964) 369-370.

[7] N. Ohta, Estimating absorption bands of component dyes by means of principal component analysis, Anal. Chem. 45 (3) (1973) 553-557.

[8] L.T. Maloney, B.A. Wandell, Color constancy: a method for recovering surface spectral reflectance, J. Opt. Soc. Am. A 3 (1) (1986) 29-33.

[9] L.T. Maloney, Evaluation of linear models of surface spectral reflectance with small numbers of parameters, J. Opt. Soc. Am. A 3 (10) (1986) 1673-1683.

[10] J.P.S. Parkkinen, et al. Pattern recognition approach to color measurement and discrimination, Acta Polytech. Scand. Appl. Phys. Ser. No. 149 1 (1985) 171-174.

[11] J.P.S. Parkkinen, Subspace methods in two machine vision problems, (PhD thesis), Kuopio, Finland, 1989.

[12] E. Oja, Subspace Methods of Pattern Recognition, Research Studies Press, Letchworth, England, 1983.

[13] H.J. Caulfield, P.F. Mueller, Direct optical computation of linear discriminants for color recognition, Opt. Eng. 23 (1) (1984) 16-19.

[14] J. Parkkinen, et al., Color analysis by learning subspaces and optical processing, in: Proceedings of the IEEE Annual International Conference on Neural Networks, vol. 2, July 24-27, 1988, IEEE, San Diego, CA, USA, 1988, pp. 421-427.
[15] T. Jaaskelainen, et al., Color classification by vector subspace method and its optical implementation using liquid crystal spatial light modulator, Opt. Commun. 89 (1) (1992) 23-29.
[16] H. Hotelling, Analysis of a complex of statistical variables into principal components, J. Educ. Psychol. 24 (1933) 498-520.
[17] I. Jolliffe, Principal Component Analysis, second ed., Springer-Verlag, Heidelberg, Germany, 2002.
[18] E. Oja, A simplified neuron model as a principal component analyzer, J. Math. Biol. 15 (1982) 267-273.
[19] E. Oja, Principal components, minor components and linear neural networks, Neural Netw. 5 (6) (1992) 927-935.
[20] A. Hyvärinen, et al., Independent Component Analysis, John Wiley & Sons, New York, NY, USA, 2001.
[21] K. Karhunen, Über lineare Methoden in der Wahrscheinlichkeitsrechnung, Ann. Acad. Sci. Fenn. A I (37) (1947) 1-79.
[22] J. Romero, et al., Linear bases for representation of natural and artificial illuminants, J. Opt. Soc. Am. A 14 (5) (1997) 1007-1014.
[23] Munsell Book of Color, Matte Finish Collection, Munsell Color, Baltimore, USA, 1976.
[24] University of Eastern Finland, Spectral Color Research Group, Color spectra database, November 12, 2014, Available from: https://www.uef.fi/spectral/spectral-database.
[25] M. Flinkman, et al. Eigenvectors of optimal color spectra, J. Opt. Soc. Am. A 30 (9) (2013) 1806-1813.
[26] H. Laamanen, et al., Weighted compression of spectral color information, J. Opt. Soc. Am. A 25 (6) (2008) 1383-1388.
[27] R. Lenz, T.H. Bui, Statistical properties of color-signal space, J. Opt. Soc. Am. A 22 (5) (2005) 820-827.
[28] R. Piche, Nonnegative color spectrum analysis filters from principal component, J. Opt. Soc. Am. A 19 (10) (2002) 1946-1950.
[29] X. Zhang, H. Xu, Reconstructing spectral reflectance by dividing spectral space and extending the principal components in principal component analysis, J. Opt. Soc. Am. A 25 (2) (2008) 371-378.
[30] M. Barakzehi, et al., Reconstruction of total radiance spectra of fluorescent samples by means of nonlinear principal component analysis, J. Opt. Soc. Am. A 30 (9) (2013) 1862-1870.
[31] R.V. Hogg, A.T. Craig, Introduction to Mathematical Statistics, fourth ed., Macmillan Publishing Co. Inc., New York, NY, USA, 1978.
[32] T.W. Lee, et al., Color opponency constitutes a sparse representation for the chromatic structure of natural scenes, Adv. Neural Inf. Process. Syst. 13 (2001) 866-872.
[33] V. Heikkinen, Kernel methods for estimation and classification of data from spectral imaging (PhD thesis), Joensuu, Finland, 2011.
[34] G. Wyszecki, W.S. Stiles, Color Science, second ed., John Wiley & Sons, New York, NY, USA, 2000.
[35] A. Calcagni, et al., Multispectral retinal image analysis: a novel non-invasive tool for retinal imaging, Eye 25 (12) (2011) 1562-1569.

[36] S. Toyoka, N. Hayasaka, Two-dimensional spectral analysis using broad-band filters, Opt. Commun. 137 (1-3) (1997) 22-26.
[37] M. Hauta-Kasari, Computational techniques for spectral image analysis, (PhD thesis), Lappeenranta, Finland, 1999.
[38] H. Laamanen, Spectral color and spectral color image analysis, (PhD thesis), Joensuu, Finland, 2007.
[39] J.S. Tyo, et al., Principal-components-based display strategy for spectral imagery, IEEE Trans. Geosci. Remote Sens. 41 (3) (2003) 708-718.
[40] C. Liu, et al., Nondestructive determination of transgenic *Bacillus thuringiensis* rice seeds (*Oryza sativa* L.) using multispectral imaging and chemometric methods, Food Chem. 153 (2014) 87-93.
[41] R.M. Cavalli, et al. Detection of anomalies produced by buried archaeological structures using nonlinear principal component analysis applied to airborne hyperspectral image, IEEE J. Sel. Top. Appl. Earth Obs. Remote Sens. 6 (2) (2013) 659-669.
[42] J.M. Kainerstorfer, et al., Principal component model of multispectral data for near real-time skin chromophore mapping, J. Biomed. Opt. 15 (4) (2010) 046007.
[43] N. Tsumura, H. Haneishi, Y. Miyake, Independent component analysis of spectral absorbance image in human skin, Opt. Rev. 7 (6) (2000) 479-482.
[44] J. Kinnunen, et al., Optical spectral reflectance of human articular cartilage – relationships with tissue structure, composition and mechanical properties, Biomed. Opt. Express 2 (5) (2011) 1394-1402.
[45] T.A. Jones, S.A. Christopher, Multispectral analysis of aerosols over oceans using principal components, IEEE Trans. Geosci. Remote Sens. 46 (9) (2008) 2659-2665.
[46] U. Amato, et al., Statistical cloud detection from SEVIRI multispectral images, Remote Sens. Environ. 112 (3) (2008) 750-766.
[47] B. Penna, et al., Transform coding techniques for lossy hyperspectral data compression, IEEE trans. Geosci. Remote Sens. 45 (5) (2007) 1408-1421.
[48] S. Baronti, et al., Principal component analysis of visible and near-infrared multispectral images of works of art, Chemom. Intell. Lab. Syst. 39 (1) (1997) 103-114.
[49] S. Kukkonen, et al., Color features for quality control in ceramic tile industry, Opt. Eng. 40 (2) (2001) 170-177.
[50] J. Hernández-Andrés, et al., Color and spectral analysis of daylight in southern Europe, J. Opt. Soc. Am. A 18 (6) (2001) 1325-1335.
[51] R. Josth, et al., Real-time PCA calculation for spectral imaging (using SIMD and GP-GPU), J. Real-Time Image Process. 7 (2) (2012) 95-103.

CHAPTER

From pattern recognition methods to machine vision applications 11

Heikki Kälviäinen

Machine Vision and Pattern Recognition Laboratory, Department of Mathematics and Physics, Lappeenranta University of Technology, Lappeenranta, Finland

11.1 INTRODUCTION

This chapter considers scientific challenges in developing machine vision applications based on pattern recognition methods. The focus is on the research carried out at the Machine Vision and Pattern Recognition (MVPR) Laboratory of Lappeenranta University of Technology (LUT) [1]. The goal of machine vision solutions is to create useful and significant value-added applications, especially using digital image processing and analysis, such as machine vision systems for the process industry, and medical image analysis for efficient health care of eye diseases. The chapter also considers modeling human experience and ground truth. It is important to recognize relevant phenomena, to measure them, and to understand how humans define their experience based on the measured data. This knowledge enables us to develop intelligent and robust methods for practical applications with smart feature selection and parameter sensitivity analysis. The problem contains two levels: (i) either the experts can relatively accurately define the ground truth among themselves, or (ii) the ground truth is based on representative observers' (experts and/or laymen) mean opinions with variation. The task becomes more difficult when visual evaluation is based more on humans' subjective opinions than on well-defined objective details agreed by experts, especially in the case of modeling visual image quality. This leads to a very challenging task of modeling the connection between a human observer (psychometric data) and a machine vision system (physical data). An interesting question is how much image content affects perception, and thus how much there is a need for robust object recognition to find salient features. Real-world applications for industrial machine vision, psychometric quality assessment, medical image processing, and traffic sign condition monitoring are introduced. Developed annotation tools and expert databases are also discussed.

The chapter is organized as follows: Section 11.2 studies relationships between machine vision and human vision. Moreover, fundamental steps in modeling of industrial phenomena based on visual information are introduced. Methods for feature extraction and classification are briefly discussed. Visual inspection and

E. Bingham, S. Kaski, J. Laaksonen and J. Lampinen (Eds) Advances in Independent Component Analysis and Learning Machines.
http://dx.doi.org/10.1016/B978-0-12-802806-3.00011-7

computer vision are considered in Section 11.3 where we introduce several industrial applications. Section 11.4 discusses medical image processing and analysis for retinal image diagnosis, and biomolecular vision is considered in Section 11.5. Conclusions are given in Section 11.6 including pointers to future research.

11.2 FROM HUMAN VISION TO MACHINE VISION

The main objective in developing digital image processing and analysis applications is to replace human vision by machine vision in suitable tasks. A machine vision solution automates actions done manually. Many such tasks are monotonous, perhaps even dangerous. As compared to human vision, machine vision can be faster, more accurate, less expensive, and yet easy to use. A machine vision system can replace human vision (e.g., in industrial automation), assist human vision (e.g., in medical diagnoses), and perform tasks that are impossible for the human vision system (e.g., spectral imaging, or video analyses containing a lot of information). In the industry, there is often a need to fully automate some task as shown in Figure 11.1.

There is a fundamental challenge that a machine "sees" differently from a human being but it should nevertheless understand similarly to a human being. For a human being it takes only a couple of seconds to understand what is in Figure 11.2(a). However, a machine sees only a set of local pixels as in Figure 11.2(b). Together these sets constitute more global information as shown in Figure 11.2(c).

In industrial machine vision, the following steps are important: first, to recognize phenomena that are significant to the process, then, to measure optimally, and finally, to understand the measurements correctly. It is important to observe who knows the ground truth, and why and how. We must be sure that modeling expert knowledge is indeed possible. It must be considered whether there are challenges to

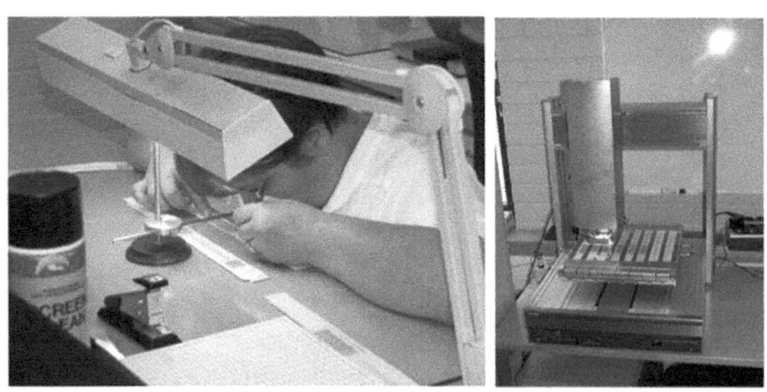

(a) (b)

FIGURE 11.1

Human vision vs. machine vision: (a) manually annotated expert knowledge and (b) a fully automatic machine vision system.

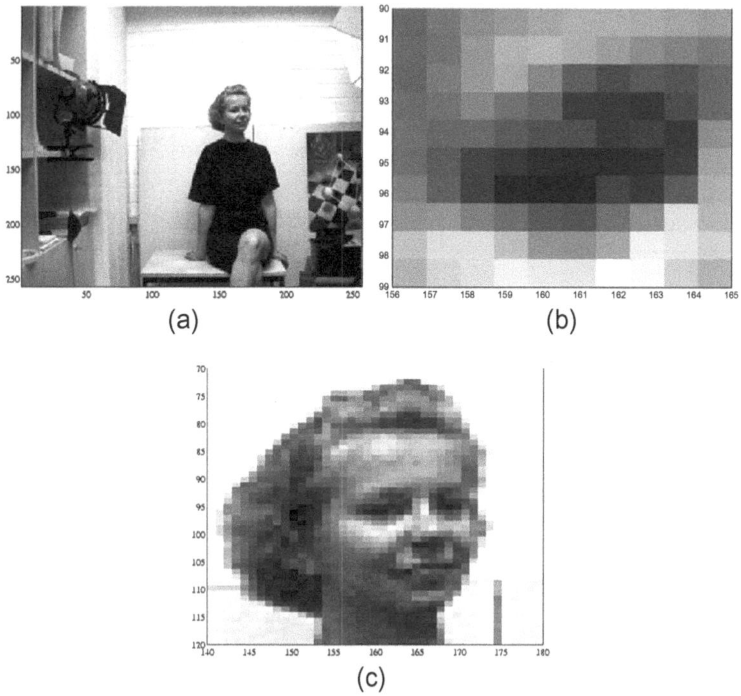

FIGURE 11.2

Human vision vs. machine vision as digital information: (a) full image; (b) local pixels; and (c) more global information [2].

be expected in imaging and whether multimodal information is needed. To understand the measurements, it must be known whether the categorization of the obtained data is clear, that is, object classes are known, and no more clustering is needed. Usually there is a trade-off between accuracy and computation time. In general, the final goal is to build solutions for overall quality management and process control. An example of a machine vision system is shown in Figure 11.3 where a solution for pulping has been built according to these three design principles. The behavior of pulp suspension is studied inside a container using the detected bubbles and the total volume and size distributions of bubbles to make this process stage more efficient.

The fundamental steps of machine visions are also shown in Figure 11.3. To facilitate these steps in general, a lot of research methods exist: randomized hough transform (RHT) for geometric primitives detection [4–13], Gabor filtering for object detection [14–18], Gaussian mixture models for object classification [19,20], SOM- and PCA-based image compression and representation of spectral images [21–23], surface analysis for 2D and 3D images [24–26], unsupervised methods for visual object categorization (VOC) [27–31], tracking methods for computer vision [32–34],

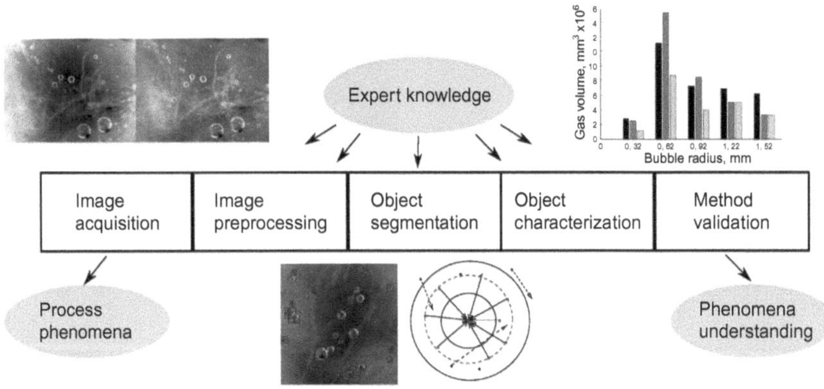

FIGURE 11.3

Steps in a machine vision system for pulping [3].

probabilistic methods for image quality assessment [35–38], and pattern recognition methods for nonvisual information [39].

11.3 VISUAL INSPECTION AND COMPUTATIONAL VISION

Visual inspection enables image-based quality control and resource-efficient production. Imaging, image processing, and image analysis methods have thus been developed for industrial machine vision. In practice, this requires an analyzing laboratory, as well as online and inline measurements of the processes for the characterization of phenomena as shown in Figure 11.3. In this section, we discuss how computational vision can be used in the forest and printing industries. Modeling and predicting human experience can be made based on full reference (FR), reduced reference (RR), and nonreference (NR) visual quality metrics. We discuss modeling the visual quality experience with Bayesian networks (evolve-estimate-simulate) and using human-computer interaction (HCI) based on multimodal data. At the end of the section, we also briefly discuss applying computational vision to face biometrics and traffic sign condition analysis.

11.3.1 PROCESS CONTROL FOR PULPING AND PAPERMAKING

Our main objective is to develop novel resource-efficient and environmentally sound imaging, image processing, and image analysis methods for machine vision applications, especially for overall quality assessment and control. In pulping and papermaking, the goal is resource-efficient and environmentally sound production on a required quality level, using less capital, raw material, water, and energy, following a slogan "less is more."

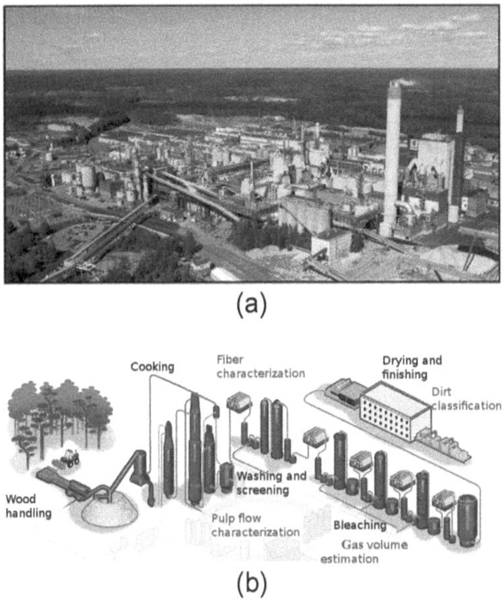

FIGURE 11.4

Pulping and papermaking: (a) industrial environment and (b) pulping process [3], modified from [40].

Machine vision can provide quality control along the whole manufacturing line of paper and board products. The scope is from pulping to papermaking which can be understood as a part of the following production chain: forest → wood → pulp → paper → print. This industrial environment is very complex as seen in Figure 11.4, providing large amounts of rich data.

The motivation of this study comes from the necessity to predict the quality of printing on paper or board, especially in the case of images. Printed materials should look good enough to a consumer; an advertisement should get positive attention and a high-quality journal should be easy to read. Thus, a paper manufacturer should know which kind of quality it offers to a printing house. The quality should not be too high or too low but just sufficient for a known purpose. This requires quality assessment before printing and after printing. In both cases, the visual quality assessment is usually done manually or semiautomatically observing either the manufacturing processes or test prints.

Pulping is the first step in the chain to be considered. It would be interesting to know more about raw material at the pulping stage, also called the wet stage. One often focuses on the following process steps [3]: (i) fiber characterization in pulp suspension, (ii) gas volume estimation at the bleaching stage of pulping, (iii) pulp flow characterization, and (iv) dirt particle classification in dried pulp sheets. In Figure 11.3, we show an example of gas volume estimation based on the detection of bubbles in pulp suspension. In this research, a novel framework for bubble detection titled as concentric circular arrangements (CCA) was proposed [41]. The CCAs are recovered in a hypothesize-optimize-verify framework. The hypothesis generation is based on sampling from the partially linked components of the nonmaximum suppressed responses of oriented ridge filters, and is followed by the CCA parameter estimation. Parameter optimization is carried out by minimizing a novel cost-function. Besides bubbles, it is important to detect and characterize fiber segments since fibers affect very much the quality of paper. In [42], one starts with an edge detection algorithm after which the task of object detection becomes a problem of edge linking. A state-of-the-art local linking approach called tensor voting is used to estimate the edge point saliency describing the likelihood of a point belonging to a curve, and to extract the end points and junction points of these curves. Another interesting new approach is the framework for dirt particle detection and classification in pulp sheets as well as for the generation of the semisynthetic ground truth [25]. Pulp sheets are used as raw material for papermaking so their purity affects the quality. To classify the dirt particles, a set of features is computed for each image segment. Sequential feature selection is employed to determine a close-to-optimal set of features to be used in classification.

While making paper and board, interesting questions include whether we can in general measure and control the quality on a paper web online during manufacturing, and whether we can predict the quality and get the desired quality by overall quality management. Thus, imaging methods for paper characterization (low and high resolutions in real-time) and methods for management of paper characteristics by vision-based control have been proposed especially for paper web surface analysis [24].

Before printing, the quality of paper and board products can be tested either manually or by a machine vision system as shown in Figure 11.1. Methods for the evaluation of the unevenness of printing measured in an image (Mottling) [43], the detection of missing dots in test slips for testing printability (Heliotest) [44], and the detection of surface defects for testing runnability (IGT Picking) [45] have been suggested.

11.3.2 IMAGE QUALITY ASSESSMENT AND VISUAL OBJECT CATEGORIZATION

The quality of printing is a function of the selected paper type and the printer used in printing. Similarly, the quality of a digital image on a display depends on display devices used. The main question is how to model the connection between human perception and physical measurements: which image we prefer and why when seeing the same image produced with different media as shown in Figure 11.5.

This leads us to consider modeling human experience and ground truth. It is important to recognize relevant phenomena, to measure them, and to understand how humans define their experience based on the measured data. This knowledge enables us to develop intelligent and robust methods for practical applications, with smart feature selection and parameter sensitivity analysis. The problem contains

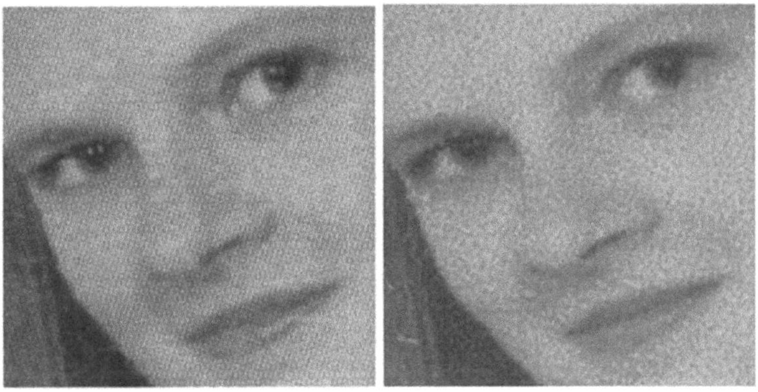

FIGURE 11.5

Which looks better and why? Two close-ups from the same image produced by different media.

two levels: (i) either the experts can relatively accurately define the ground truth among themselves, or (ii) the ground truth is based on representative observers' (experts and/or laymen) mean opinions with variation. The task becomes more difficult when visual evaluation is based more on laymen's subjective opinions than on well-defined objective details agreed by experts, especially in the case of modeling visual image quality. This leads to a very challenging task of modeling the connection between a human observer (psychometric data) and a machine vision system (physical data). An interesting question is how much image content affects perception, and thus how much there is a need for robust object recognition to find salient features.

Methods based on the overall visual quality index (VQI) of the whole image have been proposed in [35–37]. Tuytelaars et al. have given a survey on VOC [31], and partly inspired by their article, methods based on the regions of interest in an image using VOC have been proposed in [27–30], enabling a content-based VQI where the regions of the image are weighted. The former methods based on the overall VQI connect the physical measurements to experienced overall quality by subjective quality attributes using a Bayes network as shown in Figure 11.6. Psychometric data were generated as mean opinions among the panelists. Based on these data, hypotheses can be verified as shown in Figure 11.7(a) where the experience of quality decreases when mottling increases. Effects of different quality characteristics are shown in Figure 11.7(b) as Cumulative Match Score (CMS). In the figure, the nth bin of the cumulative histogram tells how many times (percent of all samples) the ground truth match (human evaluation) is contained in the set of n closest samples in the second evaluation space (computational). See [35–37] for further information.

The latter methods are based on VOC where areas of images do not have an uniform impact on the quality. Different areas, that is, objects, are considered to affect the quality differently. An unsupervised VOC (UVOC) approach [27,29] is

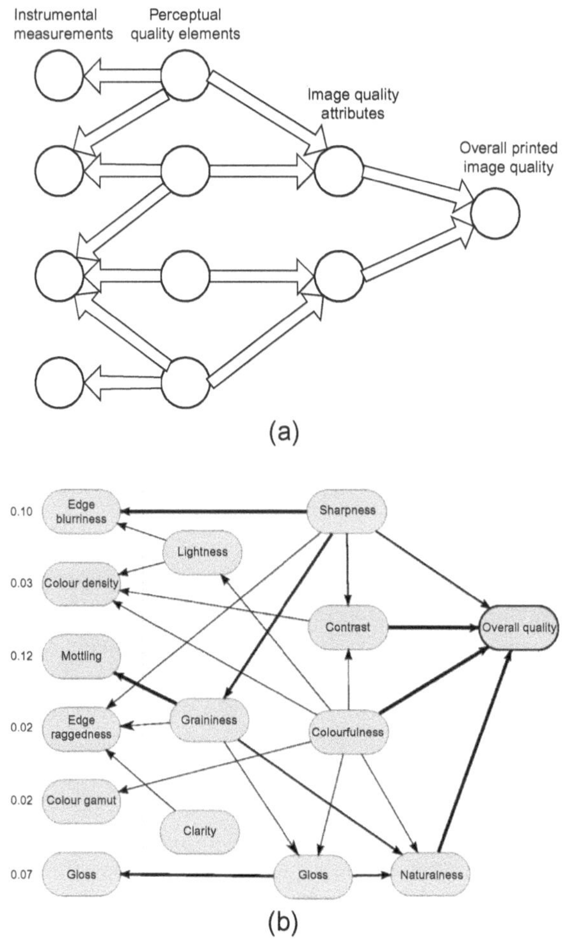

FIGURE 11.6

Connecting physical measurements and subjective quality attributes.

considered in Figure 11.8 where unsupervised categorization means that object categories are not known beforehand, and thus the most important ones are detected without supervision. Usually there are lots of objects to be categorized. We have generated the Randomized Caltech-101 image set [28] (with the known ground truth) and the Abstract image set [30], including human opinions as "the ground truth" [46] to test the quality of our approach. In our approach, the objects are categorized in an unsupervised manner, that is, there are no attached class labels or a predefined number of clusters but the objects are clustered automatically based on their characteristics. This suits computing a content-based VQI since important features are localized but there is no need to classify them. The bag-of-features approach has been applied

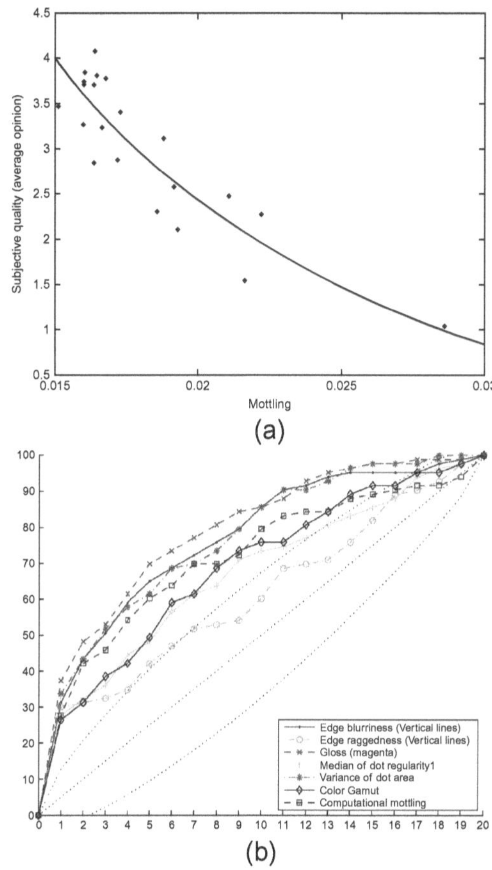

FIGURE 11.7

Using psychometric data: (a) verifying rules, i.e., hypotheses, and (b) Cumulative Match Score of each quality characteristic.

to UVOC as shown in Figure 11.8(b). In the first row, detected local features are drawn with rectangles. In the second row, detected local features are described by computing the gradients in eight directions which are illustrated with arrows. In the third row, visual vocabulary is built by using a self-organizing map (SOM). In the fourth row, codebook histograms are shown. In the fifth row, images are categorized using self-organization. The results are shown in Figure 11.8(c).

The study on the connection between human perception and physical measurements was extended to observations of hand gestures, especially finger movements, for example, as shown in Figure 11.9. The goal is to understand experimental aspects of HCI with touch and gesture interfaces. This is achieved by measuring psychological, behavioral, and environmental interaction variables with novel

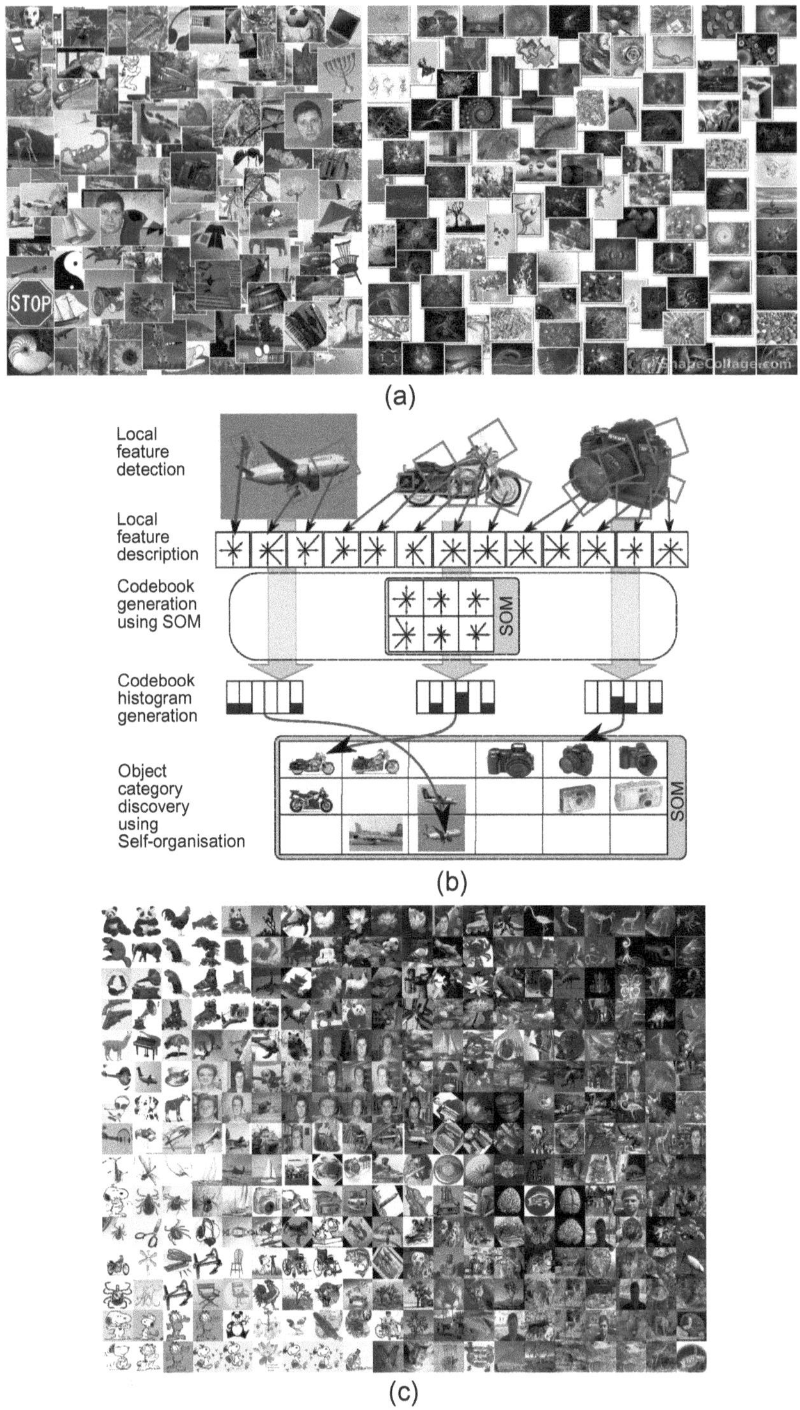

FIGURE 11.8

Unsupervised visual object categorization using SOM: (a) Randomized Caltech-101 image set and Abstract image set; (b) UVOC framework; and (c) results of the image categorization on Caltech-101 images. Only one image is shown for each node [46].

FIGURE 11.9

How do we behave while using fingers for gesture interfaces?

methodologies and mathematically modeling the interaction experience. A comprehensive comparison of the state-of-the-art methods is given in [47] and the new datasets are published in [48].

11.3.3 FACE DETECTION AND RECOGNITION

The face is one of the main features in many biometrics applications, for example, in automatic passport control. Before the face can be recognized it must be detected in a digital image. It is very important that this localization succeeds accurately. MVPR has developed feature-based face detection methods which use Gabor filtering [14,15,17,18]. The proposed framework is illustrated in Figure 11.10. Gabor filtering as a feature-based method generates many labeled candidates (eye, ear, nose, mouth). The number of candidates is cleverly reduced by taking advantage of physical restrictions of the face. The obtained detection rates are very accurate as compared to the other state-of-the-art methods. To improve the accuracy further, 3D face models are generated as shown in Figure 11.10, approaching the optimal recognition rate of 100%. Naturally, a 3D biometric passport still needs improved technology to be feasible.

11.3.4 TRAFFIC SIGN CONDITION ANALYSIS

Roadways are full of traffic signs which need to be correctly placed and in good condition. These signs are usually inspected manually without automated assisting systems and this inspection happens less frequently, once in 5-7 years. There are five condition categories from 1 (worst) to 5 (best). Thus, automatic traffic sign inventory and condition analysis using machine vision and pattern recognition methods could be very useful for more efficient road maintenance, improving processes, decreasing maintenance costs, and guaranteeing the quality of traffic signs in real-time enabling also intelligent driving systems. Machine vision-based inventory of traffic signs consists of detection (HOG, LUV Color, channel features, and AdaBoost), classification (HOG and LDA features, and KNN and random forest), localization (Kalman filtering and GPS estimation), and condition analysis of traffic signs (color segmentation, Canny filtering, and KNN) [49]. Images are produced by a camera which is attached to a moving maintenance vehicle.

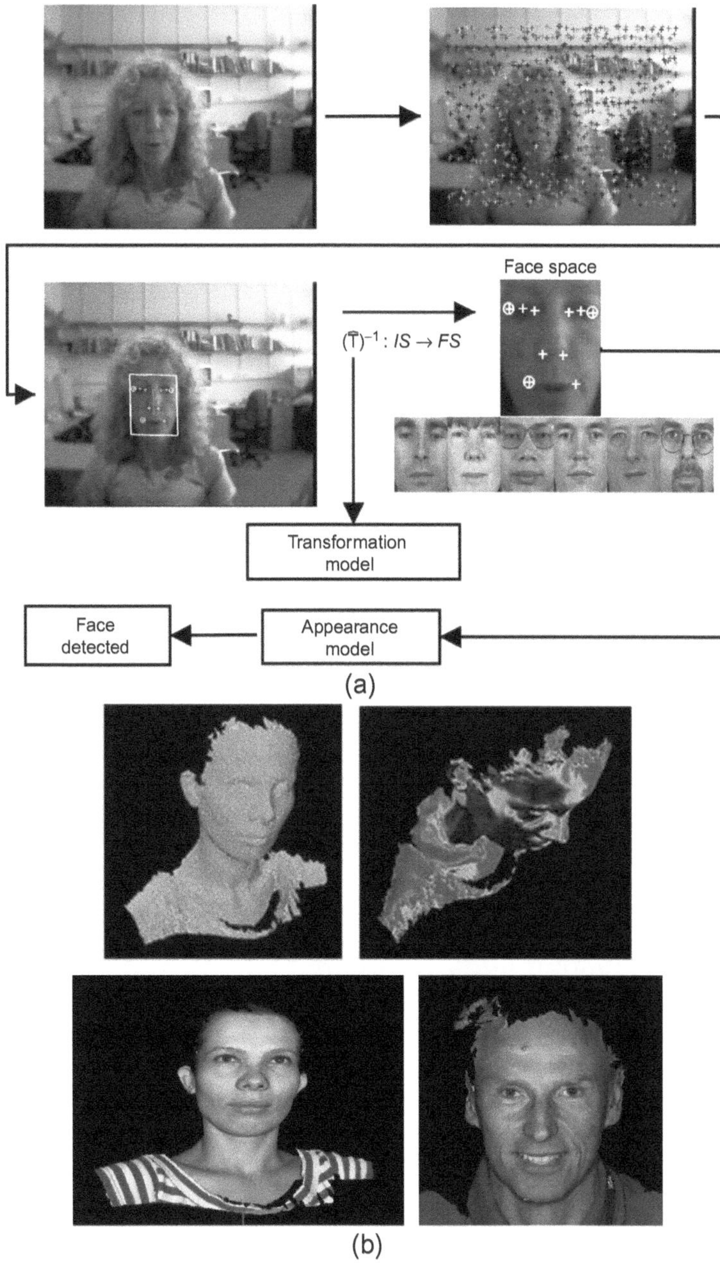

FIGURE 11.10

Face detection and recognition: (a) 2D framework and (b) 3D modeling.

We have proposed a machine vision system including the annotation tool shown in Figure 11.11. An expert can use the tool for marking his/her visual opinion about the condition of a traffic sign. Expert knowledge can be modeled based on these markings. The performance of the developed machine vision system performance has been estimated with three datasets, two of which have been collected by MVPR.

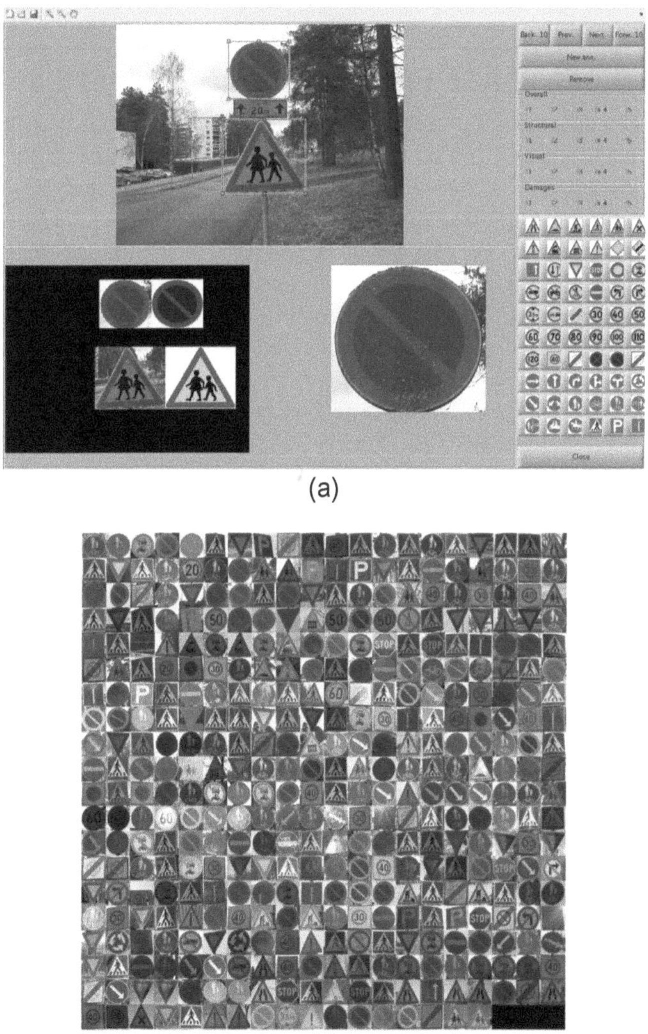

(a)

(b)

FIGURE 11.11

Traffic sign condition analysis: (a) annotation tool and (b) Lappeenranta road signs database [49].

Images of traffic signs taken in the downtown area of Lappeenranta, Finland, are shown in Figure 11.11 and all of them are annotated by an expert. Another dataset was collected during winter with a video camera attached to a moving maintenance car. Based on the experiments [49], almost all traffic signs can be detected, classified, and located automatically using machine vision. Their condition can be analyzed using computational vision quite as accurately as manual evaluation by an expert [49].

11.4 MEDICAL IMAGE PROCESSING AND ANALYSIS

Digital image processing and analysis does not need to replace human vision but it can significantly assist it. This is the reason why medical imaging is one of the most important application areas of computer vision and image analysis. This section considers an important application field, namely the processing of microscopy images for the detection of lesions of diabetic retinopathy from fundus images [50,51]. The goal is to demonstrate how lesions in a retina caused by diabetic retinopathy can be detected from color fundus images by using machine vision methods [52–54]. Diabetes is a metabolic disorder characterized by an impaired control of blood glucose level. Two types exist: Type 1 is characterized by lack of insulin (mostly children and young persons) while Type 2 is typically characterized by an insulin resistance (mostly middle-aged and elderly people) [55].

The automated diagnosis consists of two steps: in the first step, it is decided whether an eye needs further analysis (too many lesions visible) or not (the eye is healthy enough). In the second step, fundus images selected for further analysis are automatically diagnosed. The developed system saves both the resources of medical experts and costs in health care. It offers a tool for healthcare providers to improve the quality of life of diabetes patients, for example, by more robust screening. This is important since the number of diabetes patients is increasing especially in the developed countries [56,57]. Without proper treatment it can lead to a decrease of vision or even blindness: it is the leading cause of blindness in the working age population. Thus the early detection of retinal complications is crucial. Photographic detection and follow-up of retinal changes with a retinal camera is the most versatile way to monitor the condition of the retina. The whole automated process is described in Figure 11.12.

As a case study, the framework and the tools (Figure 11.13) were utilized to establish the DiaRetDB1 V2.1 database for benchmarking diabetic retinopathy detection algorithms, originally published in [58,59], and later discussed in [54]. The database contains a set of retinal images, the ground truth based on information from multiple experts, and a baseline algorithm for the detection of retinopathy lesions. The main contributions can be summarized as follows [54]: (1) an image annotation tool for medical experts, (2) a public retinal image database with expert annotations, (3) a solid evaluation framework for image analysis system development and comparison, and (4) image-based and pixel-based evaluation methods.

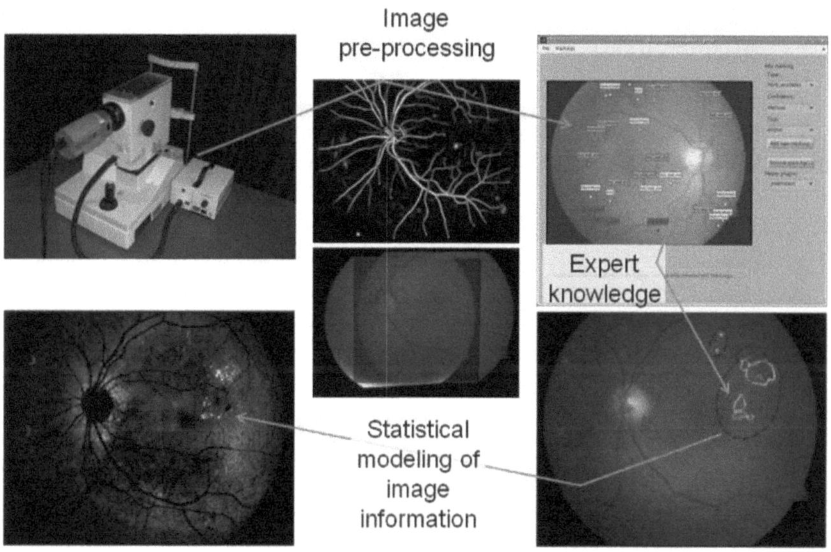

FIGURE 11.12

Medical image processing of retina images.

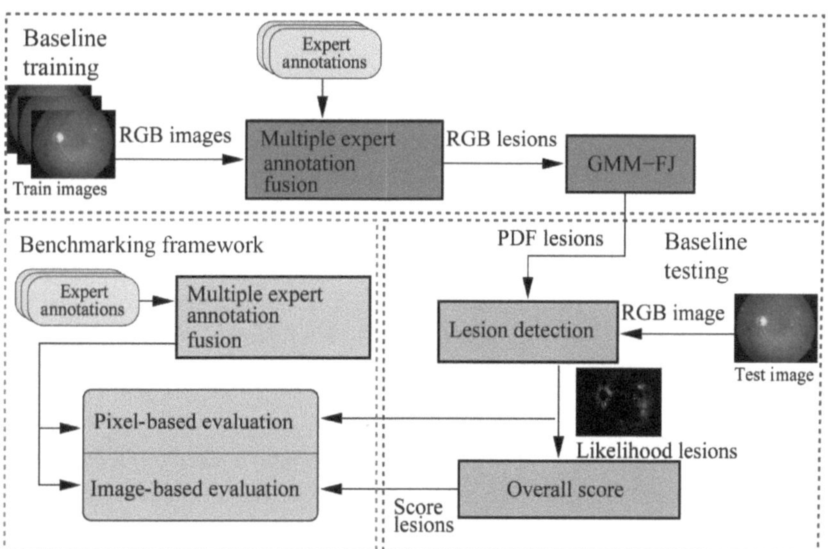

FIGURE 11.13

A framework for constructing benchmark databases and protocols [52].

11.4.1 IMAGE ANNOTATION TOOLS FOR GENERATING THE GROUND TRUTH

Image annotation is a process of collecting metadata for digital images, for example, by annotating identifiers of lesions indicative of diabetic retinopathy in eye fundus images. By using this metadata, benchmarking and developing image analysis algorithms are possible. During the research, medical experts annotated diabetic lesions in several eye fundus images by drawing a perimeter around each lesion and assigning their subjective certainty about the annotation. A software tool for medical image annotation (see the upper right corner of Figure 11.12) was provided helping to collect class label, spatial span, and expert's confidence on lesions. The tool also appropriately combines the manual segmentations done by multiple experts.

11.4.2 PUBLIC RETINA IMAGE DATABASES

Public image databases are the important resource in the development and comparison of automatic eye fundus image analysis that support the technology transfer from research laboratories to clinical practice. Two medical image databases have been published with the accompanied ground truth: DiaRetDB0 and DiaRetDB1 [54,58–60]. The work on DiaRetDB0 provided us with essential information on how diabetic retinopathy data should be collected, stored, annotated, and distributed. DiaRetDB1 was a continuation to establish a better database for algorithm evaluation. DiaRetDB1 contains retinal images selected by experienced ophthalmologists. The lesion types of interest were selected by medical doctors: microaneurysms (distensions in the capillary), hemorrhages (caused by ruptured or permeable capillaries), hard exudates (leaking lipid formations), soft exudates (microinfarcts), and neovascularization (new fragile blood vessels). These lesions are signs of mild, moderate, and severe diabetic retinopathy, and they provide evidence also for early diagnosis. An example of the database is shown in Figure 11.14.

The images were annotated by four independent and experienced medical doctors inspecting similar images in their regular work. Expert knowledge is combined and it can be presented as shown in the lower right corner of Figure 11.12. It is also possible to establish private patient databases where the full medical history of each patient is stored.

The images and ground truth are publicly available in the Internet [50]. The images are in PNG format, and the ground truth annotations follow the XML format. Moreover, we provide a DiaRetDB1 kit containing full MATLAB functionality (M-files) for reading and writing the images and ground truth, fusing expert annotations, and generating image-based evaluation scores. The whole pipeline from images to evaluation results (including the Strawman algorithm) can be tested using the provided functionality. The annotation software (MATLAB files and executables) is also available upon request.

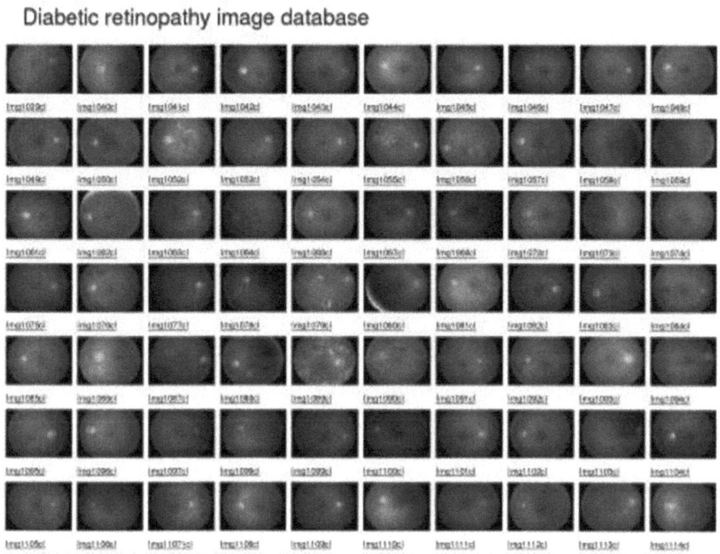

FIGURE 11.14

Retina image database.

11.4.3 BENCHMARKING FRAMEWORK FOR DEVELOPMENT AND COMPARISON

Performance evaluation practices for developing medical image analysis methods would be desirable. In particular, how to establish and share databases of medical images with verified ground truth and solid evaluation protocols. Such databases support the development of better algorithms, execution of profound method comparisons, and consequently, technology transfer from research laboratories to clinical practice. For this purpose, a framework consisting of reusable methods and tools for the laborious task of constructing a benchmark database was proposed in [54]. A benchmarking framework was developed to provide guidelines on how to construct benchmarking databases for eye fundus images. The guidelines comprise three mandatory components required in benchmarking: (1) true patient images, (2) ground truth from experts, and (3) an evaluation protocol. Following the benchmarking framework shown in Figure 11.13, a researcher can develop and compare own solutions using the public DiaRetDB1 diabetic retinopathy image database.

11.4.4 IMAGE-BASED AND PIXEL-BASED METHODS

The automatic eye fundus image analysis algorithms support two diagnostic steps: screening and diagnosing the diabetic retinopathy. Preprocessing may be needed as shown in the upper middle part of Figure 11.12 where blood vessel segments are detected to eliminate them from other diagnoses. Moreover, positioning of a single image or the registration of images may be needed, especially in the case of spectral images where several images of narrow bandwidth must be aligned to each other. In the first step, it is decided whether an eye needs further examinations (too many lesions visible) or not (the eye is healthy enough). In the second step, it is decided whether the eye requires medical action or not, that is, the state of the disease is determined. The methods that support the former step are called image- or subject-based methods, and the ones supporting the latter step are called pixel- or lesion-based methods. In the image-based approach, each image is assigned with a label (e.g., diabetic retinopathy present or absent) or a probability of the disease being present, whereas the pixel-based approach assigns the label or probability of lesion being present for each pixel. Correspondingly, lesion-based methods assign a label or probability of lesion for a group of connected pixels representing a lesion. See an example of a probability distribution in the lower left corner of Figure 11.12. The methods are based on suitable features (e.g., color, texture, shapes) and Bayesian reasoning. The results can be analyzed using ROC curves as shown in Figure 11.15. See [52,54] for further information. Based on the current method development it can be concluded that automatic image analysis under human supervision is now possible. Its applications include medical diagnosis assistance in screening, fundus image sorting according to severity or certainty of the disease, a semiautomatic tool to aid remote diagnosis, quality control of diagnosis work, and patient-specific image databases.

11.5 BIOMOLECULAR VISION

Molecular computing is a relatively new field of science where novel computing approaches are searched from the domain of molecules and atoms. More specifically, the aim is to understand how to control molecular reactions for information processing. Despite the fact that Moore's "law" still holds, these studies are motivated by the increasing technical difficulties to further develop the CMOS transistors as the building blocks of computing devices. Due to these difficulties the microelectronics industry has already pushed multicore and other parallel architectures to the market.

Biomolecules offer several advantages over synthetic ones. In their natural environment, their functionality and robustness is usually close to optimal due to evolutionary steps during the development of their structure and function. Therefore, many things can be learned from nature by studying the biomolecules and their interactions. Most of the studies concerning information processing using biomolecules have focused on DNA and photoactive biomolecules, for example, rhodopsins, chloroplasts, photosynthetic reaction centers and light-harvesting complexes, and retinal proteins. Bacteriorhodopsin (BR) is a retinal protein which has been intensively studied and proposed for various applications.

In an ongoing project biomolecules and their usage in technical applications and information processing are studied, and our specific goal is to understand especially the photoelectric functionality of BR and its applicability in implementing

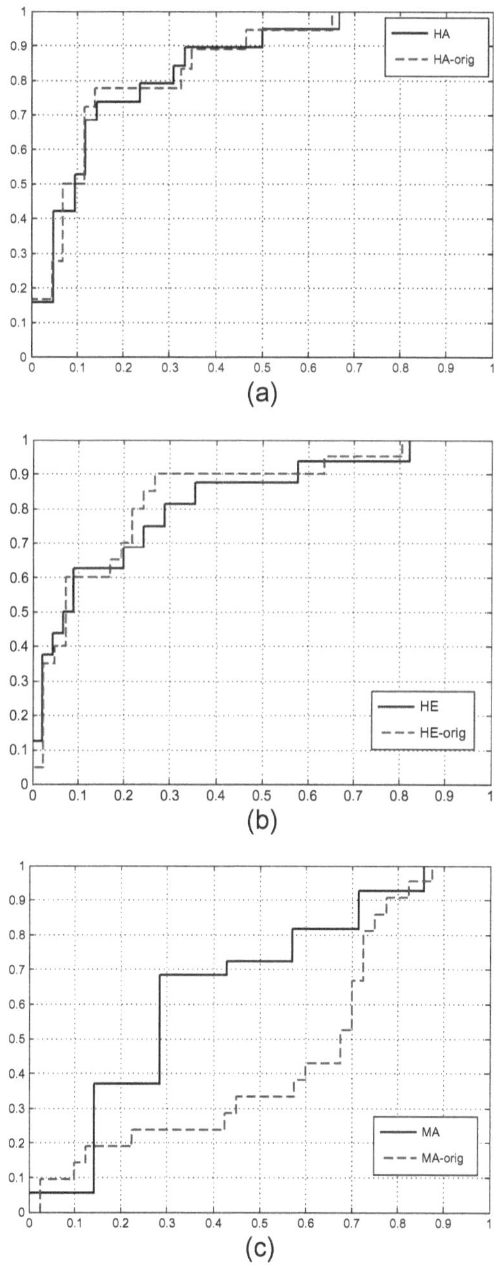

FIGURE 11.15

ROC curves for the DiaRetDB1 baseline method using the original and revised (max) method to generate the training and testing data: (a) hemorrhage; (b) hard exudate; (c) microaneurysm; and

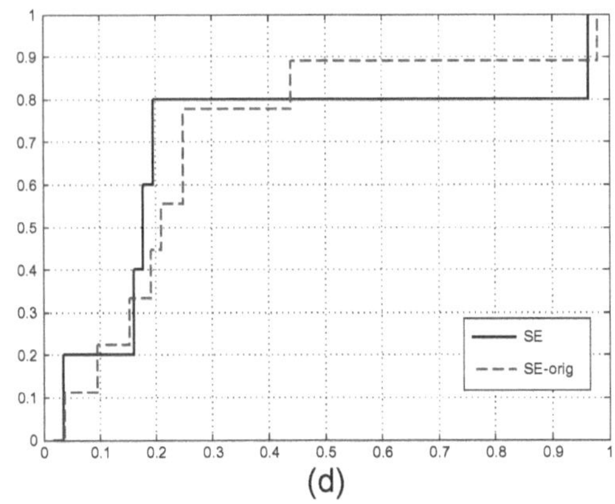

FIGURE 11.15, CONT'D

(d) soft exudate [61].

FIGURE 11.16

Color-sensitive digital cameras based on three types of BR: (a) camera matrix based on wild-type BR and (b) camera matrix based on three types of BR.

a color-sensitive artificial retina [62]. So far, our most important results are as follows: (1) cultivation of the Archaea, *Halobacterium salinarum* as the source of BR, and preparation of BR-in-polyvinylalcohol thick films, (2) single-element optoelectronic sensors based on wild-type BR and its variants (Figure 11.16), (3) color-sensitive digital camera based on three types of BR (Figure 11.16), (4) models for color vision systems based on, for example, BR, and (5) simulation environment for the reduced photocycle of BR.

11.6 CONCLUSIONS

This chapter considered scientific challenges in developing machine vision applications based on pattern recognition methods. The goal of such solutions is to create useful and significant value-added applications, especially using digital image processing and analysis, such as machine vision systems for the process industry, and medical image analysis for efficient health care of eye diseases. The chapter gave an overview on research carried out in the MVPR Laboratory of Lappeenranta University of Technology, especially in the fields of visual inspection, computational vision, medical image processing and analysis, and biomolecular vision. Several real-world applications discussed in this chapter have shown that digital image processing and analysis can be utilized robustly.

Development of advanced methods and technologies will offer great possibilities in future. In the short-term future, the research field will focus more on real-time recognition and tracking, three-dimensional computer vision, and multimodal systems with many sensors (image, speech, touch, etc.). It can be predicted, adapting from [63], that after some decades vehicles will travel completely without human interaction, airplanes will fly totally unmanned, humans will stay young "forever," a football team of robots can play better than a team of humans, and robots and humans may even have mixed relationships. This vision includes demanding challenges for software and hardware development in digital image processing and analysis.

ACKNOWLEDGMENTS

This chapter is a summary of research done by several researchers. The author would like to thank all these researchers. The author would especially like to thank Prof. Erkki Oja for establishing machine vision and pattern recognition research in Lappeenranta University of Technology (LUT) in the 1980s. Prof. Oja's legacy was actively continued by Prof. Jussi Parkkinen in the 1990s. The professors of the next generation, Professors Ville Kyrki, Joni-Kristian Kämärainen, and Lasse Lensu, have also greatly contributed to the research carried out in the Machine Vision and Pattern Recognition Laboratory (MVPR) at LUT. Besides many other talented researchers at MVPR such as Associate Professor Arto Kaarna, too many to name all of them, there have been many national and international collaborators to thank, especially Prof. Josef Kittler and the late Prof. Maria Petrou from CVSSP at University of Surrey. Besides CVSSP, the research groups of Prof. Sanna-Katriina Asikainen, Prof. Heikki Handroos, Prof. Jarmo Partanen, Prof. Kaisu Puumalainen, and Prof. Antti Salminen from LUT, Prof. Alan Bovik from University of Texas at Austin, Prof. Pasi Fränti, Prof. Markku Hauta-Kasari, and Prof. Jussi Parkkinen from University of Eastern Finland, Dr. Jari Käyhkö from Mikkeli University of Applied Sciences, Prof. Nahum Kiryati from Tel Aviv University, Prof. Danica Kragic from Swedish Royal Institute of Technology, Prof. Jivri Matas from Czech Technical University, Prof. Göte Nyman from Helsinki University, Prof. Pirkko Oittinen from Aalto University, Prof. Risto Ritala from Tampere University of Technology, Prof. Hannu Uusitalo from University of Tampere, Prof. Lei Xu from Chinese University of Hong Kong, and Prof. Pavel Zemcík from Brno University of Technology are very much appreciated. Finnish Academy of Finland, Finnish Funding Agency for Innovation (Tekes), European Union, and participating companies are acknowledged for significant financial support.

REFERENCES

[1] Machine Vision and Pattern Recognition Laboratory (MVPR), Electronic material (Online), Available from: http://www2.it.lut.fi/mvpr/.

[2] H. Kälviäinen, Motion detection using sets of moving pixels, Pattern Recogn. Image Anal. 13 (3) (2003) 394-408.

[3] N. Strokina, Machine vision methods for process measurements in pulping (PhD thesis), Lappeenranta University of Technology, 2013.

[4] L. Xu, E. Oja, P. Kultanen, A new curve detection method: randomized hough transform (RHT), Pattern Recogn. Lett. 11 (5) (1990) 331-338.

[5] H. Kälviäinen, P. Hirvonen, L. Xu, E. Oja, Probabilistic and non-probabilistic hough transforms: overview and comparisons, Image Vis. Comput. 13 (4) (1995) 239-252.

[6] H. Kälviäinen, Motion detection using the randomized hough transform (RHT): exploiting gradient information and detecting multiple moving objects, IEE Proc. 143 (6) (1996) 361-369.

[7] P. Bosdogianni, H. Kälviäinen, M. Petrou, J. Kittler, Robust unmixing of large sets of mixed pixels, Pattern Recogn. Lett. 18 (1997) 415-424.

[8] V. Kyrki, H. Kälviäinen, Combination of local and global line extraction, J. Real-Time Imag. 6 (2) (2000) 79-91.

[9] N. Kiryati, H. Kälviäinen, S. Alaoutinen, Randomized or probabilistic hough transform: unified performance evaluation, Pattern Recogn. Lett. 21 (13) (2000) 1157-1164.

[10] P. Fränti, E. Ageenko, S. Kukkonen, H. Kälviäinen, Using hough transform for context-based image compression in hybrid raster/vector applications, J. Electron. Imag. 11 (2) (2002) 236-245.

[11] L. Xu, A unified perspective and new results on RHT computing, mixture based learning, and multi-learner based problem solving, Pattern Recogn. 40 (2007) 2129-2153.

[12] J. Matas, C. Galambos, J. Kittler, Robust detection of lines using progressive probabilistic hough transform, Comput. Vis. Image Und. 78 (2000) 119-137.

[13] M. Dubská, A. Herout, R. Juránek, J. Sochor, Fully automatic roadside camera calibration for traffic surveillance, IEEE Trans. Intell. Transp. Syst. 2014 (1) (2014) 1-10.

[14] V. Kyrki, J.-K. Kamarainen, H. Kälviäinen, Simple Gabor feature space for invariant object recognition, Pattern Recogn. Lett. 25 (3) (2004) 311-318.

[15] M. Hamouz, J. Kittler, J.-K. Kamarainen, P. Paalanen, H. Kälviäinen, J. Matas, Feature-based affine-invariant localization of faces, IEEE Trans. Pattern Anal. Mach. Intell. (PAMI) 27 (9) (2005) 1490-1495.

[16] J. Ilonen, J.-K. Kamarainen, T. Lindh, J. Ahola, H. Kälviäinen, J. Partanen, Diagnosis tool for motor condition monitoring, IEEE Trans. Ind. Appl. 41 (4) (2005) 963-971.

[17] J.-K. Kamarainen, V. Kyrki, H. Kälviäinen, Invariance properties of Gabor filter based features – overview and applications, IEEE Trans. Image Process. 15 (5) (2006) 1088-1099.

[18] J. Ilonen, J.-K. Kamarainen, P. Paalanen, M. Hamouz, J. Kittler, H. Kälviäinen, Image feature localization by multiple hypothesis testing of Gabor features, IEEE Trans. Image Process. 17 (3) (2008) 311-325.

[19] P. Paalanen, J.-K. Kamarainen, J. Ilonen, H. Kälviäinen, Feature representation and discrimination based on Gaussian mixture model probability densities – practices and algorithms, Pattern Recogn. 39 (7) (2006) 1346-1358.

[20] Z. Chen, T. Ellis, A self-adaptive Gaussian mixture model, Comput. Vis. Image Und. 122 (2014) 35-46.
[21] A. Kaarna, P. Zemcik, H. Kälviäinen, J. Parkkinen, Compression of multispectral remote sensing images using clustering and spectral reduction, IEEE Trans. Geosci. Remote Sens. 28 (2) (2000) 1073-1082.
[22] S. Kukkonen, H. Kälviäinen, J. Parkkinen, Color features for quality control in ceramic tile industry, Opt. Eng. 40 (2) (2001) 170-177.
[23] R. Josth, J. Antikainen, J. Havel, A. Herout, P. Zemcik, M. Hauta-Kasari, Real-time PCA calculation for spectral imaging (using SIMD and GP-GPU), J. Real-Time Image Process. 7 (2012) 95-103.
[24] T. Kuparinen, V. Kyrki, Optimal reconstruction of approximate planar surfaces using photometric stereo, IEEE Trans. Pattern Anal. Mach. Intell. 31 (12) (2009) 2282-2289.
[25] N. Strokina, A. Mankki, T. Eerola, L. Lensu, J. Käyhkö, H. Klviäinen, Framework for developing image-based dirt particle classifiers for dry pulp sheets, Mach. Vis. Appl. 24 (4) (2013) 869-881.
[26] M. Pietikäinen, A. Hadid, G. Zhao, T. Ahonen, Computer Vision Using Local Binary Patterns, vol. 40, Springer, Berlin, 2011.
[27] T. Kinnunen, J.-K. Kämäräinen, L. Lensu, H. Kälviäinen, Unsupervised visual object categorization via self-organization, in: International Conference on Pattern Recognition (ICPR2010), 2010.
[28] T. Kinnunen, J.K. Kamarainen, L. Lensu, J. Lankinen, H. Kälviäinen, Making visual object categorization more challenging: randomized caltech 101 data set, in: International Conference on Pattern Recognition (ICPR2010), 2010.
[29] T. Kinnunen, J.-K. Kamarainen, L. Lensu, H. Kälviäinen, Unsupervised object discovery via self-organization, Pattern Recogn. Lett. 33 (16) (2012) 2102-2112.
[30] M. Laine-Hernandez, T. Kinnunen, J.-K. Kamarainen, L. Lensu, H. Kälviäinen, P. Oittinen, Visual saliency and categorization of abstract images, in: 21st International Conf. on Pattern Recognition (ICPR2012), Tsukuba Science City, Japan, 2012.
[31] T. Tuytelaars, C.H. Lampert, M.B. Blaschko, W. Buntine, Unsupervised object discovery: a comparison, Int. J. Comput. Vis. 88 (2) (2010) 284-302.
[32] O. Alkkiomäki, V. Kyrki, H. Kälviäinen, Y. Liu, H. Handroos, Complementing visual tracking of moving targets by fusion of tactile sensing, Robot. Auton. Syst. 57 (11) (2009) 1129-1139.
[33] H. Fennander, V. Kyrki, A. Fellman, A. Salminen, H. Kälviäinen, Visual measurement and tracking in hybrid welding, Mach. Vis. Appl. 20 (2) (2009) 103-118.
[34] V. Kyrki, D. Kragic, Tracking rigid objects using integration of model-based and model-free cues, Mach. Vis. Appl. 22 (2) (2011) 323-335.
[35] T. Eerola, J.-K. Kamarainen, L. Lensu, T. Leisti, R. Halonen, H. Kälviäinen, G. Nyman, P. Oittinen, Full reference printed image quality: measurement framework and statistical evaluation, J. Imag. Sci. Technol. 54 (1) (2010) 010201, IS&T 2011 Charles E. Ives Journal Award.
[36] T. Eerola, L. Lensu, J.-K. Kamarainen, T. Leisti, R. Ritala, G. Nyman, H. Kälviäinen, Bayesian network model of overall print quality: construction and structural optimization, Pattern Recogn. Lett. 32 (11) (2011) 1558-1566.
[37] T. Eerola, L. Lensu, H. Kälviäinen, A.C. Bovik, Study of no-reference image quality assessment algorithms on printed images, J. Electron. Imag. 23 (6) 2014 061106-1-061106-12.

[38] W. Xue, X. Mou, L. Zhang, A.C. Bovik, X. Feng, Blind image quality assessment using joint statistics of gradient magnitude and Laplacian features, IEEE Trans. Image Process. 23 (2014) 4850-4862.
[39] J. Ilonen, J.-K. Kamarainen, K. Puumalainen, S. Sundqvist, H. Kälviäinen, Toward automatic forecasts for diffusion of innovations, Technol. Forecast. Soc. Change 73 (2) (2006) 182-198.
[40] Knowpulp, Electronic material (Online), Available from: http://www.knowpulp.com/english/index.htm.
[41] N. Strokina, J. Matas, T. Eerola, L. Lensu, H. Kälviäinen, Detection of bubbles as concentric circular arrangements, in: Proceedings of the 21st International Conference on Pattern Recognition (ICPR), 2012, pp. 2655-2659.
[42] N. Strokina, T. Kurakina, T. Eerola, L. Lensu, H. Kälviäinen, Detection of curvilinear structures by tensor voting applied to fiber characterization, in: Proceedings of the 18th Scandinavian Conference on Image Analysis, SCIA, 2013, pp. 22-33.
[43] A. Sadovnikov, Computational evaluation of print unevenness according to human vision (PhD thesis), Lappeenranta University of Technology, 2010.
[44] J. Vartiainen, A. Sadovnikov, J.-K. Kamarainen, L. Lensu, H. Kälviäinen, Detection of irregularities in regular patterns, Mach. Vision Appl. 19 (4) (2008) 249-259.
[45] A. Drobchenko, J.-K. Kamarainen, L. Lensu, J. Vartiainen, H. Kälviäinen, T. Eerola, Thresholding-based detection of fine and sparse details, Front. Electr. Electron. Eng. China 6 (2) (2011) 328-338.
[46] T. Kinnunen. Bag-of-features approach to unsupervised visual object categorization (PhD thesis), Lappeenranta University of Technology, 2011.
[47] V. Hiltunen, T. Eerola, L. Lensu, H. Kälviäinen, Comparison of general object trackers for hand tracking in high-speed videos, in: Proceedings of the 22nd International Conference on Pattern Recognition (ICPR), 2014.
[48] Copex finger movements high-speed video dataset, Electronic material (Online), Available from: http://www2.it.lut.fi/project/copex/hshanddb01/index.shtm.
[49] P. Hienonen, Automatic traffic sign inventory- and condition analysis (Master's thesis), Lappeenranta University of Technology, 2014.
[50] ImageRet project: optimal detection and decision-support diagnosis of diabetic retinopathy, Electronic material (Online), Available from: http://www.it.lut.fi/project/imageret/.
[51] ReVision project: re-engineering retinal imaging with photonics and computational science, Electronic material (Online), Available from: http://www.it.lut.fi/project/revision/.
[52] T. Kauppi, Eye fundus image analysis for automatic detection of diabetic retinopathy (PhD thesis), Lappeenranta University of Technology, 2010.
[53] T. Kauppi, J.-K. Kamarainen, L. Lensu, H. Uusitalo, H. Kälviäinen, Detection and decision-support diagnosis of diabetic retinopathy using machine vision, Pattern Recogn. Image Anal. 21 (2) (2011) 140-143.
[54] T. Kauppi, J.-K. Kamarainen, L. Lensu, V. Kalesnykiene, I. Sorri, H. Uusitalo, H. Kälviäinen, Constructing benchmark databases and protocols for medical image analysis: diabetic retinopathy, Comput. Math. Methods Med. 2013 (Article ID 368514).
[55] K. Winell, A. Reunanen, Diabetes Barometer 2005, 2006, ISBN 952-486-023-6.
[56] World Health Organization, Definition, diagnosis and classification of diabetes mellitus and its complications. Part 1: diagnosis and classification of diabetes mellitus, Technical report, World Health Organization, Department of Noncommunicable Disease Surveillance, Geneva, 1999. Report of a WHO Consultation.

[57] World Health Organization and the International Diabetes Federation, Diabetes action now: an initiative of the World Health Organization and the International Diabetes Federation, 2004. ISBN 92 4 159151.

[58] T. Kauppi, V. Kalesnykiene, J.-K. Kamarainen, L. Lensu, I. Sorri, A. Raninen, R. Voutilainen, H. Uusitalo, H. Kälviäinen, J. Pietilä, The DiaRetDB1 diabetic retinopathy database and evaluation protocol, in: Proceedings of British Machine Vision Conference (BMVC), 2007, pp. 252-261.

[59] T. Kauppi, V. Kalesnykiene, J.-K. Kamarainen, L. Lensu, I. Sorri, A. Raninen, R. Voutilainen, J. Pietilä, H. Kälviäinen, H. Uusitalo, DiaRetDB1 diabetic retinopathy database and evaluation protocol, in: Proceedings of Medical Image Understanding and Analysis (MIUA), 2007, pp. 61-65.

[60] T. Kauppi, J.-K. Kamarainen, L. Lensu, V. Kalesnykiene, I. Sorri, H. Uusitalo, H. Kälviäinen, A framework for constructing benchmark databases and protocols for retinopathy in medical image analysis, in: J. Yang, F. Fang, C. Sun (Eds.), Machine Learning in Medical Imaging: Third Sino-Foreign-Interchange Workshop, IScIDE 2012, Nanjing, China, October 15-17, 2012. Revised Selected Papers, vol. 7751 of Lecture Notes in Computer Science, 2012, pp. 832-843.

[61] J.-K. Kamarainen, L. Lensu, T. Kauppi, Combining multiple image segmentations by maximizing expert agreement, in: F. Wang, D. Shen, P. Yan, K. Suzuki (Eds.), Machine Learning in Medical Imaging, Lecture Notes in Computer Science, Springer, Berlin, Heidelberg, 2012, pp. 193-200.

[62] L. Lensu, S. Parkkinen, T. Jaaskelainen, M. Frydrych, J. Parkkinen, Photoelectric properties of bacteriorhodopsin analogs for color-sensitive optoelectronic device, Opt. Mater. 27 (1) (2004) 57-62.

[63] Tiede (a scientific magazine in Finnish, a special issue), 34 (10) 2009.

CHAPTER

Advances in visual concept detection: Ten years of TRECVID

12

Ville Viitaniemi[1], Mats Sjöberg[2], Markus Koskela[2], Satoru Ishikawa[1] and Jorma Laaksonen[1]

[1]Department of Computer Science, Aalto University, Espoo, Finland, [2]Department of Computer Science, University of Helsinki, Helsinki, Finland

12.1 INTRODUCTION

Content-based multimedia information retrieval addresses the problem of finding data relevant to the users' information needs from multimedia databases. In early content-based image and video retrieval systems, the retrieval was usually based solely on querying by examples and measuring the similarity of the database objects (images, video shots) with *low-level features* automatically extracted from the objects. Generic low-level features are often, however, insufficient to discriminate content well on a conceptual level. This "semantic gap" is the fundamental problem in content-based multimedia retrieval.

In recent years, it has become common to build semantic representations of multimedia content by applying machine learning techniques for detecting *mid-level semantic concepts* (events, objects, locations, people, etc.) on the basis of the content's low-level visual and aural features [1–3]. This kind of mid-level representation at least narrows the semantic gap. In recent studies, it has been observed that, despite the far-from-perfect accuracy of concept detectors, the representation often is very useful in supporting *high-level indexing and querying* on multimedia data [4]. This is mainly because semantic concept detectors can be trained off-line with computationally more demanding supervised learning algorithms and with considerably more positive and negative training examples than what are typically available at query time. The automatic machine learning-based approach is scalable to large numbers of multimedia objects and features. The introduction of large-scale multimedia ontologies, such as LSCOM [5] and ImageNet [6] and large manually annotated data sets (e.g., [7]), have enabled generic analysis of multimedia content as well as an increase in multimedia lexicon sizes by orders of magnitude.

Through years of experimentation and evaluation of concept-detection techniques by the multimedia retrieval community, an understanding has emerged that machine learning systems for concept detection should generally be based on fusion of several low-level features extracted from the multimedia content, not just a single well-performing feature. Accepting this boundary condition of feature fusion, there

E. Bingham, S. Kaski, J. Laaksonen and J. Lampinen (Eds) Advances in Independent Component Analysis and Learning Machines.
http://dx.doi.org/10.1016/B978-0-12-802806-3.00012-9

still remain many design choices in implementing a concept-detection system. Typically such systems are complex and consist of several sub-modules. The modules themselves can be implemented using a multitude of alternative technologies, and there are alternative ways to combine the modules together.

Given the complex nature of concept-detection systems, it is not self-evident which factors and techniques are beneficial for concept-detection performance. Some of the techniques applied in systems exhibiting good overall concept-detection performance might be essential, whereas some other, attractive-looking techniques might just be parts of otherwise well-functioning systems, without being particularly effective themselves. This situation calls for controlled experiments where just one component of a concept-detection system is varied while other system parts are kept constant.

In this chapter, we describe the development of the concept-detection subsystem in our PicSOM multimedia analysis and retrieval framework. We discuss several alternative ways of implementing its components. As one highlight, we propose and study *N*-gram-based postprocessing techniques for taking advantage of temporal correlations that many semantic concepts exhibit between video shots. Most of the current state-of-the-art multimedia retrieval systems do not include inter-shot temporal analysis.

In the experiments of this chapter, we extensively compare the concept-detection performances the use of the component technologies leads to, employing the large-scale experimental settings of the high-level feature extraction (HLFE) and semantic indexing (SIN) tasks of the annual TRECVID video retrieval evaluation campaign. The name of the task was changed in 2010, but the content of the task remained very much the same, even though also the type and amount of video material employed was changed simultaneously. The TRECVID settings have arguably represented the multimedia research community's best effort to realistically model large-scale multimedia search tasks in a controlled benchmark setting. Yearly dozens of research groups evaluate their techniques and systems using this benchmark.

The remaining sections of this chapter are organized as follows. In Section 12.2, we describe the parts of a generic video retrieval system in order to provide context for the concept-detection subsystem, which is the main topic of this chapter. We also describe some implementation details of those parts in our PicSOM multimedia retrieval framework to the degree in which they are relevant from the concept-detection point of view. Section 12.3 contains the essential theoretical and methodological contribution of this chapter. There we describe the concept-detection techniques implemented in the PicSOM system. Section 12.4 presents empirical verifications of the proposed concept-detection algorithms in the TRECVID evaluations of years 2005, 2009, and 2014. In Section 12.5, we give our conclusions from our experiments and experiences.

12.2 PARTS OF A VIDEO RETRIEVAL SYSTEM

Figure 12.1 schematically shows the architecture of the automatic concept-detection and search subsystems in our PicSOM multimedia retrieval system. The

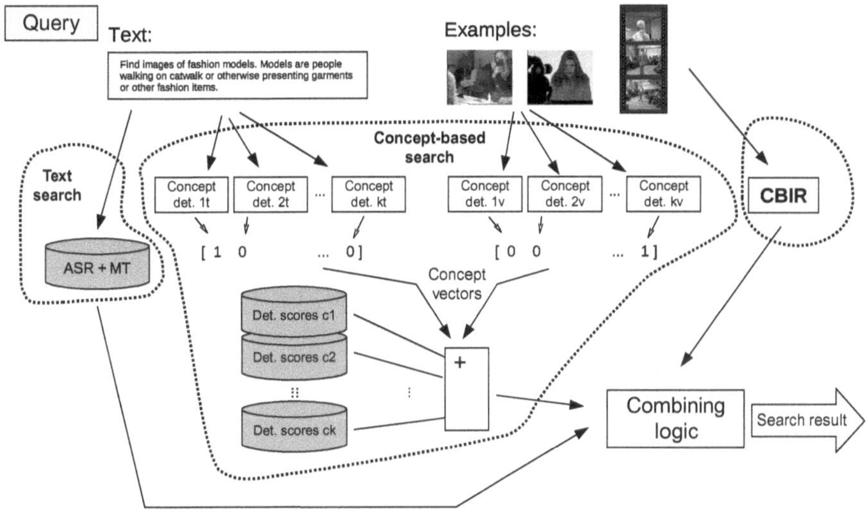

FIGURE 12.1

General architecture of the PicSOM multimedia retrieval system when applied to video search. Concept-based search is supplemented with textual and content-based (CBIR) search. The text is extracted from the video soundtracks using a combination of automatic speech recognition (ASR) and machine translation (MT).

implementation of the concept-detection system, seen in the center of the figure, is the focus of this chapter. In the search phase, the outputs of the concept-detection system can be supplemented with outputs of the interactive content-based information retrieval (CBIR) and textual search modules also depicted in the illustration.

The operation of a video search system generally consists of two phases. In the first phase, the system is *prepared* for a video corpus. The corpus is divided into an annotated training part and an unannotated testing part, on which video retrieval is going to be performed in the second *search* phase.

In the preparing phase, the whole video corpus is first segmented into shots and the annotations are associated with the shots. A number of low-level visual, audio and textual feature descriptors are extracted from each shot and content-based indices prepared based on the features. In systems that rely on automatic detection of concepts, the annotated part can then be used to train shot-wise detectors for the concepts that have been specified in the annotations. The detectors apply supervised learning techniques to form a mapping between low-level shot features and the annotation concepts, earlier often referred as *high-level features*, and more recently as *visual semantic concepts*. The preparing phase is allowed to be relatively time-consuming as it is intended to be performed off-line prior to the actual on-line use of the retrieval system.

After the preparation phase, the retrieval system is ready to be used for video retrieval in the search phase. In this phase, the system is queried with a textual phrase, combined with image and video examples of the desired query topic. The result of a query is a list of video shots, ranked in the order of decreasing predicted likelihood to match the query. The system operation in the search phase is intended to be sufficiently fast to enable the retrieval needs of a real user to be satisfied while the user is waiting, typically in a couple of seconds. The example images and video shots will require preprocessing, feature extraction, and classification that cannot be performed during the preparing phase, but will inevitably need to be done while the user is waiting for the output.

As description and evaluation of concept-detection techniques forms the essence of this chapter, the detailed discussion of those techniques are postponed to Section 12.3. In the remainder of this section, we discuss the other parts of a video retrieval system. The preparation phase parts are described to the extent in which they are relevant for the subsequent concept-detection experiments. The description of the video search phase in turn motivates and emphasizes the need for well-functioning semantic concept detectors.

12.2.1 SHOT SEGMENTATION AND KEYFRAME SELECTION

The first task of the preparing phase for a comprehensive video retrieval system is to segment the video corpus temporally into sequential basic units. The PicSOM multimedia retrieval system implements two shot boundary detection techniques, based on global visual feature evolution [8] and interest point tracking statistics [9]. However, for the experiments of this chapter with TRECVID video material, we employ the openly available master definition of shots [10] so that our results are comparable with those by other groups that have performed TRECVID tasks.

Another preparation phase task is the extraction of one or more keyframes from each video shot. The keyframes are needed both for extracting visual features to describe the content of the shot and for presenting them to the users of the system as still replacements for the dynamic video content. The most straightforward keyframe selection method is to use the center-most frame of each shot. Better results can be obtained by selecting the keyframe on the basis of the content of the shot, by comparing the frames with their neighbors and the calculated average of the shot [11]. In recent years, the organizers of the TRECVID semantic indexing task have provided also all i-frames of the MPEG-4 compressed video streams, and it has been computationally feasible to use all of them as keyframes.

12.2.2 LOW-LEVEL FEATURES

Automatic extraction of low-level features is the foundation of large-scale content-based multimedia processing. Using pixel values of video or image data directly in search and retrieval is typically neither sensible nor feasible. Effective features combined with an appropriate distance or similarity measure facilitates the use of the statistical vector space model approach, which is the basis of most current multimedia analysis methods. In many cases, a single well-chosen keyframe can compactly express the most central visual characteristics of that shot. Consequently, one can use still-image features, often originally developed for image-only retrieval systems, as a way to compare video shots.

In the following sections, we briefly go through different modalities of features that can be extracted to represent different relevant and complementary aspects of the underlying video data. Feature types that are used in the concept-detection experiments of this chapter are described in more detail as the nature and quality of the extracted features critically determine the maximum level of performance that a concept-detection system based on those features can achieve.

12.2.2.1 Global image features

Many of the classical image features are global, that is, calculated from all pixels of the image, thus representing characteristics of the image as a whole. An increasingly popular alternative has been to calculate features separately for smaller image segments; for example, calculating each block in a grid or pyramid structure placed over the image. It is also possible to use automatic segmentation, where the image is split into visually homogeneous segments, for which features are calculated separately [12].

The PicSOM system uses a wide range of image features. In TRECVID 2005, many of PicSOM's global image features were based on the standardized MPEG-7 descriptors [13]. We used both the implementations of the MPEG-7 XM reference software and our own more efficient implementations of the following MPEG-7 features: *Color Layout*, *Color Structure*, *Dominant Color*, *Scalable Color*, *Edge Histogram*, and *Region Shape*. Furthermore, PicSOM implements some nonstandard image features developed in-house: *Average Color*, *Color Moments*, *Texture Neighborhood*, *Edge Histogram*, *Edge Co-occurrence* [14], *Edge Fourier*, and *Centrist* [15]. These have been calculated either globally or for five spatial zones (center, top, bottom, left, and right) of the image. In the case of zoning, the final image-wise feature vector has been obtained as a concatenation of the zone-wise features.

12.2.2.2 BoV image features

Until very recently, the field of image analysis has been dominated by the approach of characterizing images by describing the statistics of their local feature descriptors. The local descriptors can be calculated for visually salient *interest points* [16]. For instance, the points can be edge or corner points where the image content changes substantially. Another strategy is to densely sample image area and calculate local descriptors for the sample of image locations. Histograms of robust, scale-invariant local descriptors – such as Scale-Invariant Feature Transform (SIFT) [17] and the Speeded Up Robust Features (SURF) [18] – and later their Fisher vector encodings provided the state-of-the-art image descriptors between 2008 and 2013.

Histograms of localized features are also called *bag-of-visual-words* (BoV) features in analogy to the traditional bag-of-words approach in textual information retrieval. In this interpretation, each histogram bin – representing a specific local pattern – is seen as a "visual word" in the vocabulary of all the histogram bins. The BoV features can be enhanced by calculating the histograms for different subdivisions of the image [19]. Another improvement to the BoV methodology is to use soft-assignment in histogram generation as demonstrated, for example, in [20].

In addition to the BoV encoding, other approaches include sparse coding of the local descriptors [21], supervector encoding [22], vector of locally aggregated descriptors (VLAD) [23], and the Fisher vector [24]. The Fisher vector encoding can arguably be considered as the current state of the art in local feature-based image classification. By measuring the deviation of a sample from a GMM-based generative model in the SIFT descriptor space, one ends up, however, with very high-dimensional image signatures.

The BoV features used in the PicSOM system in TRECVID evaluation of year 2009 were based on the *SIFT* local descriptors and the opponent color space version of the *color SIFT* descriptor [25]. We have employed two different strategies for selecting the points from which the local descriptors are extracted: the Harris-Laplace interest point detector and dense sampling of images. The codebooks have been generated with k-means and Self-Organizing Map (SOM) clustering algorithms.

In 2014, the PicSOM system also used densely sampled SIFT descriptors encoded with VLAD and Fisher vectors. The codebooks were generated using k-means with 512 cluster centers and a 128-component GMM, respectively.

12.2.2.3 Deep convolutional network features

A recent major development in image classification has been the use of deep convolutional neural networks (CNNs), with excellent results [26–28]. The convolutional networks based on the structure of Krizhevsky et al. [26] typically contain five or more convolutional layers, followed by two fully connected layers, and the output layer. However, one drawback with CNNs is that they require huge amounts of training data and delicate tuning of the training parameters. It has, however, been observed that CNNs trained with one visual dataset can function as highly discriminative features even for considerably different data domains and tasks [29,30]. We can therefore employ CNNs trained with external data as feature extractors in a standard concept-detection framework.

In 2014, the PicSOM system included a total of 24 CNN features extracted with four different CNN networks [30]. We use the activations of the first fully connected layers of each network as our features. In addition, we use the spatial pyramid pooling proposed in [28] with two-scale levels. The first level corresponds to the full image, and the second level consists of nine regions with scale of two. On each scale, the CNN activations of the regions are averaged, and the activations of the different scales are concatenated.

12.2.2.4 Video features

In many cases, the static visual properties of a video keyframe are not enough to describe the salient features of the full scene. The dynamic properties may also make the computational learning problem easier. It has been reported in various publications that using video features beyond the single keyframe approach can improve the results [31–33]. For the experiments reported in this chapter for the

TRECVID 2005 and 2009 evaluations, we extracted video features by temporally extending some of the still-image features described in the previous sections. When calculating these features, the video shot is first divided into nonoverlapping temporal subshots of equal lengths. A feature vector is calculated separately for each frame and all the frame feature vectors averaged within the subshots to form feature vectors are finally concatenated to form one shot-wise feature vector.

12.2.2.5 Audio features

Most video shots include a sound track, containing for example human speech, music, or different environment sounds. The general-level characteristics of the sound track can be described either globally or the track can be segmented into separately described parts. A popular approach for the description is to calculate the mel-scaled cepstral coefficients (MFCC) [34]. Besides coarse general-level description of audio, speech can often be automatically recognized and thus handled as text, as will be described in the following section. Depending on the video analysis and retrieval task at hand, analyzing music and environment sounds may or may not be beneficial. For the experiments of this chapter audio features were used only in the TRECVID 2005 evaluation.

12.2.2.6 Textual features

Video material often includes textual data or metadata that can facilitate text-based indexing and retrieval. Textual data for video shots may originate, for example, from speech recognition, closed captions, subtitles, or video OCR. As text-based information retrieval methodology is very mature and text indices can provide fast and accurate results [35,36], an effective video retrieval system will generally benefit from a text search component when responding to the high semantic-level queries from the user. For detecting mid-level semantic concepts, however, textual features are not always useful. The textual information in the TRECVID corpora used before year 2010 was obtained through an ASR and MT process. Experiences had proven that textual features extracted from that material performed poorly in detecting visual concepts.

In the experiments of this chapter, textual features were used in the TRECVID evaluation of year 2005, but not in the evaluation of year 2009. Since year 2010, textual data have not been provided in the TRECVID semantic indexing task.

12.2.3 SEARCH PHASE

The ultimate goal of video retrieval is to find relevant video content for a specific information need of the user. The conventional approach has been to rely on textual descriptions, keywords, and other metadata to achieve this functionality, but this requires manual annotation and does not usually scale well to large and dynamic video collections. In some applications, such as YouTube, the text-based approach works reasonably well, but it fails when there is no metadata available or when the metadata cannot adequately capture the essential content of the video material.

Content-based video retrieval, on the other hand, utilizes techniques from related research fields, such as image and audio processing, computer vision, and machine learning, to automatically index the video material. Content-based queries are typically based on a small number of provided examples (i.e., *query-by-example*). The material of a video collection is ranked based on its similarity to the examples according to low-level features [37–39]. In recent works, the content-based techniques are commonly combined with separately pretrained detectors for various semantic concepts (*query-by-concepts*) [3,4]. It has been empirically observed that visual concept lexicons or ontologies are an integral part of effective content-based video retrieval systems.

Concept-based video retrieval can operate in either automatic or interactive mode. In *automatic concept-based video retrieval*, no user interaction is needed after a query has been presented to the retrieval system. In the automatic mode, the fundamental challenge is mapping the user's information need into the space of available concepts in the used concept ontology [40]. The basic approach is to select a small number of concept detectors as active and weight them based either on the performance of the detectors or their estimated suitability for the current query. Negative or complementary concepts are not typically used. In [40], the methods for automatic selection of concepts were divided into three categories: *text-based*, *visual-example-based*, and *results-based methods*. Text-based methods use lexical analysis of the textual query and resources such as WordNet [41] to map query words into concepts. Methods based on visual examples measure the similarity between the provided example objects and the concept detectors to identify suitable concepts. Results-based methods perform an initial retrieval step and analyze the results to determine the concepts that are then incorporated into the actual retrieval algorithm.

In addition to automatic retrieval, *interactive concept-based retrieval* constitutes a parallel paradigm. Interactive video retrieval systems include the user in the loop at all stages of the retrieval session and therefore call for sophisticated and flexible user interfaces. A global database visualization tool providing an overview of the database as well as a localized point-of-interest with increased level of detail are typically needed. Relevance feedback can also be used to steer the interactive query toward video material the user considers relevant [42]. Semantic concept detection has generally been recognized as an important component also in interactive video retrieval [4], and interactive video retrieval systems (e.g., [43]) typically use concept detectors as a starting point for the interactive search functionality.

12.3 CONCEPT DETECTION IN PicSOM

After having extracted low-level video features from each shot, supervised learning techniques can be applied in order to learn the associations between the low-level features and the concepts in the annotations of the video corpus. The PicSOM multimedia retrieval system includes a supervised concept-detection subsystem trained in the preparing phase of the video corpus. Figure 12.2 illustrates the overall architecture

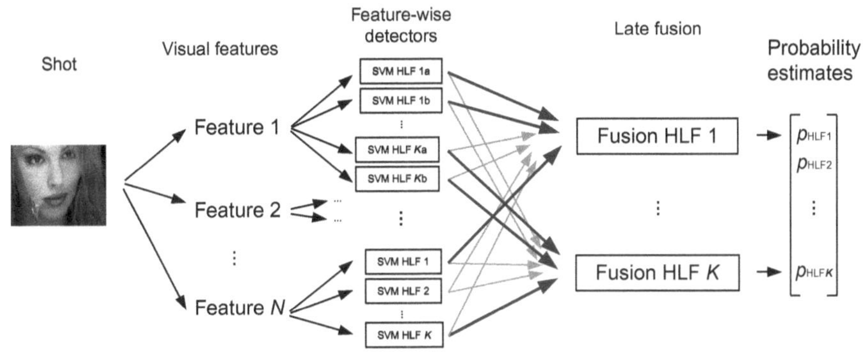

FIGURE 12.2

Fusion-based shot-wise concept-detection module in the PicSOM system. K denotes the number of concepts that are to be detected. The dark thick arrows between the feature-wise detector and fusion stages are intra-concept connections; the shaded thin arrows represent cross-concept links.

of this system. All the K concepts are first detected from each shot, based on the shot's low-level features, K being the number of concepts that have been annotated in the training part of the video corpus. This step results in a K-dimensional vector of detection scores. After the shot-level concept detection, the scores are readjusted in a postprocessing step according to the score vectors of temporally neighboring shots, based on the estimated likelihood of observing particular temporal concept patterns.

12.3.1 SHOT-LEVEL CONCEPT DETECTION

The shot-level concept-detection task is in the PicSOM system addressed with a well-established fusion-based architecture. The fusion-based approach is common also in other well-performing state-of-the-art image and video analysis systems (e.g., [32,44]). In our approach, dozens of supervised probabilistic detectors are first trained for each concept, based on the different shot-wise low-level features, detailed in Section 12.2.2, and their early fusion combinations. The feature-wise detector outcomes are then fused in a postclassifier fusion (also called late fusion) step. The outlined shot-level detection architecture contains a number of components that can be implemented in several alternative ways. In the following, we describe the techniques implemented in the PicSOM system during the various stages of its development.

Given the extracted shot-wise features, the first stage in our fusion algorithm is the feature-wise supervised detection of concepts. Each concept and feature is treated symmetrically, that is, every concept is detected with the same algorithms.

12.3.1.1 Self-organizing maps

Historically, the SOM-based detectors have always been part of the PicSOM throughout the system's existence. The early emphasis was on interactive CBIR, where the rapidness of the SOM approach in learning new category definitions is essential for a satisfactory user experience. Also much of the early work where the PicSOM system has been used in off-line category detection tasks used SOM-based detectors. One of the first evaluations of this approach took place in the TRECVID 2005 evaluation as will be described in Section 12.4.

The construction of the SOM-based detectors begins with quantizing the feature spaces using the TS-SOM [45] algorithm, a tree-structured variant of the SOM [46]. In the subsequent learning algorithm, the bottom levels of TS-SOMs define the quantization and the upper levels act as an index structure for rapid search. Typically, TS-SOMs from two to four stacked levels have been used, the bottom levels measuring from 16×16 to 256×256 map units, respectively. Figure 12.3 shows an example of a TS-SOM quantization of a feature space based on the color and texture distribution of image segments.

The TS-SOM preparation step needs to be performed only once for each feature type in an image collection. After that, generating a classifier for any binary partitioning of the training images is very fast. Any partitioning is characterized by the division of the training images into positive and negative examples. The classifier for

FIGURE 12.3

A TS-SOM partitioning of the feature space defined by color and texture distribution of image segments. From [47].

the partitioning is created by subtracting the proportion of negative examples that fall into each bottom-level TS-SOM unit, that is, quantization bin, from the corresponding proportion of positive examples. This way a classification score is assigned to each quantization bin. After this initial scoring, the scores are low-pass filtered on the two-dimensional TS-SOM grid surface, taking advantage of the topology-preserving characteristic of the SOM clustering and efficiently emphasizing the differences between the feature space regions where positive and negative examples are well separated, or occur mixed with each other.

When the preparation step is complete, a detection score is associated with each quantization bin of the feature space. Assigning a feature-wise detection score to an independent test image is then simple: the extracted feature vector of the image is quantized using the same quantization scheme and the image receives the detection score of the quantization bin into which its feature vector is mapped.

12.3.1.2 Nonlinear support vector machines

After the SOM, a nonlinear Support Vector Machine (SVM) [48] algorithm was used as the supervised detection algorithm in PicSOM. The SVM implementation used is an adaptation of the C-SVC classifier of the LIBSVM software library [49].

We have used the radial basis function (RBF) SVM kernel

$$g_{\mathrm{RBF}}(\mathbf{x},\mathbf{x}') = \exp\left(-\gamma \|\mathbf{x}-\mathbf{x}'\|^2\right) \tag{12.1}$$

for all the shot-wise features and also have the option to use the exponential χ^2 kernel

$$g_{\chi^2}(\mathbf{x},\mathbf{x}') = \exp\left(-\gamma \sum_{i=1}^{d} \frac{(x_i - x_i')^2}{x_i + x_i'}\right) \tag{12.2}$$

for d-dimensional histogram-like visual features.

The free parameters of the SVMs are selected with an approximate 10-fold cross-validation search procedure that consists of a heuristic line search to identify a promising parameter region, followed by a grid search in that region. To speed up the computation, the data set is radically downsampled for the parameter search phase. Further speed-up is gained by optimizing the C-SVC cost function only very approximately during the search. For the final detectors we also downsample the data set, but less radically than in the parameter search phase. Usually there are much fewer annotated example shots of a concept (positive examples) than there are example shots not exhibiting that concept (negative examples). In these cases, we usually retain all the positive examples and just limit the number of negative examples.

12.3.1.3 Linear support vector machines

There have been numerous approaches to reduce the computational complexity from the level of standard nonlinear SVMs. Such approaches include using approximate SVM solvers [50,51], reducing the number of support vectors [52,53], and replacing the nonlinear SVMs with linear classifiers [54]. It is also possible to speed up SVMs by using GPUs [55]. Using linear classifiers is particularly appealing, as both the training and classification time requirements can be several orders of magnitude smaller than with nonlinear SVMs. Recent algorithms for training large-scale linear classifiers include the stochastic sub-gradient descent in Pegasos [56] and the dual coordinate descent algorithm in LIBLINEAR [57]. As a practical example, in our current implementation and the TRECVID data used in experiments reported in this chapter, evaluating a linear classifier for a single image (excluding feature extraction) takes only a fraction of a millisecond whereas nonlinear SVMs require 100-200 ms per image. In PicSOM, we have focused on two approaches, homogeneous kernel maps and power mean SVM.

Nonlinear kernel classifiers can be considered as linear classifiers in a feature space for which there exists a corresponding implicit feature map $\Psi : \mathbb{R}^d \rightarrow \mathbb{R}^D$. Therefore, one approach is to perform an explicit (either exact or approximate) feature mapping to convert the nonlinear problem into a linear one and use a standard linear solver. With an exact feature map this is straightforward:

$$K(\mathbf{x}_i, \mathbf{x}_j) = \langle \Psi(\mathbf{x}_i), \Psi(\mathbf{x}_j) \rangle. \tag{12.3}$$

The exact mapping approach can work in certain cases, but, in general, the dimensionality D of the feature map Ψ can be high or even infinite, as is the case, for example, with the RBF kernel. Therefore, a more practical approach is to approximate the nonlinear kernel. One approach is to try to find a mapping function $\hat{\Psi} : \mathbb{R}^d \rightarrow \mathbb{R}^r$ so that

$$\langle \hat{\Psi}(\mathbf{x}_i), \hat{\Psi}(\mathbf{x}_j) \rangle \approx K(\mathbf{x}_i, \mathbf{x}_j). \tag{12.4}$$

In the general case, finding such mappings is difficult, but it has turned out that with additive kernels this is possible. A kernel is *additive* if it can be represented as a sum of feature-component-wise one-dimensional functions, that is, if it can be written as

$$K(\mathbf{x}, \mathbf{z}) = \sum_{i=1}^{d} k_i(x_i, z_i), \tag{12.5}$$

where $\mathbf{x} = [x_1, \ldots, x_d]^T, \mathbf{z} = [z_1, \ldots, z_d]^T \in \mathbb{R}_+^d$. Common additive kernels include the intersection kernel

$$k_{\text{int}}(x_i, z_i) = \min(x_i, z_i), \tag{12.6}$$

the χ^2 kernel

$$k_{\chi^2}(x_i, z_i) = -\frac{(x_i - z_i)^2}{x_i + z_i}, \tag{12.7}$$

the Bhattacharyya kernel

$$k_{\text{bha}}(x_i, z_i) = \sqrt{x_i z_i}, \tag{12.8}$$

and the Jensen-Shannon kernel

$$k_{\text{js}}(x_i, z_i) = \frac{x_i}{2} \log_2 \frac{(x_i + z_i)}{x_i} + \frac{z_i}{2} \log_2 \frac{(x_i + z_i)}{z_i}. \tag{12.9}$$

Maji et al. [58] proposed a sparse feature map for the intersection kernel, and subsequently Vedaldi and Zisserman proposed *homogeneous kernel maps* [59] for any additive homogeneous kernel. Such explicit kernel maps are convenient to use as they do not require any changes to the linear classification algorithm and are data independent. As a result, no learning is required and the kernel map can be computed on-the-fly using a look-up table.

The homogeneous kernel map of order n is a $(2n+1)$-dimensional linear approximation of an additive kernel for a scalar feature, $\hat{\Psi}_n : \mathbb{R} \rightarrow \mathbb{R}^{2n+1}$. Due to the additivity property (Eq. 12.5), one can then encode a d-dimensional feature vector as a $d(2n+1)$-dimensional linear problem using the kernel map and use any standard linear solver with it to approximate the corresponding nonlinear kernel. The complexity of evaluating the classifier is thus $O(d)$. In [59], homogeneous kernel maps are provided for many common additive kernels used in computer vision. Among them, an implementation of the homogeneous kernel map of order $n = 2$ for the intersection kernel has been adopted for the experiments described in this chapter.

12.3.2 FUSION ALGORITHMS

The PicSOM system includes several alternative algorithms for the fusion of feature-wise concept detectors. As a baseline approach we form the geometric mean of all the detector outcomes for each processed video frame. Besides this unsupervised fusion approach, we also implement several supervised fusion methods that make use of the cross-validated detector outcomes for the training set.

One supervised technique is SVM-based fusion employing RBF kernels, another Bayesian binary regression (BBR) [60]. The other implemented alternatives are variations of the scheme where the basic fusion mechanism is still the geometric mean, but the mean is calculated only over a subset of the detector outcomes, selected by a sequential forward-backward search (SFBS).

In addition to basic SFBS, we implement the idea of partitioning the training set into multiple folds. In our implementation, we have used a fixed number of sixfolds. The SFBS algorithm is run several times, each time leaving onefold outside the training set. The final fusion outcome is the geometric mean of the fold-wise geometric means. For later reference, we denote this fusion algorithm multifold-SFBS.

We also consider reserving a part of the training set for validation and early stopping the search based on the performance in this validation set. This early stopping can be combined with both the basic SFBS and multifold-SFBS algorithms. For the basic SFBS, one sixth of the training data is used as a validation set. In the case of multifold-SFBS, the left-out fold for each fold-wise run is re-used as the validation set.

In addition to the fusion mechanisms used to combine the outputs of multiple detectors for a single video frame, one needs to fuse the frame-wise results to shot-wise detection scores, provided that there is more than one keyframe in a shot. In PicSOM, we have employed for this purpose a simple maximum pooling approach which we have found to perform better than, for example, arithmetic and geometric average pooling techniques.

12.3.3 TEMPORAL POSTPROCESSING

For temporal postprocessing of the fusion outcomes, the PicSOM system implements techniques first described in [61]. The techniques operate on a stream of K-tuples corresponding the concept-detector outputs for the sequential video shots, where K is the number of the detected concepts. The methods thus ignore the absolute timing and duration of the video shots, preserving only their ordering.

Methodologically, our temporal postprocessing is based on N-gram modeling performed for each concept individually. In the following, $c_n \in \{0, 1\}$ is an indicator variable of the occurrence of the concept to be detected at time instant n and $s_n \in \mathbf{R}$ is the output of the corresponding concept detector. H_n denotes the recursive prediction history known at time instant n, extending $N - 1$ steps backwards in time:

$$H_n = \left\{\hat{p}(c_{n-i}|s_{n-i}, H_{n-i})\right\}_{i=1}^{N-1}. \tag{12.10}$$

Using this notation, we can write the recursive N-gram model as

$$\hat{p}\,(c_n|s_n, H_n) \propto \hat{p}(s_n|c_n)\hat{p}\,(c_n|H_n) \tag{12.11}$$

if we assume the conditional independence of s_n and H_n given c_n, that is,

$$\hat{p}(s_n|c_n, H_n) = \hat{p}(s_n|c_n). \tag{12.12}$$

Then the recursive model can be written as

$$\hat{p}\,(c_n|H_n) = \sum_{c_{n-1}} \cdots \sum_{c_{n-N+1}} p_0(c_n|c_{n-1}, \ldots, c_{n-N+1}) \prod_{i=1}^{N-1} \hat{p}\,(c_{n-i}|s_{n-i}, H_{n-i}). \tag{12.13}$$

Here p_0 is the marginalized N-gram probability that is estimated from the training data. The N-gram model is initialized in the beginning of each video by using models of lower order, for example, a bigram model is used on the second time instant. The conditional distributions of detector outputs $\hat{p}(s_n|c_n)$ are modeled as exponential distributions

$$\hat{p}(s_n|c_n = i) = \frac{1}{\lambda_i} \mathrm{e}^{-s_n/\lambda_i}, \quad i \in \{0, 1\}. \tag{12.14}$$

For concept-wise parameters λ we use the maximum likelihood estimates

$$\hat{\lambda}_i = \frac{\sum_{n|c_n=i} s_n}{\sum_{n|c_n=i} 1}, \quad i \in \{0, 1\}, \tag{12.15}$$

where the summation is over the shots of the training set.

In addition to this causal model, we also form the corresponding anticausal model that is obtained by reversing the time flow. The causal and anticausal models are then combined by logarithmic averaging of the model outcomes.

12.4 EXPERIMENTS

In this section, we describe the experiments we have performed in the HLFE and semantic indexing tasks of the TRECVID evaluation campaigns in 2005, 2009, and

2014, and present an analysis of the results. The experiments are based on our submissions to corresponding TRECVID evaluations [62–64], but for this chapter we have complemented the submitted results with additional experiments based on retrospective analysis of the annual results.

12.4.1 TRECVID EVALUATION CAMPAIGN

TRECVID [65] is an annual workshop series organized by the National Institute of Standards and Technology (NIST) and arguably the leading venue for evaluating research on content-based video analysis and retrieval. It started in 2001 as TREC Workshop's video track and since 2003 it has been organized as a workshop of its own. TRECVID provides the participating organizations large test collections, uniform scoring procedures, and a forum for comparing the results. Each year the TRECVID evaluation contains a set of video analysis tasks, such as HLFE or semantic indexing, video search, video summarization, event detection, and content-based copy detection.

In the experiments of this chapter, we focus on the HLFE task of TRECVID 2005 and 2009, and the semantic indexing (SIN) task of TRECVID 2014. As already stated, these tasks are basically the same from the point of view of visual analysis, only the name was changed in 2010.

The video material used in TRECVID has consisted of television news broadcasts (until 2006), documentaries, news reports, educational programs (2007-2009), and consumer videos from the Internet Archive (www.archive.org) (since 2010). The video material is divided into shots in advance and these reference shots are used as the unit of concept detection [10]. To obtain training data for the HLFE/SIN task, a collaborative annotation effort has been organized [7] annually.

Due to the size of the test corpora, it has been infeasible within the resources of the TRECVID initiative to perform an exhaustive examination in order to determine the topic-wise ground truth. Therefore, a pooling technique has been used instead. First, a pool of possibly relevant shots is obtained by gathering the sets of shots returned by the participating teams. These sets are then merged, duplicate shots are removed, and the relevance of only this subset of shots is assessed manually.

The main performance measure in the HLFE/SIN task has been first the *inferred average precision* (infAP) [66] and later the *extended inferred average precision* (xinfAP) [67], which approximate the standard *average precision* very closely, but require only a subset of the pooled results to be evaluated manually. The mean of the concept-wise precision values over a set of queries, *mean (extended) inferred average precision* (MIAP and MXIAP), is then used to provide an overview of the results.

12.4.2 EXPERIMENTS 2005

In our TRECVID experiments in 2005 [62], we were using the SOM as the shot-level concept-detection method. The main emphasis of our experiments was on evaluating the performance of the features, both visual and textual, for each concept. This was feasible because only 10 concepts were used. The visual features that were available that time were mostly global features, but some histogram-based texture features were also already available.

12.4.2.1 Data

In 2005, the TRECVID HLFE task data consisted of 170 h of video data recorded from TV news broadcasts in Lebanon, China, and the United States. ASR and MT results in English were provided by the organizers. One-half of the data was given for the development of the systems and the other half was reserved for testing.

12.4.2.2 Video features

On the video shot level, we used the MPEG-7 [68] *Motion Activity* descriptor (MA) and temporal versions of three still image features. The temporal image features were calculated by dividing each video first into five nonoverlapping parts with equal lengths. All the frames of these five subshots were then extracted, and each frame divided spatially into five separate zones: the upper, the lower, the left-hand side, the right-hand side, and the central zone. A feature vectors were calculated separately for each zone, and were then concatenated to form a vector depicting the whole frame. All the frame-wise feature vectors of a subshot were then averaged and these average vectors were concatenated to form the feature vector of the video clip. Several different video features were calculated using this method by varying the feature that is calculated for the zones of the frames. *Average Color* (AC), *Color Moments* (CM), and *Texture Neighborhood* (TN) features were the three zone features that were used.

The AC feature vector is a three-element vector that contains the average RGB values of all the pixels within the zone. The CM feature is calculated by separating the HSV color channels from the zone. Then the values of the color channels are treated as probability distributions, and the first three moments (mean, variance, and skewness) are calculated for each distribution. The feature vector contains the three moment values for the three color channels.

The Texture Neighborhood feature is calculated from the Y (luminance) component of the YIQ color representation of the zone pixels. The 8-neighborhood of each inner pixel is examined, and a probability estimate is calculated for the probabilities that the neighbor pixel in each surrounding relative position is brighter than the central pixel. The feature vector contains these eight probability estimates.

12.4.2.3 Image features

For the keyframe indices we used a set of six standard MPEG-7 [68] descriptors, namely, *Color Layout* (CL), *Color Structure* (CS), *Dominant Color* (DC), *Scalable Color* (SC), *Edge Histogram* (EH), and *Homogeneous Texture* (HT). The descriptors were extracted globally from every keyframe in the collection, that is, no segmentation or zoning was used.

12.4.2.4 Audio features

The mel-scaled cepstral coefficient, or shortly *Mel Cepstrum* (CE), is the discrete cosine transform (DCT) applied to the logarithm of the mel-scaled filter bank energies. The number of coefficients taken is 12, and these are organized as vector. Finally the total power of the signal is appended to the vector giving a feature vector of length 13.

12.4.2.5 Text features

Unlike the other features, an inverted file instead of an SOM index was used for the ASR/MT output. For the HLFE task, the text features were constructed by gathering concept-dependent lists of most informative terms. Let us denote the number of shots in the development set associated with concept c as N_c and assume that of these shots, $n_{c,t}$ contain the term t in the ASR/MT output. After preprocessing and stemming, the following measure is applied for term t regarding the concept c:

$$S_c(t) = \frac{n_{c,t}}{N_c} - \frac{n_{\text{all},t}}{N_{\text{all}}}. \tag{12.16}$$

For every concept, we recorded the 10 and 100 most informative terms and use them as alternative text features.

12.4.2.6 Feature selection

The set of used features was selected for each concept separately. For this purpose, we applied an SFS-type feature selection scheme, in which we began with an empty set of features and computed a criterion value for each of the potential features. If adding the feature with the highest value improved the overall result, the feature was used in the task and the process continued. Otherwise the selection process was stopped. As the optimization criterion we used the average precision at 2000 returned items with twofold cross-validation on the development set.

12.4.2.7 Results

Table 12.1 lists the sets of selected features for each of the 10 studied concepts. As can be seen, the selection process typically resulted in four to seven features to be

Table 12.1 Features Used in TRECVID 2005 in the High-Level Feature Extraction Task for Each Concept

	Video				Image						Audio	Text	
High-level feature	**MA**	**AC**	**CM**	**TN**	**CL**	**CS**	**DC**	**SC**	**EH**	**HT**	**CE**	**10**	**100**
Walking/running	×	×							×				
Explosion/fire			×		×					×			×
Maps		×	×			×		×	×	×			×
Flag-US	×		×	×					×			×	
Building		×	×				×		×	×			×
Waterscape/waterfront	×	×	×	×	×				×			×	
Mountain	×	×		×					×	×	×		
Prisoner										×			
Sports	×	×	×		×				×			×	
Car	×	×	×	×	×				×	×	×		×

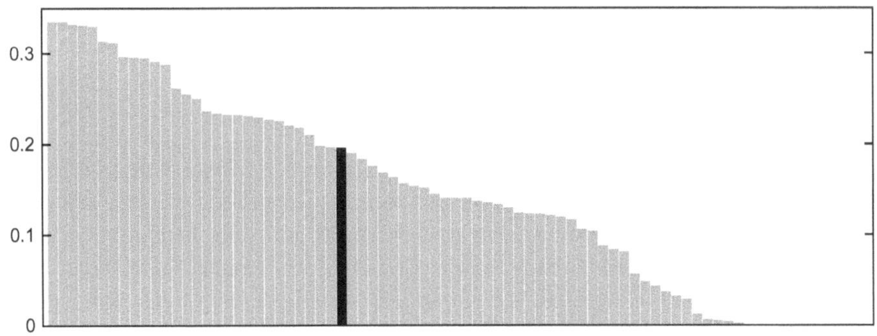

FIGURE 12.4

Mean average precision values for all runs submitted to the TRECVID 2005 high-level feature extraction task, our run highlighted.

selected and fused. The *Prisoner* concept was a notable exception as adding any second feature, including the text features, beside Homogeneous Texture resulted in performance degradation. In general, video features seem to be included in the feature set more often than still image features. Among the image features, Edge Histogram and Homogeneous Texture were used more than the color-based features. Audio and text features were beneficial only for a subset of the concepts.

Figure 12.4 shows the PicSOM Team's placement among the submissions evaluated by NIST. As this was our first participation in the evaluation, we were able to submit only one result. Our performance can be regarded as mediocre. In 2005, most of the other participants were already using SVMs in their systems and it was therefore decided that also the PicSOM system should start to use them as the principal classifier technology for concept detection.

12.4.3 EXPERIMENTS 2009

In 2009, we had followed the example given by other successful groups in the TRECVID evaluations and started to use BoV features and nonlinear SVMs in the PicSOM system. The aim of our experiments in TRECVID 2009 was to evaluate the advantage that could be obtained with the SIFT and Color SIFT BoV features compared to the global features used earlier. Other research questions were the benefits that could result from late fusion of the detector outputs and from applying temporal postprocessing to the shot-wise detection results.

12.4.3.1 Data

The video data for the system development in TRECVID 2009 consisted of approximately 100 h of documentaries, news reports, and educational programs from Dutch TV; 280 h of similar video data were used for evaluation. Table 12.2 lists all the concepts detected in TRECVID 2009.

In the experiments reported in the following subsections, the shot-wise feature sets that we have used as a starting point consist solely of various combinations of

Table 12.2 The 20 Concepts Detected in TRECVID 2009 High-Level Feature Extraction Task

Classroom	Person playing a musical instrument	Hand
Chair	Person playing soccer	People dancing
Infant	Cityscape	Nighttime
Traffic	Person riding a bicycle	Boat or ship
Doorway	Telephone	Female human face closeup
Airplane flying	Person eating	Singing
Bus	Demonstration or protest	

visual features, that is, keyframe image and video features. Audio and text features have not been used.

12.4.3.2 Shot-wise features

As a preparation for the postclassifier fusion, we trained a number of individual SVM detectors, each based on a single shot-level feature. This lets us compare different shot-level features in terms of their detection accuracies, although the individual detectors are only used as components of the final fusion-based detection subsystem.

The best individual feature performances we observed resulted from histograms of local image features collected according to the BoV paradigm, that is, variants of SIFT and Color SIFT features. Table 12.3 compares different BoV feature variants in terms of MIAP [66]. As expected, Color SIFT outperforms normal SIFT. Dense sampling is a more effective approach than interest point detection. The soft histogram technique and spatial pyramids improve the performance of the BoV features as well. These results hold on average, but concept-wise differences are large. It does not seem likely that all the differences would result from statistical fluctuations. Table 12.4 lists the most accurate non-BoV features, which can be seen to be clearly inferior in performance to the BoV features.

Combining the features with early fusion did produce more accurate detectors than the individual image features alone, the best early fusion combination having MIAP 0.0601. In the whole concept-detection system, the detector results based on single features and their early fusion combinations are further processed using late

Table 12.3 Concept-Detection Accuracy Based on Various BoV Image Features in TRECVID 2009

Feature	Sampling	Histograms	Spatial Partitioning	MIAP
Color SIFT	Dense	Soft histograms	Spatial pyramid	0.1166
Color SIFT	Dense	Soft histograms	Global	0.1031
Color SIFT	Interest points	Soft histograms	Spatial pyramid	0.1014
Color SIFT	Interest points	Soft histograms	Global	0.0961
Color SIFT	Dense	Hard histograms	Global	0.0988
SIFT	Interest points	Hard histograms	Global	0.0832

Table 12.4 A Selection of Feature-Wise Concept-Detection Accuracies in TRECVID 2009

Feature	Type	MIAP
Edge Histogram	Video	0.0625
Color Moments	Image	0.0438
MPEG-7 Edge Histogram	Image	0.0417
Edge Histogram	Image	0.0403
Color Layout	Video	0.0340
Color Layout	Image	0.0309
Scalable Color	Image	0.0330
Edge Fourier	Image	0.0290
MPEG-7 Color Structure	Image	0.0263

fusion. In some of our previous investigations, also the overall system performance has benefited from early fusion [69]. However, in the experiments with the 2009 data, the use of early fusion did not improve the overall system performance when also late fusion stage was included.

12.4.3.3 Fusion algorithms

We performed a preliminary evaluation of the various postclassifier fusion algorithms in a setting where the annotated part of the video corpus was further partitioned to a training and validation part in 2:1 proportions. In this preliminary experiment, SVM- and BBR-based fusion algorithms were significantly and consistently outperformed by geometric mean-based fusion algorithms, both by the unsupervised basic version and by the supervised SFBS variants. Moreover, the SVM and BBR fusion mechanisms are computationally much more costly. Consequently, the remaining evaluation with the full data set was constrained to the variants of geometric mean fusion.

Figure 12.5(a) compares different geometric mean-based fusion algorithms with the whole video corpus and four different sets D1-D4 of detectors to be fused. These sets result from different sets of shot-wise features, different SVM training parameters and different cross-concept strategies. The number of fused detectors ranges between 77 (D1) and 26 (D4). We can see that the geometric mean of all detectors (the leftmost bar) is always inferior to methods where the set of detectors is selected with SFBS. This has not always been the case in our earlier experiments as SFBS easily overfits to the training data. The figure also shows that multifold-SFBS performs better than the basic SFBS. Early stopping has no essential effect on the average performance. It, however, seems to increase the variance of the results. These experiments thus confirm that early stopping is not a suitable way of regularizing SFBS.

The results of this section – when compared with the MIAP values of the best individual features in Section 12.4.3.2 – can be used to confirm the observation that fusion of features usually outperforms individual features, even if the best

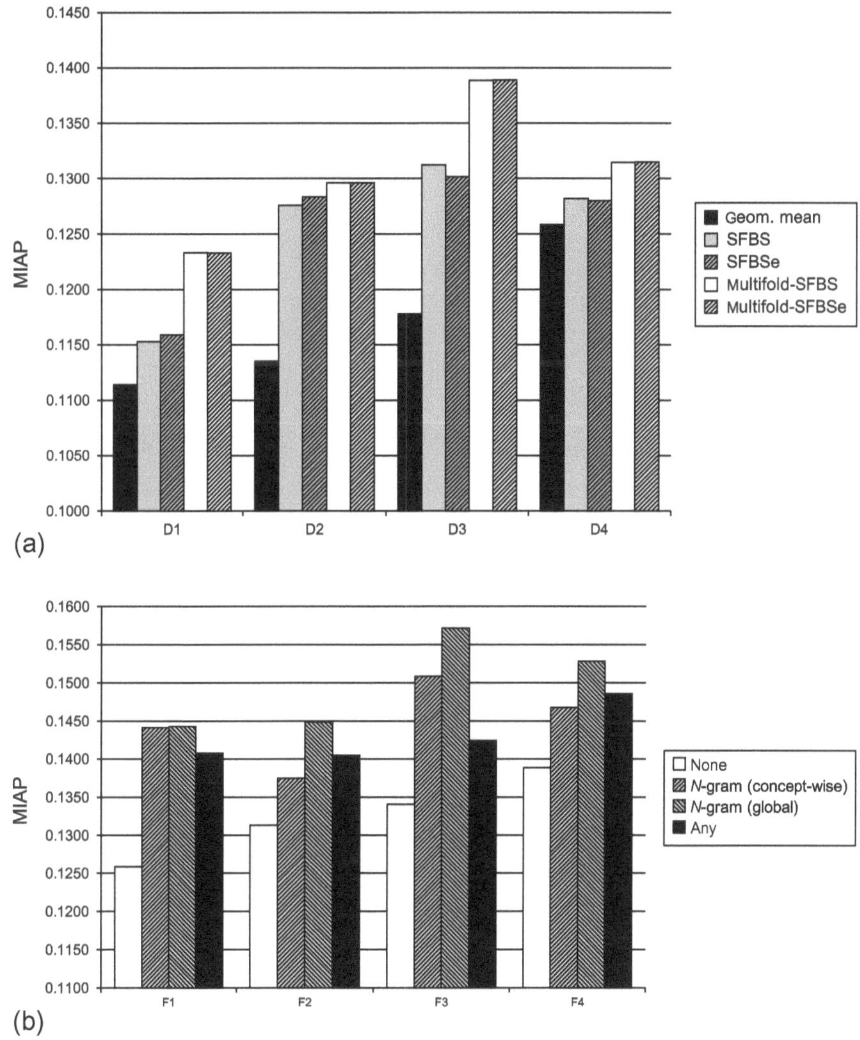

FIGURE 12.5

(a) Comparison of algorithms for selecting detectors for geometric mean fusion for four different sets of detectors D1-D4. The SFBSe and multifold-SFBSe bars with diagonal hatching correspond to algorithms with early stopping. (b) The effect of applying temporal postprocessing on four different shot-wise fusion-based detectors. The bars with diagonal hatching correspond to the *N*-gram technique with two different strategies for order selection.

individual features are clearly better than the worst-performing fused features. With a good fusion algorithm, benefit can be obtained from individually rather badly performing features. In one experiment, we picked approximately 75% of the best features for fusion, thus leaving just the worst performing 25% of the features outside. Still, with the multifold-SFBS fusion algorithm the fusion accuracy improved when the worst 25% were returned to the feature set. With a less-developed fusion algorithm, the saturation point is reached earlier where further addition of features no longer improves the fusion result. An example of this behavior can be seen in Figure 12.5(a) when comparing sets of detectors D3 and D4. Here set D3 is a superset of D4 having almost three times as many detectors. When the geometric mean fusion is used, better performance is obtained by using the smaller set D4, whereas with the more advanced SFBS fusion algorithms the situation reverses: benefit can be obtained from the extra detectors in D3.

12.4.3.4 Temporal postprocessing

Figure 12.5(b) shows the effect of temporal postprocessing for a selection of shot-wise fusion-based detectors F1-F4. The detectors employ different sets of shot-wise features and fusion algorithms. From the figure we can observe that the *N*-gram postprocessing (bars with diagonal hatching) improves MIAP markedly over the baseline with no postprocessing (white bars). We evaluated two strategies for choosing the order of *N*-gram models. In one strategy, the *N*-gram order was selected for each concept separately based on a validation experiment performed with 2:1 split of the training data. The other strategy was to choose the order globally, that is select the order of *N*-grams that resulted in the best mean performance over all the concepts in the validation experiment. As the results show, the global order-selection approach works somewhat better. In almost all the cases the global selection resulted in the selection of order eight, the maximum order that was considered. Generally, the mean performance seems to increase rapidly with increasing *N*-gram order at first. Gradually, the performance starts to saturate and eventually begins to degrade slowly when the order is further increased.

The postprocessing methods marked with identifier "any" (solid dark bars) refer to the concept-wise selection of the postprocessing method from a larger pool of methods according to the best performance in the 2:1 validation experiment. In addition to the *N*-gram methods, this pool included clustering-based techniques that take advantage of temporal and instantaneous inter-concept correlations. Those techniques turned out to be useful in an experiment with data sets of TRECVID 2005-2007 [61], although in that case the baseline detectors were based on less powerful SOM detectors instead of nonlinear SVMs. As can be observed from the figure, in these TRECVID 2009 concept-detection experiments, the inter-concept methods did not bring any improvement over *N*-grams.

12.4.3.5 The best 2009 PicSOM system and its performance

In the above sections, we have investigated many alternative techniques for implementing semantic concept detection from videos. In this section, we collect the results together and describe the best-performing concept-detection module of a video retrieval system that we can assemble from the discussed components. We compare the performance of such a module with that of the state-of-the-art systems that participated in the TRECVID HLFE tasks in year 2009.

Our experiments have shown that with our fusion algorithms, the PicSOM system can benefit from all the shot-wise visual features we have extracted. For concept detection, we thus train one nonlinear SVM detector based on each shot-wise feature. In SVM training, we have to make a compromise between accuracy and training time. The detectors from all the features are fused together with the multifold-SFBS postclassifier fusion algorithm. The concept detection is finalized with an N-gram temporal postprocessing stage where we use the same N-gram order (eight) for all the concepts.

Figure 12.6 shows the MIAP concept-detection performance of the PicSOM system in the TRECVID HLFE tasks of year 2009 in comparison with the best-performing systems of that year. The baseline system PicSOM B closely

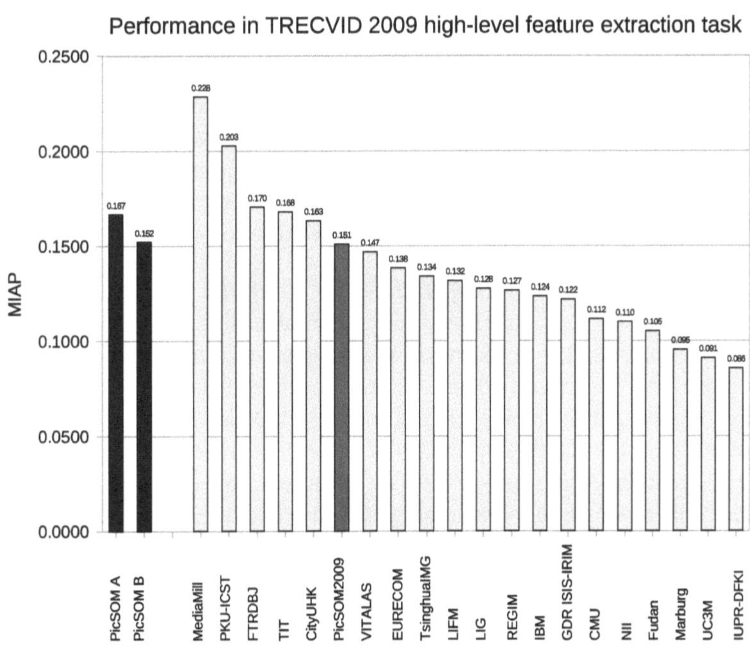

FIGURE 12.6

The MIAP performance in TRECVID 2009 high-level feature extraction task compared with the systems submitted by the best groups to the evaluation. The dark bars correspond to the PicSOM system discussed here, not any submitted system. Note that the figures show only the best-performing end of the distribution, all the systems are significantly more accurate than median MIAP 0.049 of the submissions.

resembles our official submission in the evaluation. The PicSOM A result has been obtained after the official evaluation by using a somewhat more comprehensive set of low-level visual features and more elaborate SVM training. It can be seen that the TRECVID 2009 performance has been further improved, and while not being absolutely the best, PicSOM's HLFE performance compared well with the state-of-the-art systems of that time.

12.4.4 EXPERIMENTS 2014

In the experiments of TRECVID 2014, the used PicSOM system was again based on late fusion of a large variety of supervised detectors trained for each concept. We augmented the set of used features with CNN activation features (see Section 12.2.2.3) and dense SIFT descriptors encoded with Fisher vectors and VLAD encoding (Section 12.2.2.2). As classifiers for the CNN features, we utilized linear SVMs with homogeneous kernel maps [59] of order $d = 2$ to approximate the intersection kernel. For Fisher vectors and VLAD, the classifiers were trained using linear SVMs due to the high dimensionality of the vectors and consequent computational complexity.

12.4.4.1 Data

In 2014, the TRECVID semantic indexing task used Internet Archive video data that consisted of 800 h for development and 200 h for evaluation. The development set contained 28,123 videos with average 1 min 40 s length whereas there were 2373 videos with average 5 min length in the test set. In the training material, keyframes were extracted and used from each shot, and in the test material the i-frames provided by NIST were utilized. Submissions were requested for 60 semantic concepts.

12.4.4.2 Hard negative mining in detector training

A concept-wise, two-class classifier generally produces false positives on negative examples that are similar to the positive examples according to the used feature space. Therefore, to acquire more relevant negative examples, we performed n rounds of hard negative mining [70]. The final classifier for a given feature was obtained by fusing the classifier trained with the original, randomly sampled negatives and the n classifiers using mined relevant negatives. We observed in preliminary experiments that a single round of mining hard negatives already brought the greatest improvement. We therefore used the value $n = 1$ in the following experiments.

12.4.4.3 Submitted runs and results

Table 12.5 shows an overview of our submitted runs, where the four columns in the middle refer to the used features: global features, BoV features, Fisher vectors + VLAD, and CNN features. The next column indicates whether hard negative mining was used, and the rightmost column lists the corresponding mean extended inferred average precision (MXIAP) [67] values.

Run 1 is intended to match our best submission in TRECVID 2013, that is, to use the same features, classifiers, and method of fusion [71]. In Run 2, the Fisher

Table 12.5 An Overview of Our Runs Submitted for the TRECVID 2014 Evaluation

	Features				Hard Neg.	
ID	Glob.	BoV	FV	CNN	Mining	MXIAP
1	•	•				0.1951
2	•	•	•	•		0.2722
3				•	•	0.2843
4	•	•	•	•	•	0.2880

vector and VLAD features and the set of 24 CNN features were included and the global image features discarded. Run 3 uses only the CNN features, together with hard negative mining, and Run 4 combines the characteristics of Runs 2 and 3, that is, all SIFT-based and CNN features with hard negative mining.

The most striking observation on the results is the notable increase of performance compared to our last year's submissions. This is mostly due to the extended set of features, in particular the CNN activation features. By comparing Runs 1 and 2, we observe a 40% increase on MXIAP induced by the different feature sets.

Second, the mining of hard negatives further improved the results, as can be observed by comparing Runs 2 and 4, the latter including the mining step and obtaining the highest MXIAP among our runs, 0.2880 (a 6% increase). The solid performance of the CNN features can furthermore be observed from Run 3, which contains only the CNN features but still almost reaches the MXIAP value of Run 4.

Figure 12.7 shows all runs submitted to the TRECVID 2014 semantic indexing task, our runs highlighted. In total, there were 75 submissions, and only the MediaMill group of the University of Amsterdam submitted runs that were superior to the two best PicSOM runs in their MXIAP results.

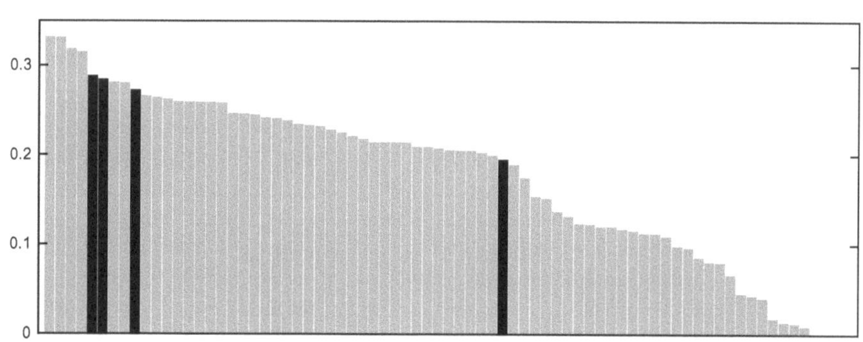

FIGURE 12.7

MXIAP values for all submissions to the TRECVID 2014 semantic indexing task, our runs highlighted.

12.5 CONCLUSIONS

In this chapter, we have described the concept-detection architecture of our PicSOM multimedia retrieval system and proposed and evaluated several alternative techniques for implementing its components. The presented experiments started with TRECVID 2005, where we used the SOMs as the detector algorithm for mostly global image and video features.

In TRECVID 2009, the nonlinear SVMs had replaced SOMs as the detectors, and SIFT and Color SIFT-based BoV features were shown to be superior in performance compared to the global descriptors. In that study, we also demonstrated the usefulness of feature selection and evolved late fusion, as well as that of temporal postprocessing of shot-wise detection results. Using the proposed techniques, the performance of the PicSOM concept-detection subsystem compares favorably with other state-of-the-art systems of that time.

With our recent experiments in TRECVID 2014, we have shown that the top performance obtained in many image classification tasks with deep CNNs can be carried over to semantic video indexing tasks. For the reasons of computational complexity, we used linear SVM detectors with homogeneous kernel maps to approximate the intersection kernel. Combined with the hard negative mining technique in detector training, the PicSOM group ranked second among 21 participants to the semantic indexing task.

As a whole, this chapter has shown an overview of the gradual evolution of the PicSOM multimedia retrieval system since our first participation in TRECVID's visual concept-detection evaluations in 2005. This evolution has concerned the used features, which have developed from global to BoV-based and further to CNN-based, the applied detector algorithms, which have changed from the SOM to nonlinear and linear SVMs, and also various fusion and postprocessing techniques. As a general trend, the PicSOM team's performance and ranking in the evaluation results has been steadily improving – being now the second in this highly competitive and appreciated evaluation.

REFERENCES

[1] M.R. Naphade, T.S. Huang, Extracting semantics from audiovisual content: the final frontier in multimedia retrieval, IEEE Trans. Neural Netw. 13 (4) (2002) 793-810.

[2] M. Koskela, J. Laaksonen, Semantic concept detection from news videos with self-organizing maps, in: I. Maglogiannis, K. Karpouzis, M. Bramer (Eds.), Proceedings of 3rd IFIP Conference on Artificial Intelligence Applications and Innovations, IFIP, Springer, Athens, Greece, 2006, pp. 591-599.

[3] C.G.M. Snoek, M. Worring, Concept-based video retrieval, Found. Trends Inform. Retriev. 4 (2) (2009) 215-322.

[4] A.G. Hauptmann, M.G. Christel, R. Yan, Video retrieval based on semantic concepts, in: Proceedings of the IEEE 96, vol. 4 (April), 2008, pp. 602-622.

[5] M. Naphade, J.R. Smith, J. Tevsić, S.-F. Chang, W. Hsu, L. Kennedy, A. Hauptmann, J. Curtis, Large-scale concept ontology for multimedia, IEEE MultiMedia 13 (3) (2006) 86-91.

[6] J. Deng, W. Dong, R. Socher, L.-J. Li, K. Li, L. Fei-Fei, ImageNet: a large-scale hierarchical image database, in: CVPR, 2009.

[7] S. Ayache, G. Quénot, Video corpus annotation using active learning, in: Proceedings of 30th European Conference on Information Retrieval (ECIR'08), Glasgow, UK, 2008, pp. 187-198.

[8] H. Muurinen, J. Laaksonen, Video segmentation and shot boundary detection using self-organizing maps, in: Proceedings of 15th Scandinavian Conference on Image Analysis (SCIA 2007), Aalborg, Denmark, 2007, pp. 770-779.

[9] M. Koskela, M. Sjöberg, V. Viitaniemi, J. Laaksonen, PicSOM experiments in TRECVID 2008, in: Proceedings of the TRECVID 2008 Workshop, Gaithersburg, MD, USA, 2008, Available from: http://www-nlpir.nist.gov/projects/tvpubs/tv.pubs.org.html.

[10] C. Petersohn, Fraunhofer HHI at TRECVID 2004: shot boundary detection system, in: TREC Video Retrieval Evaluation Online Proceedings, TRECVID, 2004.

[11] M. Sjöberg, M. Koskela, M. Chechev, J. Laaksonen, PicSOM experiments in TRECVID 2010, in: Proceedings of the TRECVID 2010 Workshop, Gaithersburg, MD, USA, 2010, Available from: http://www-nlpir.nist.gov/projects/tvpubs/tv.pubs.org.html.

[12] K. Barnard, N.V. Shirahatti, A method for comparing content based image retrieval methods, in: Proceedings of SPIE Internet Imaging IV, vol. 5018, Santa Clara, CA, USA, 2003, pp. 1-8.

[13] B.S. Manjunath, P. Salembier, T. Sikora (Eds.), Introduction to MPEG-7: Multimedia Content Description Interface, John Wiley & Sons Ltd., London, 2002.

[14] S. Brandt, J. Laaksonen, E. Oja, Statistical shape features for content-based image retrieval, J. Math. Imag. Vis. 17 (2) (2002) 187-198.

[15] M. Sjöberg, S. Ishikawa, M. Koskela, J. Laaksonen, E. Oja, PicSOM experiments in TRECVID 2011, in: Proceedings of the TRECVID 2011 Workshop, Gaithersburg, MD, USA, 2011.

[16] L. Amsaleg, P. Gros, Content-based retrieval using local descriptors: problems and issues from a database perspective, Pattern Anal. Appl. 4 (2/3) (2001) 108-124.

[17] D.G. Lowe, Distinctive image features from scale-invariant keypoints, Int. J. Comput. Vis. 60 (2) (2004) 91-110.

[18] H. Bay, T. Tuytelaars, L.V. Gool. SURF: speeded up robust features, in: Proc. ECCV 2006, 2006.

[19] S. Lazebnik, C. Schmid, J. Ponce, Beyond bags of features: spatial pyramid matching for recognizing natural scene categories, in: Proceedings of the IEEE Conference on Computer Vision and Pattern Recognition, vol. 2, 2006, pp. 2169-2178.

[20] J.C. van Gemert, C.J. Veenman, A.W.M. Smeulders, J.M. Geusebroek, Visual word ambiguity, IEEE Trans. Pattern Anal. Mach. Intell. 32 (7) (2010) 1271-1283.

[21] J. Yang, K. Yu, Y. Gong, T. Huang, Linear spatial pyramid matching using sparse coding for image classification, in: IEEE Conference on Computer Vision and Pattern Recognition. CVPR 2009, 2009, pp. 1794-1801.

[22] X. Zhou, K. Yu, T. Zhang, T. Huang, Image classification using super-vector coding of local image descriptors, in: Proceedings of European Conference on Computer Vision (ECCV 2010), 2010.

[23] H. Jegou, M. Douze, C. Schmid, P. Perez, Aggregating local descriptors into a compact image representation, in: 2010 IEEE Conference on Computer Vision and Pattern Recognition (CVPR), 2010.
[24] F. Perronnin, C. Dance, Fisher kernels on visual vocabularies for image categorization, in: IEEE Conference on Computer Vision and Pattern Recognition, 2007, CVPR'07, 2007, pp. 1-8.
[25] K.E.A. van de Sande, T. Gevers, C.G.M. Snoek, Evaluating color descriptors for object and scene recognition, IEEE Trans. Pattern Anal. Mach. Intell. 32 (9) (2010) 1582-1596.
[26] A. Krizhevsky, I. Sutskever, G. Hinton, ImageNet classification with deep convolutional neural networks, in: NIPS, 2012.
[27] M. Zeiler, R. Fergus, Visualizing and understanding convolutional networks, 2013, arXiv:1311.2901.
[28] Y. Gong, L. Wang, R. Guo, S. Lazebnik, Multi-scale orderless pooling of deep convolutional activation features, 2014, arXiv.org:1403.1840.
[29] J. Donahue, Y. Jia, O. Vinyals, J. Hoffman, N. Zhang, E. Tzeng, T. Darrell, DeCAF: a deep convolutional activation feature for generic visual recognition, in: ICML 2014, 2014.
[30] M. Koskela, J. Laaksonen, Convolutional network features for scene recognition, in: Proceedings of the 22nd International Conference on Multimedia, Orlando, Florida, 2014.
[31] M. Sjöberg, H. Muurinen, J. Laaksonen, M. Koskela, PicSOM experiments in TRECVID 2006, in: Proc. of the TRECVID 2006 Workshop, Gaithersburg, MD, USA, 2006.
[32] C.G.M. Snoek, K.E.A. van de Sande, O. de Rooij, B. Huurnink, J.C. van Gemert, J.R.R. Uijlings, et al., The MediaMill TRECVID 2008 semantic video search engine, in: Proceedings of the TRECVID Workshop, 2008.
[33] N. Inoue, S. Hao, T. Saito, K. Shinoda, I. Kim, C. Lee, TITGT at TRECVID 2009 Workshop, in: Proceedings of the TRECVID 2009 Workshop, Gaithersburg, MD, USA, 2009.
[34] S.B. Davis, P. Mermelstein, Comparison of parametric representations for monosyllabic word recognition in continuously spoken sentences, in: A. Waibel, K. Lee (Eds.), Readings in Speech Recognition, Morgan Kaufmann Publishers, Inc., San Francisco, CA, USA, 1990, pp. 65-74.
[35] G. Salton, M.J. McGill, Introduction to Modern Information Retrieval. Computer Science Series. McGraw-Hill, New York, 1983.
[36] R. Baeza-Yates, B. Ribeiro-Neto, Modern Information Retrieval, Addison-Wesley, Boston, MA, 1999.
[37] S.W. Smoliar, H. Zhang, Content-based video indexing and retrieval, IEEE MultiMedia 1 (2) (1994) 62-72.
[38] N. Dimitrova, H.-J. Zhang, B. Shahraray, I. Sezan, T. Huang, A. Zakhor, Applications of video-content analysis and retrieval, IEEE MultiMedia 9 (3) (2002) 42-55.
[39] A.F. Smeaton, P. Wilkins, M. Worring, O. de Rooij, T.-S. Chua, H. Lua, Content-based video retrieval: three example systems from TRECVid, Int. J. Imag. Syst. Technol. 18 (2/3) (2008) 195-201.
[40] A.P. Natsev, A. Haubold, J. Tevsić, L. Xie, R. Yan, Semantic concept-based query expansion and re-ranking for multimedia retrieval, in: Proceedings of ACM Multimedia (ACM MM'07), Augsburg, Germany, 2007, pp. 991-1000.
[41] C. Fellbaum (Ed.), WordNet: An Electronic Lexical Database, The MIT Press, Cambridge, MA, 1998.

[42] M. Koskela, Interactive image retrieval using self-organizing maps (PhD thesis), Laboratory of Computer and Information Science, Helsinki University of Technology, 2003, Available from: http://lib.hut.fi/Diss/2003/isbn9512267659/.

[43] O. de Rooij, C.G.M. Snoek, M. Worring, Balancing thread based navigation for targeted video search, in: Proceedings of the International Conference on Image and Video Retrieval (CIVR 2008), Niagara Falls, Canada, 2008, pp. 485-494.

[44] C.-W. Ngo, Y.-G. Jiang, X.-Y. Wei, W. Zhao, Y. Liu, J. Wang, S. Zhu, S.-F. Chang, VIREO/DVMM at TRECVID 2009: high-level feature extraction, automatic video search, and content-based copy detection, in: Proceedings of the TRECVID Workshop, 2009, pp. 415-432.

[45] P. Koikkalainen, E. Oja, Self-organizing hierarchical feature maps, in: Proceedings of International Joint Conference on Neural Networks, vol. II, San Diego, CA, USA, 1990, pp. 279-284.

[46] T. Kohonen, Self-Organizing Maps, Third ed., Springer Series in Information Sciences, vol. 30. Springer-Verlag, Berlin, 2001.

[47] V. Viitaniemi, J. Laaksonen, Use of image regions in context-adaptive image classification, in: Y. Avrithis, S. Staab, N. O'Connor (Eds.), Proceedings of the 1st International Conference on Semantic and Digital Media Technologies (SAMT 2006), Lecture Notes in Computer Science. Springer, Athens, Greece, 2006, pp. 169-183.

[48] C. Cortes, V. Vapnik, Support-vector networks, Mach. Learn. 20 (3) (1995) 273-297.

[49] C.-C. Chang, C.-J. Lin, LIBSVM: a library for support vector machines, Software, 2001, Available from: http://www.csie.ntu.edu.tw/~cjlin/libsvm.

[50] A. Bordes, S. Ertekin, J. Weston, L. Bottou, Fast kernel classifiers with online and active learning, J. Mach. Learn. Res. 6 (2005) 1579-1619.

[51] J. Wu, Power mean SVM for large scale visual classification, in: Proceedings of the IEEE Int'l Conference on Computer Vision and Pattern Recognition (CVPR 2012), Providence, USA, 2012.

[52] C.J.C. Burges, B. Schölkopf, Improving the accuracy and speed of support vector learning machines, in: Proc. NIPS, 1997.

[53] T. Downs, K.E. Gates, A. Masters. Exact simplification of support vector solutions. J. Mach. Learn. Res. 2 (2002) 293-297.

[54] G.-X. Yuan, C.-H. Ho, C.-J. Lin, Recent advances of large-scale linear classification, in: Proceedings of the IEEE 100, vol. 9, 2012, pp. 2584-2603.

[55] K. van de Sande, T. Gevers, C. Snoek, Empowering visual categorization with the GPU, IEEE Trans. Multim. 13 (1) (2011) 60-70.

[56] S. Shalev-Shwartz, Y. Singer, N. Srebro, Pegasos: primal estimated sub-gradient solver for SVM, in: Proceedings of the 24th International Conference on Machine Learning, ICML'07, ACM, New York, NY, USA, 2007, pp. 807-814.

[57] C.-J. Hsieh, K.-W. Chang, C.-J. Lin, S.S. Keerthi, S. Sundararajan, A dual coordinate descent method for large-scale linear SVM, in: Proceedings of 25th International Conference on Machine Learning (ICML 2008), Helsinki, Finland, 2008.

[58] S. Maji, A. Berg, J. Malik, Classification using intersection kernel support vector machines is efficient, in: Proceedings of IEEE Conf. on Computer Vision and Pattern Recognition (CVPR 2008), 2008, pp. 1-8.

[59] A. Vedaldi, A. Zisserman, Efficient additive kernels via explicit feature maps, in: Proceedings of the IEEE Conf. on Computer Vision and Pattern Recognition (CVPR 2010), 2010.

[60] A. Genkin, D.D. Lewis, D. Madigan, BBR: Bayesian logistic regression software, Available from: http://www.stat.rutgers.edu/~madigan/BBR/.
[61] V. Viitaniemi, M. Sjöberg, M. Koskela, J. Laaksonen, Exploiting temporal and inter-concept co-occurrence structure to detect high-level features in broadcast videos, in: Proceedings of 9th International Workshop on Image Analysis for Multimedia Interactive Services (WIAMIS 2008), Klagenfurt, Austria, 2008, pp. 12-15.
[62] M. Koskela, J. Laaksonen, M. Sjöberg, H. Muurinen, PicSOM experiments in TRECVID 2005, in: Proceedings of the TRECVID 2005 Workshop, Gaithersburg, MD, USA, 2005, pp. 262-270, Available from: http://www-nlpir.nist.gov/projects/tvpubs/tv.pubs.org.html.
[63] M. Sjöberg, V. Viitaniemi, M. Koskela, J. Laaksonen, PicSOM experiments in TRECVID 2009, in: Proceedings of the TRECVID 2009 Workshop, Gaithersburg, MD, USA, 2009, Available from: http://www-nlpir.nist.gov/projects/tvpubs/tv.pubs.org.html.
[64] S. Ishikawa, M. Koskela, M. Sjöberg, R.M. Anwer, J. Laaksonen, E. Oja, PicSOM experiments in TRECVID 2014, in: Proceedings of the TRECVID 2014 Workshop, Orlando, FL, USA, 2014.
[65] A.F. Smeaton, P. Over, W. Kraaij, Evaluation campaigns and TRECVid, in: MIR'06: Proceedings of the 8th ACM International Workshop on Multimedia Information Retrieval, ACM Press, New York, NY, USA, 2006, pp. 321-330.
[66] E. Yilmaz, J.A. Aslam, Estimating average precision with incomplete and imperfect judgments, in: Proceedings of 15th International Conference on Information and Knowledge Management (CIKM 2006), Arlington, VA, USA, 2006.
[67] E. Yilmaz, E. Kanoulas, J.A. Aslam, A simple and efficient sampling method for estimating AP and NDCG, in: Proceedings of the 31st Annual International ACM SIGIR Conference on Research and Development in Information Retrieval (SIGIR'08), 2008, pp. 603-610.
[68] ISO/IEC, Information Technology – Multimedia Content Description Interface – Part 3: Visual, 15938-3:2002(E), 2002.
[69] V. Viitaniemi, J. Laaksonen, Improving the accuracy of global feature fusion based image categorization, in: B. Falcidieno, M. Spagnuolo, Y.S. Avrithis, I. Kompatsiaris, P. Buitelaar (Eds.), Proceedings of the 2nd International Conference on Semantic and Digital Media Technologies (SAMT 2007), Lecture Notes in Computer Science, vol. 4669. Springer, Geneva, Italy, 2007, pp. 1-14.
[70] X. Li, C.G.M. Snoek, M. Worring, D.C. Koelma, A.W.M. Smeulders, Bootstrapping visual categorization with relevant negatives, IEEE Trans. Multim. 15 (4) (2013) 933-945.
[71] S. Ishikawa, M. Koskela, M. Sjöberg, J. Laaksonen, E. Oja, E. Amid, K. Palomäki, A. Mesaros, M. Kurimo, PicSOM experiments in TRECVID 2013, in: Proceedings of the TRECVID 2013 Workshop, Gaithersburg, MD, USA, 2013.

CHAPTER

On the applicability of latent variable modeling to research system data

13

Ella Bingham[1] and Heikki Mannila[2]*

[1]Helsinki Institute for Information Technology, Aalto University and University of Helsinki, Helsinki, Finland, [2]Department of Computer Science, Aalto University, Espoo, Finland

13.1 INTRODUCTION

Science policy can be vaguely defined as the part of public policy that concerns the organization of universities and research institutes and the funding of science [1–3]. OECD countries use about 0.5-1% of their GDP in public funding of research. As in many areas of public policy, the impact of decisions in science policy is hard to measure. The research system is a complicated one, consisting of universities, research institutes, and other organizations. Furthermore, the research system is strongly connected to many parts of the society: educational policies, industrial structure of the country, etc., influence the impact of research in many ways. The research system is not closed: decisions made elsewhere have a strong effect on the system. The time span from decisions to results in science policy is always years and sometimes decades. For these reasons it is quite difficult to separate the effect of individual science policy decisions from the effects of other changes that have taken place during the same period.

"Science of science policy" [2] aims at applying the methods of science to science policy, or "to provide a scientifically rigorous, quantitative basis for science policy" [4]. This area is still at its infancy, but there are some interesting developments; see, for example, [1–3].

Even the simple case of trying to understand why one research group is more productive than another can be quite hard. Rough measures for quantitative indicators of the inputs to the group (funding, person-years, etc.) and also outputs (publications) can be found. These do not capture all the aspects that influence the productivity of the group. There are important variables such as the skill or competence in the group, the current state of the research carried in the group (is there a long-term research effort close to completion), etc., which are typically not observed. Also,

*Currently on leave at the Academy of Finland.

E. Bingham, S. Kaski, J. Laaksonen and J. Lampinen (Eds) Advances in Independent Component Analysis and Learning Machines.
http://dx.doi.org/10.1016/B978-0-12-802806-3.00013-0

the environment of the group can have an effect: a group in one university can perhaps function quite differently from another group with similar inputs at another university, if the policies and collaboration possibilities of the universities differ from each other.

Professor Erkki Oja has, in addition to his distinguished scientific career, also been very influential in Finnish science policy: he was for a long time the chairperson of the Research Council for Natural Sciences and Engineering at the Academy of Finland. Thus it is interesting to study how the type of methods he has been developing might be applicable to the analysis of data about the research system.

13.2 PROBLEM SETTING

In this chapter, we address a small part of the problem of analyzing the structure of a research system from the viewpoint of selecting suitable computational models. We consider whether tensor methods are appropriate for modeling the personnel structure and publication results of different scientific disciplines in different universities.

Part of the motivation for this small study comes from the difficulties in using scientific disciplines as a background variable in analyzing the science system. The use of a disciplinary classification is necessary for the study of structure of such a system. At the same time it has to be understood that such a classification easily causes some artifacts. For example, the discipline of the department in which a researcher works and the discipline of her/his publications can be different: a biochemist can publish in journals in clinical medicine, a computer scientist in statistics or mathematics journals, etc. Multidisciplinarity is in some areas more of a norm than an exception. A possible way of handling these problems would be to use a more coarse-grained classification, but this might lose a lot of valuable information. Furthermore, such an aggregation might miss interesting combinations of disciplines.

Schematically, the data we consider have information about personnel and publications of different disciplines at different universities in Finland. Thus we have for each measured variable (e.g., number of professors) a matrix with rows corresponding to the university and columns to the scientific discipline. As the measured variables interact with both the universities and the disciplines, it is useful to view the data as a tensor.

Our aim in this preliminary study is to investigate the application of latent variable methods to the study of such data sets. Given the data of the above form about the universities in Finland, we study whether tensor methods can be used to find structure within the disciplines and within the universities. Given the small size of the data, the expectation is that the potential structure should somehow correspond to the preconceptions on the roles and profiles of the different universities and scientific disciplines. Specifically, as our data are from Finland, we expect the group of four to five technical universities to show up, as well as the group of the largest nontechnical universities (whose sets of disciplines have large overlap). Similarly, for the disciplines we expect to see the grouping into traditional subjects (medicine, biology, engineering, arts, letters, etc.).

In more detail, the data set we consider comes from the Vipunen database of the Ministry of Education and Culture and the Finnish National Board of Education [5]. The classification for the scientific disciplines used has 54 different disciplines, and there are 14 universities. The observed variables are the person-years for four different job categories: stage I (corresponding to PhD students), stage II (postdocs), stage III (lecturers, senior researchers), and stage IV (professor level) in 2012. We use the percentages of each category I to IV and the total sum of all person-years as our variables. Additionally, we have data about the number of publications from the year 2013.[a] Thus we have altogether six measured variables, and the data can be viewed as a tensor $\mathbf{X}$ of size $6 \times 54 \times 14$. The entry x_{ntk} of $\mathbf{X}$ is the value of the nth measured variable of the discipline t in university k. Obviously these variables only give a very narrow view of the research system; the goal of this study is to evaluate the applicability of tensor methods on this type of data.

13.3 METHODS

13.3.1 PARALLEL FACTOR ANALYSIS

Our aim is to get insights into the data at hand, instead of creating a predictive model, and therefore we use a nonprobabilistic model. We apply a Parallel Factor Analysis (PARAFAC, [6]) decomposition to the three-way data. It is a generalization of the singular value decomposition (SVD) to tensors. PARAFAC is also known as Canonical Decomposition (CANDECOMP, [7]) and more recently as Canonical Polyadic Decomposition (CPD, [8]). PARAFAC is a major improvement to classical Gaussian factor analysis or principal component analysis (PCA) in the sense that it can uniquely estimate the components also for Gaussian data, whereas PCA cannot. For a pleasant survey of unsupervised multi-way data analysis, see [9]. De Lathauwer et al. present a general multi-way SVD in [10].

We assume that our data can be represented by a multilinear factor model such that the scientific disciplines obey some commonalities with respect to the variables (person-years and publications) across the universities, and, on the other hand, the scientific disciplines have varying presence at each university.

A three-way PARAFAC model provides a good way of treating the scientific disciplines and universities in a symmetric manner: resorting to a two-way model would require us to assume that all disciplines behave independently; or alternatively to neglect the information on the university.

An underlying idea at PARAFAC is that the same factors are present in each sample under different conditions, but scaled depending on the conditions. A PARAFAC model decomposes a tensor of size $N \times T \times K$ as a linear combination of R rank-1 tensors $\mathbf{A}_{N\times R} = [a_{nr}]$, $\mathbf{B}_{T\times R} = [b_{tr}]$, and $\mathbf{C}_{K\times R} = [c_{kr}]$, as follows:

$$x_{ntk} = \sum_{r=1}^{R} a_{nr} b_{tr} c_{kr} + e_{ntk}. \tag{13.1}$$

Uniqueness of component matrices is obtained by requiring that factors in different dimensions can only interact factorwise: factor r in the first dimension (**A**) can only interact with the rth factor in the second (**B**) and third (**C**) dimensions. With this restriction the component matrices are determined uniquely up to a permutation and scaling of columns.

As noted by Acar and Yener [9] the model is multilinear, that is, linear in each dimension (variables, disciplines, and universities), and factors extracted from each dimension are linear combinations of the variables in that dimension.

We use a logarithmic transformation of the data as our original variables are of somewhat different scales. We estimate the PARAFAC model by least squares fitting, using the N-Way Toolbox for MATLAB [11]. We choose R, the number of components, using the core consistency test as suggested by [12].

13.3.2 INDEPENDENT COMPONENT ANALYSIS

We next consider one specific latent variable method, namely independent component analysis (ICA) [13]. ICA for tensor data has been studied especially in the case of neuroimaging [14,15] and also in atmospheric sciences [16]. Hyvärinen [17] suggests estimating ICA for three-way data by collapsing the data into an ordinary matrix. In our notation, let $\mathbf{X}_k$ be a data matrix of size $N \times T$, containing the measured variables n of all scientific disciplines t of university k. We can concatenate all $\mathbf{X}_k$ column-wise into a matrix $\chi = (\mathbf{X}_1, \ldots, \mathbf{X}_K)$ for which an ordinary ICA model $\chi = \mathbf{A}(\mathbf{S}_1, \ldots, \mathbf{S}_K)$ can be fitted. Here $(\mathbf{S}_1, \ldots, \mathbf{S}_K)$ is a matrix containing university-specific independent components given in matrices $\mathbf{S}_k$ that are concatenated column-wise. The mixing matrix **A** is common to all universities k. In essence, the estimation would now neglect the information on one of the dimensions.

Hyvärinen [17] also gives an interesting interpretation of PARAFAC in terms of ICA: one can write the PARAFAC model (Eq. 13.1) in an equivalent manner as follows. For each observation (here, university) k write the observed data matrix $\mathbf{X}_k$ as

$$\mathbf{X}_k = \mathbf{A}\mathbf{D}_k\mathbf{B}^T + \mathbf{E}_k, \tag{13.2}$$

where $\mathbf{D}_k$ is a diagonal matrix whose diagonal elements are the kth row of the third component matrix **C**, and $\mathbf{E}_k$ contains the error terms. The matrices **A** and **B** are the same for all k. The above notation is equivalent to an ICA model where **A** is the mixing matrix and $\mathbf{B}^T$ contains the independent components, common to all data sets or "subjects" (here, universities) k, and $\mathbf{D}_k$ gives subject-specific scaling factors. The same ICA model thus holds for all subjects, save for the scaling factors. The matrices $\mathbf{D}_k$ must be linearly independent, and hence the data matrices $\mathbf{X}_k$ must be sufficiently different with respect to the scalings for different k [17].

13.4 RESULTS

The core consistency test [12] suggests using $R = 2$ components in PARAFAC, which is reasonable taking into account the small dimensionality of the data.

The results are shown in Figures 13.1–13.3. We can nicely visualize the estimated factors in terms of measured variables, disciplines, and universities. For the purposes of visualization, the rows of the matrices have been reordered using the so-called barycentric algorithm [18,19]. This method is basically an eigenvalue approach for finding similar rows and columns.

Figure 13.1 shows the component matrix $\mathbf{A}_{6\times 2}$ containing the loadings of the six measured variables. We see that the first factor captures especially the number of publications and the percentage of lecturers and senior researchers (stage III in the universities' career system), while the second factor is strong in the percentage

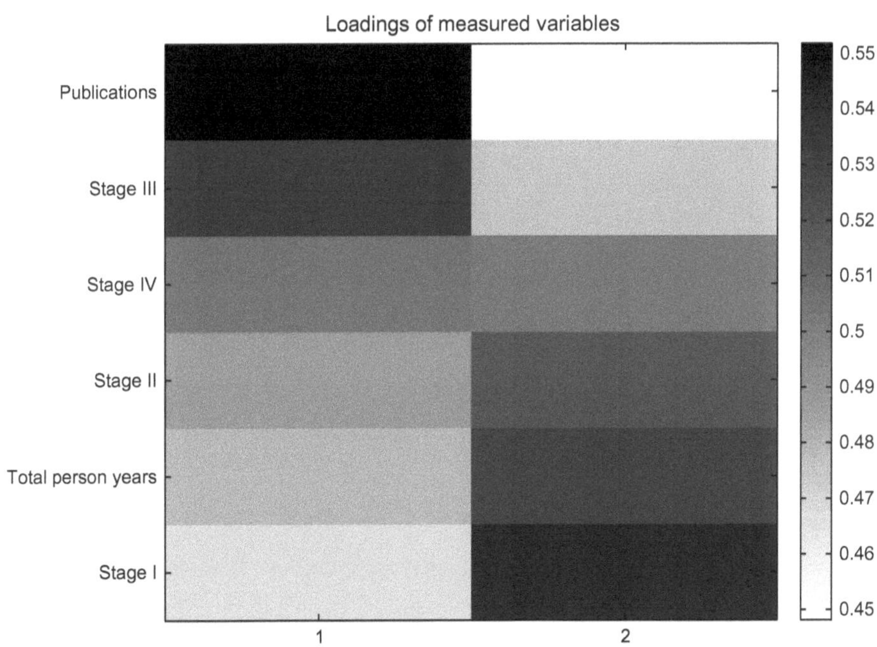

FIGURE 13.1

Visualization of the two PARAFAC factors in terms of the measured variables (y-axis). Note the small range (0.45-0.55) of the grayscale coding.

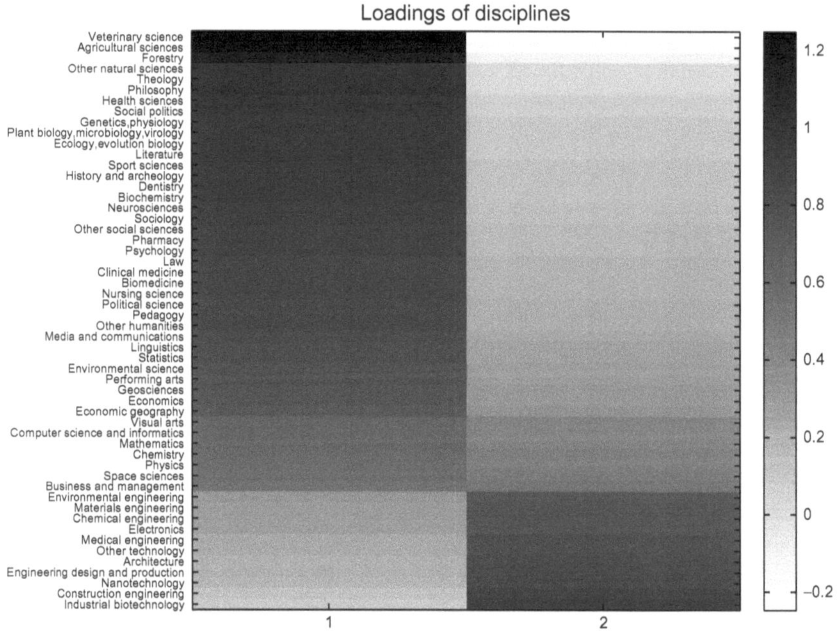

FIGURE 13.2

Visualization of the two PARAFAC factors in terms of the scientific disciplines (*y*-axis).

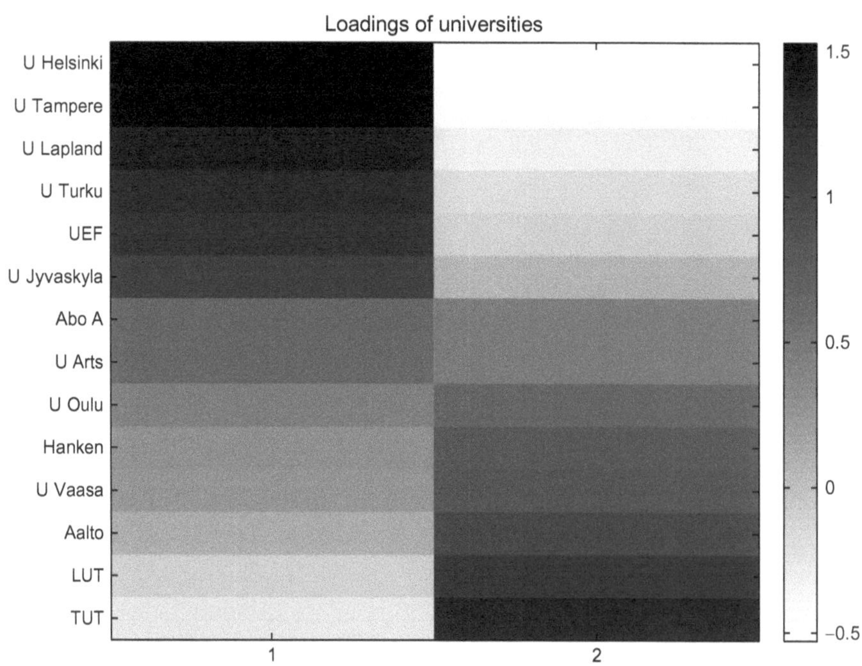

FIGURE 13.3

Visualization of the two PARAFAC factors in terms of the universities (*y*-axis).

of doctoral students (stage I), the total number of person-years and the percentage of postdocs (stage II). The proportion of professors (stage IV) is equally strongly captured by both factors.

Figure 13.2 shows the component matrix $\mathbf{B}_{54\times 2}$ of scientific disciplines. The first factor is dominated by medical sciences, agriculture and forestry, biosciences, and social sciences such as theology, philosophy, sociology, arts and letters, and so on. In contrast, the second factor is dominated by engineering disciplines and business economics.

The component matrix $\mathbf{C}_{14\times 2}$ containing the factors with respect to all Finnish universities is shown in Figure 13.3. The first factor is dominated by universities whose palette of disciplines is known to be wide, ranging from natural sciences to humanities, such as Universities of Helsinki, Tampere, Lapland, Turku, Eastern Finland, and Jyväskylä. The second factor captures universities whose focus is mainly in technical disciplines and economics: especially Tampere University of Technology, Lappeenranta University of Technology, and Aalto University. These observations are in line with those seen in Figure 13.2 regarding scientific disciplines.

Looking at Figures 13.1–13.3 we indeed see the expected structure of "generalist" universities with their palette of medical sciences, biosciences, and arts and letters being captured by one factor, and technical universities with their engineering and business disciplines captured by the other factor. Some of the differences are due to the different group structures and publication traditions in different disciplines. We emphasize that the purpose of this experiment is to evaluate whether PARAFAC is applicable to the analysis of these types of data, not to yield conclusions about the Finnish research system.

Randomization

The results shown above indicate that the PARAFAC method indeed finds intuitive structure from the data. However, the results can, of course, be misleading: some data analysis methods are able to find meaningful-looking structure in a data set that does not contain any structure.

To test the whether the signal found can be considered to be real, we performed a simple experiment. We used the randomization method of Ojala et al. [20]. This approach randomizes the entries in a matrix while approximately maintaining the row and column sums. In the method, one selects at random four entries in a data matrix, tests whether both the diagonal elements and the antidiagonal elements are close to each other, respectively, and if so, rotates the values in the four entries. That is, given a parameter ϵ and the values a, b, c, d in the data matrix with $|a - d| < \epsilon$ and $|b - c| < \epsilon$, the randomization operation works as follows:

$$\begin{pmatrix} a & b \\ c & d \end{pmatrix} \rightarrow \begin{pmatrix} c & a \\ d & b \end{pmatrix}. \tag{13.3}$$

This operation maintains the sums of rows and columns in the data approximately.

The aim in randomization is to see whether the chosen data analysis method also finds structure in a randomized data set which is constructed so that it should not contain interesting structure. If the data analysis method finds structure in a randomized data set, this casts doubt on the validity of the results on the original data set, too.

We estimated the PARAFAC model both on original data and on randomized data. We noticed that the explained variation is clearly smaller in randomized data, and the component matrices do not have interesting structures.

Independent component analysis

We also performed a quick check in which we estimated an ordinary ICA model by collapsing the tensor data into an ordinary matrix as described in Section 13.3.2 and suggested in [17], using the assumption that information on the university can be neglected. Using the FastICA toolbox [13,21] we estimated two independent components. However, the groupings of variables are not as informative as shown above in the case of PARAFAC.

Hyvärinen [17] also suggests collapsing the tensor in another direction and again using ordinary ICA on the obtained matrix. Due to the small dimensionality of our original data, this operation was not feasible.

13.5 DISCUSSION AND FURTHER WORK

ICA has proved to be a very strong method for analyzing a variety of different types of data. While ICA is a versatile method, generalizing it to tensor type of data is not straightforward. Tensor methods seem to be a viable approach to achieving some of the good properties of ICA in a higher-dimensional setting.

The aim of this study was to evaluate the potential of tensor methods in the analysis of input-output data on universities and the different disciplines. The results indicate that simple, yet interesting and meaningful structure can be found: we were able to distinguish two more or less opposite patterns that the universities and the scientific disciplines follow. A tensor representation of the data is a useful one, as this allows studying the relationships of both universities on the one hand, and scientific disciplines on the other hand, with respect to measured variables.

As mentioned earlier, the small scale of the data implies that we can easily check whether the results correspond to the existing data on the universities and the scientific disciplines. Indeed the results are more or less the expected ones.

We have chosen PARAFAC due to its simplicity. There exist other data analysis methods for multiway data. Should we prefer to neglect the multiway structure and flatten the data into an ordinary matrix, we could resort to, for example, the Tucker1 method [22,23] or basically any other two-way analysis method such as SVD. As an example we briefly discussed using ordinary ICA on our data. Methods that truly maintain the multiway structure of the data include several different extensions of PARAFAC (see Acar and Yener [9] for a clear presentation) and the Tucker family of models. De Lathauwer et al. [10] discuss SVD and its interpretations in multiway data.

In this note, we reported only a single randomized test case. A traditional hypothesis testing approach does not seem to be very useful for our current small data set. However, an interesting question is how to quantify the strength or weakness of the signal discovered by PARAFAC. One approach to addressing this question is to do some sort of sensitivity analysis of the results with respect to small variations in the data. One could, for example, use the randomization approach, and test what is the smallest distance (in some appropriate metric) between the original and randomized data such that the signal is no longer present in the data.

Analysis methods such as described above can be a useful auxiliary tool in the study of a research system. However, it has to be emphasized that any method based on solely looking at raw counts of personnel, publications, citations, etc., misses a lot of information that is only available through more detailed studies such as those provided by peer review panels.

ACKNOWLEDGMENT

We thank Ricardo Vigário for valuable comments.

NOTES

a. Publications at least on level 1 of the Publication Forum [24].

REFERENCES

[1] P.E. Stephan, The economics of science, in: B.H. Hall, N. Rosenberg (Eds.), Handbook of the Economics of Innovation, Handbooks in Economics, Chapter 5. Elsevier, Amsterdam, 2010.

[2] K. Fealing, The Science of Science Policy, Stanford University Press, Stanford, 2011.

[3] A. Scharnhorst, K. Börner, P. van den Besselaar, Models of Science Dynamics: Encounters Between Complexity Theory and Information Sciences, Springer, Berlin, 2012.

[4] Science of Science Policy, Available from: http://www.scienceofsciencepolicy.net/about.

[5] Ministry of Education and Culture Finland and the Finnish National Board of Education, Reporting portal Vipunen, Available from: http://vipunen.csc.fi.

[6] R.A. Harshman, Foundations of the PARAFAC procedure: models and conditions for an "explanatory" multi-modal factor analysis, in: UCLA Working Papers in Phonetics, 16, 1970, 84 pp. (University Microfilms, Ann Arbor, No. 10085).

[7] J.D. Carroll, J.-J. Chang, Analysis of individual differences in multidimensional scaling via an n-way generalization of "Eckart-Young" decomposition, Psychometrika 35 (3) (1970) 283-319.

[8] L. Sorber, M. Van Barel, L. De Lathauwer, Optimization-based algorithms for tensor decompositions: canonical polyadic decomposition, decomposition in rank-(L_r,L_r,1) terms, and a new generalization, SIAM J. Optim. 23 (2) (2013) 695-720.

[9] E. Acar, B. Yener, Unsupervised multiway data analysis: a literature survey, IEEE Trans. Knowl. Data Eng. 21 (1) (2009) 6-20.

[10] L. De Lathauwer, B. De Moor, J. Vandewalle, A multilinear singular value decomposition, SIAM J. Matrix Anal. Appl. 21 (2000) 1253-1278.

[11] C.A. Andersson, R. Bro, The N-way toolbox for MATLAB, Chemometr. Intell. Lab. Syst. 52 (2000) 1-4, Available from: http://www.models.life.ku.dk/source/nwaytoolbox/.

[12] R. Bro, H.A.L. Kiers, A new efficient method for determining the number of components in PARAFAC models, J. Chemometr. 17 (5) (2003) 274-286.

[13] A. Hyvärinen, J. Karhunen, E. Oja, Independent Component Analysis, Wiley, New York, 2001.

[14] C.F. Beckmann, S.M. Smith, Tensorial extensions of independent component analysis for multisubject fMRI analysis, Neuroimage 25 (1) (2005) 294-311.

[15] V.D. Calhoun, J. Liu, T. Adali, A review of group ICA for fMRI data and ICA for joint inference of imaging, genetic, and ERP data, NeuroImage, 45 (2009) S163-S172.

[16] S. Unkel, A. Hannachi, N.T. Trendafilov, I.T. Jolliffe, Independent component analysis for three-way data with an application from atmospheric science, J. Agric. Biol. Environ. Stat. 16 (3) (2011) 319-338.

[17] A. Hyvärinen, Independent component analysis: recent advances, Philos. Trans. R. Soc. A Math. Phys. Eng. Sci. 371 (1984) (2013) 20110534

[18] K. Sugiyama, S. Tagawa, M. Toda, Methods for visual understanding of hierarchical system structures, IEEE Trans. Syst. Man Cyb. 11 (2) (1981) 109-125.

[19] E. Mäkinen, H. Siirtola, The barycenter heuristic and the reorderable matrix, Informatica 29 (3) (2005) 357-363.

[20] M. Ojala, N. Vuokko, A. Kallio, N. Haiminen, H. Mannila, Randomization methods for assessing data analysis results on real-valued matrices, Stat. Anal. Data Min. 2 (4) (2009) 209-230.

[21] H. Gävert, J. Hurri, J. Särelä, A. Hyvärinen, The FastICA package for MATLAB, Available from: http://research.ics.aalto.fi/ica/fastica/.

[22] L.R. Tucker, The extension of factor analysis to three-dimensional matrices, in: H. Gulliksen, N. Frederiksen (Eds.), Contributions to Mathematical Psychology, Holt, Rinehart and Winston, New York, 1964, pp. 110-127.

[23] L.R. Tucker, Some mathematical notes on three-mode factor analysis, Psychometrika 31 (3) (1966) 279-311.

[24] O. Auranen, J. Pölönen, Classification of scientific publication channels: final report of the Publication Forum Project (2010-2012), Web publications 1, Federation of Finnish Learned Societies, 2010, Available from: http://www.tsv.fi/files/yleinen/publication_forum_project_final_report.pdf, http://www.tsv.fi/julkaisufoorumi/english.php?lang=en.

Index

Note: Page numbers followed by *f* indicate figures and *t* indicate tables.

E

F

G

H

U

V

W

Kolibri 360 GmbH
Ferdinand-Jühlke-Straße 7
99095 Erfurt, Deutschland
produktsicherheit@kolibri360.de